Unsteady Flow in Open Channels

From river floods and tides, to systems for drainage and distribution of water, flows in open channels affect the daily lives of many people. Practitioners in water engineering rely on a thorough understanding of these flows in order to safeguard our habitat, while at the same time sustaining the water environment.

Unsteady Flow in Open Channels provides a coherent approach to the analyses and modelling of various types of unsteady flow, highlighting their similarities and differences. It presents a unified framework, using the relative roles of inertia and resistance to classify the different types of unsteady flow in environmental water systems. The link between analytical approaches and numerical modelling is emphasized – in particular, demonstrating how high-level computer languages, such as Python, can be used to solve advanced problems.

Every major topic in the book is accompanied by worked examples illustrating the theoretical concepts. Practical examples, showcasing inspiring research and engineering applications from the past and present, provide insight into how the theory developed. The book is also supplemented by a range of online resources, available at www.cambridge.org/battjes, including problem sets and computer codes. A solutions manual is also available for instructors. This book is intended for students and professionals working in the field of environmental water systems, in areas such as coasts, rivers, harbours, drainage, and irrigation canals.

Jurjen Battjes is Emeritus Professor at Delft University of Technology. Now retired, Professor Battjes had a long career in university teaching and research in the field of fluid mechanics. He taught courses in modelling, introductory fluid mechanics, unsteady flow in open channels and wind-generated waves, the latter being his major research topic. He has also been active as a consultant on numerous projects in coastal engineering, including the Deltaworks in The Netherlands. Professor Battjes has received several prizes and awards, including the International Coastal Engineering Award of the American Society of Civil Engineers.

Robert Jan Labeur is Assistant Professor at Delft University of Technology. He teaches various courses in environmental fluid mechanics. His research involves numerical modelling of environmental flows, water-borne transport processes and morphology, in particular the modelling of complex three-dimensional flows. Before working at the university, he was a consultant in the field of hydraulic and coastal engineering.

Unsteady Flow in Open Channels

JURJEN BATTJES

ROBERT JAN LABEUR

CAMBRIDGE
UNIVERSITY PRESS

University Printing House, Cambridge CB2 8BS, United Kingdom

One Liberty Plaza, 20th Floor, New York, NY 10006, USA

477 Williamstown Road, Port Melbourne, VIC 3207, Australia

314-321, 3rd Floor, Plot 3, Splendor Forum, Jasola District Centre, New Delhi - 110025, India

79 Anson Road, #06-04/06, Singapore 079906

Cambridge University Press is part of the University of Cambridge.

It furthers the University's mission by disseminating knowledge in the pursuit of education, learning and research at the highest international levels of excellence.

www.cambridge.org
Information on this title: www.cambridge.org/9781107150294
10.1017/9781316576878

First published 2017

A catalogue record for this publication is available from the British Library

Library of Congress Cataloging in Publication data
Names: Battjes, J. A. (Jurjen Anno), 1939– author. | Labeur, R. J. (Robert Jan), 1964–author.
Title: Unsteady flow in open channels / Jurjen Battjes and Robert Jan Labeur.
Description: Cambridge, United Kingdom ; New York, NY : Cambridge University Press, 2017. | Includes bibliographical references and index.
Identifiers: LCCN 2016035538 | ISBN 9781107150294 (Hardback ; alk. paper) | ISBN 1107150299 (Hardback ; alk. paper)
Subjects: LCSH: Unsteady flow (Fluid dynamics)–Mathematical models. | Open-channel flow–Mathematical models. | Water waves–Mathematical models. | Wave equation. | Fluid mechanics–Mathematical models.
Classification: LCC TA357.5.U57 B38 2017 | DDC 532/.053–dc23
LC record available at https://lccn.loc.gov/2016035538

ISBN 978-1-107-15029-4 Hardback

Additional resources for this publication at www.cambridge.org/battjes

Contents

List of Symbols

Roman Symbols

a	acceleration of fluid particle (Chapter 2)
a	height of sharp-crested weir above upstream bed level (Chapter 9)
a	height of opening under a gate (Chapter 9)
A	area of entire wetted cross section
A_c	area of wetted conveyance cross section
b	binormal coordinate
B	width of the free surface
B_c	width of the free surface of the conveyance cross section
c	velocity of propagation of a disturbance relative to the fluid ahead of it
c	concentration of dissolved or suspended matter (Chapter 10)
c_D	wind drag coefficient
c_f	boundary resistance coefficient
c_{HW}	velocity of propagation of a flood wave
C	Chézy coefficient
d	cross-sectionally averaged depth of conveyance cross section ($d = A_c/B_c$)
d_{cr}	critical depth
D	grain or stone diameter (Chapter 9)
D	inner pipe diameter (Appendix A)
E	energy level above local bed elevation (Chapter 9)
E	Young's modulus of elasticity of pipe wall material (Appendix A)
F	energy flux
Fr	Froude number
F_r	resistance force
g	acceleration of gravity
h	elevation of the free surface above reference plane
h_p	piezometric level above reference plane
H	energy level above reference plane
i	imaginary unit
i_b	bed slope
i_f	friction slope

i_s slope of free surface
k wave number in harmonic wave or oscillation ($k = 2\pi/L$) (Chapters 3 and 7)
k Nikuradse sand grain diameter (Chapter 9)
K diffusivity in flood waves (Chapter 8)
K diffusivity in transport of matter (Chapter 10)
K modulus of compressibility of water (Appendix A)
ℓ length of basin
L wave length
$\mathcal{L}$ length scale of the motion
m discharge coefficient
n Manning's n
p fluid pressure
p complex root of dispersion equation ($p = \mu + ik$) (Chapter 7)
p_{atm} atmospheric pressure at the air–water interface
P wetted perimeter of conveyance cross section
P complex propagation constant in harmonic wave propagation (Chapter 7)
q discharge per unit width ($q = Q/B_c$)
Q discharge
r radius of curvature of a streamline (Chapter 1)
r ratio of wave heights at abrupt channel transition (Chapters 4 and 7)
r amplitude response factor (Chapters 6 and 7)
R hydraulic radius ($R = A_c/P$)
$R^{\pm}$ Riemann invariants
s streamwise coordinate
S relaxation length in theory of damping of translatory waves
t time
T wave period
$\vec{T}$ three-dimensional transport vector
$\mathcal{T}$ time scale of the motion
u local particle velocity
u_* shear velocity
U cross-sectionally averaged streamwise particle velocity u_s
$\mathcal{U}$ velocity scale of the motion
V velocity of propagation relative to the bed ($V = U + c$)
W total head loss in inlet–bay system
W_{10} wind speed (10-minute average at 10 m above mean water level)
x horizontal cartesian coordinate
y horizontal cartesian coordinate
z vertical cartesian coordinate, positive upward
Z complex auxiliary length representing the discharge in harmonic wave propagation

Greek Symbols

α velocity distribution coefficient in momentum flux (Chapter 1)
α dimensionless parameter of the free-surface profile of a translatory wave (Chapter 4)
α arbitrary phase angle
β angle of bed elevation with respect to the horizontal (Chapter 1)
β velocity distribution coefficient in energy flux (Chapter 9)
β arbitrary phase angle
γ ratio of Bc-values at abrupt transition
Γ dimensionless parameter in tidal inlet–bay system
δ resistance angle in propagation of harmonic waves (Chapter 7)
δ wall thickness (Appendix A)
ϵ infinitesimal dimensionless quantity
ϵ molecular diffusivity (Chapter 10)
ϵ_t turbulence diffusivity (Chapter 10)
ζ surface elevation above mean water level
η dimensionless value of ζ
θ phase lag of bay tide behind exterior tide
κ resistance factor with dimension 1/time in harmonic wave propagation (Chapter 7)
κ Von Karman constant in theory of turbulent boundary layers (Chapter 10)
μ damping modulus in harmonic wave propagation (Chapter 7)
μ contraction coefficient (Chapter 9)
ν_t kinematic turbulence viscosity
ξ head loss coefficient
ρ water density
ρ_a air density
σ dimensionless resistance factor in harmonic wave propagation (Chapter 7)
σ streamwise standard deviation of surface elevation in flood waves (Chapter 8)
σ standard deviation of concentration of transported substance (Chapter 10)
σ Courant number (Chapter 11)
Φ Gaussian probability density function
τ relaxation time in inlet–bay system
τ_b boundary shear stress
τ_s wind shear stress at the free surface
χ coefficient for expansion loss and boundary resistance in inlet–bay system
Ψ angle between wind direction and flow direction
ω radial frequency in harmonic motion ($\omega = 2\pi/T$)
ω_0 natural (Helmholtz) frequency of inlet-bay system

Diacritical Marks

$\hat{\circ}$	circumflex	(real) amplitude of a quantity
$\tilde{\circ}$	tilde	complex amplitude of a quantity
$\bar{\circ}$	macron	(turbulence) quantity averaged over time or space
$\circ'$	prime	(turbulent) fluctuation of a quantity
$\circ'$	prime	first derivative of a function
$\circ''$	double prime	second derivative of a function

Sub- and Superscripts

subscript s	refers to *sea*
subscript b	refers to *basin* or *bay* or *bed*
subscript c	refers to *conveyance* (area or width)
subscript 0	refers to an initial or undisturbed flow state
subscript 0	refers to quantities of a harmonic wave in the absence of resistance
subscript cr	refers to *critical* flow
subscript u	refers to *uniform* flow
superscript $+$	refers to propagation in *positive* s-direction
superscript $-$	refers to propagation in *negative* s-direction

Preface

This book grew out of lectures on unsteady flow in open channels in the Civil Engineering Department of the Delft University of Technology for senior BSc students and first-year MSc students in hydraulic and coastal engineering, water resources management, hydrology and sanitary engineering. It deals with gradually varying, unsteady flows, or long waves, in natural channels such as rivers, estuaries and tidal channels, as well as man-made canals for various purposes such as drainage and irrigation systems and shipping.

Most existing books on open channel flow deal mainly with steady flows, unsteady flows typically being diverted to a single chapter. In practice, unsteady flows are the rule rather than the exception. Therefore, a unified introduction was deemed necessary, in which the unsteadiness and the important associated notion of wave propagation are essential ingredients from the start. The intended readership consists of students in the above-mentioned disciplines as well as practitioners.

Subject

The subject of this book is the mathematical modelling of gradually varying unsteady flows, or long waves, in open channels. Various classes of long waves are distinguished, depending on their origin and the associated time scale, varying from the relatively rapid translatory waves to the sluggish flood waves in lowland rivers. A chapter on steady flow summarizes some relevant results within this subclass. Transport of dissolved or suspended matter in open channels is briefly dealt with as well. Lastly, the numerical modelling of flow in conduits is covered in a separate chapter.

Pressurized flow in pipelines falls formally outside the scope of this book, but it is included nonetheless because pipe flow often is an integral part of water transport systems, and the mathematical equations describing pressurized flow and those describing open channel flow are quite similar. For these reasons, a summary is presented in Appendix A.

Aim

This book offers a unified presentation of the mathematical modelling of various classes of unsteady flows that can be expected in the context of design and operation of hydraulic engineering works in tidal areas, estuaries, rivers, canals etc., e.g. dredging or the construction and operation of control structures or dams. The engineer should be able to foresee consequences of the works being designed, both qualitatively and quantitatively. This requires insight into these flows and the ability to schematize them, quantify them through mathematical modelling and computations and interpret the results. The achievement of these objectives is the primary aim of this book.

In view of the above, the more specific aims of this book can be summarized as follows:

- to provide qualitative knowledge of various classes of unsteady flow phenomena in open channels that are important in engineering practice
- to provide insight into the relative importance of various mechanisms in the dynamics of these flows
- to explain the physics of shallow-water wave propagation
- to stimulate an attitude of making a (qualitative) problem analysis including the estimation of relevant effects
- to offer a unified, systematic overview of mathematical approximations and solution methods suited to various categories of open channel flow
- to enable the reader to develop the ability to make schematizations and to perform approximative computations for the flow phenomena considered
- to develop the ability to code algorithms for the computation of unsteady open-channel flows

Approach

A key characteristic of this book is its emphasis on the development of physical insight. This is approached through the presentation of simplified models of the principal classes of flow considered and the corresponding analytical solutions, since these show explicitly the effects of the major parameters involved.

Following the derivation of the basic one-dimensional equations for long waves (the so-called shallow-water equations), distinct classes of waves are presented and discussed. For each of these, order-of-magnitude relations between the different physical processes are derived, including the boundary resistance relative to the inertia. In fact, the latter ratio is used as the ordering principle for the remainder of the book, the successive chapters dealing with a class of flows of increasing relative resistance, going from translatory waves with negligible resistance to friction-dominated flood waves in lowland rivers.

In each of these chapters, the presentation proceeds from a qualitative discussion, via the introduction of appropriate simplifications of the momentum equation, to the development of a quantitative mathematical model. Analytical solutions are presented because these are optimally suited to the development of physical insight. The aim in these chapters is not to develop models of high quantitative accuracy. This is done in a final chapter, exclusively devoted to numerical modelling.

In view of its importance, understanding of the notion of wave propagation is developed gradually. A qualitative description of the propagation of low frictionless translatory waves in prismatic conduits prepares the way to the derivation of the corresponding general linear wave equation. Subsequently, these restrictions are gradually relaxed by considering variations in channel cross section, allowing finite wave heights, and incorporating friction. The method of characteristics is introduced next to provide added understanding as well as a formal mathematical framework for quantitative evaluations.

The practical relevance of the developed concepts is demonstrated with examples from inspiring engineering cases (such as the enclosure of the previous Zuiderzee in The Netherlands) and captivating natural phenomena (e.g. tidal bores). Field observations and some laboratory data are presented for a quantitative comparison with the theory.

Worked-out examples are presented for purposes of illustration, to provide more understanding and to aid in the ability to apply the theory. Each chapter ends with a set of Problems.

Layout

Chapter 1 opens with a brief description of the approach to the mathematical modelling of unsteady flow in open channels. This is followed by a presentation of the basic equations for these flows, which are the starting point for the analysis and modelling of various categories of long waves in the following chapters. Chapter 2 describes several characteristic long-wave phenomena qualitatively and presents a quantitative analysis of their major characteristics, making visible which processes are dominant and which ones are relatively weak in the various categories of long waves.

The notion of wave propagation with neglect of the effects of resistance is developed in Chapters 3 and 4, followed by the introduction of the powerful method of characteristics in Chapter 5.

Chapters 6–8 present suitable mathematical approximations for several classes of long waves. Corresponding solution techniques and solutions are presented as well. Harmonic motions are considered in Chapter 6 for standing oscillations in basins and in Chapter 7 for propagation with friction, mainly aimed at tidal propagation, in which resistance is important but not dominant. Flood waves in lowland rivers, in which inertial effects can be neglected relative to resistance, are the subject of Chapter 8.

Chapter 9 gives a brief summary of the modelling of steady flows, again in the order of increasing slowness, viz. rapidly varying steady flow through control structures, gradually varying steady flows (backwater curves) and finally uniform flow, which in essence is a summary of expressions for boundary resistance. Chapter 10 presents an introduction to the modelling of transport processes. Principles of numerical modelling of unsteady flows in open channels are dealt with in Chapter 11.

Lastly, Appendix A covers pressurized flow in pipes. A summary of equations is presented in Appendix B.

Prior Knowledge

The treatment of the various subjects relies on prior knowledge of basic fluid mechanics and an understanding of ordinary and partial differential equations and complex algebra.

Awareness of civil engineering will help in the understanding of the book, but this is not considered essential. The book includes some worked examples using simple computer code (written in Python) for which basic programming skills will be useful. These may also be learned, however, by following along with the examples.

Course Plan

Teaching all of the material covered in the book will take some 24–36 lectures of $1\frac{1}{2}$ teaching hours each, depending on the depth at which the various topics are treated and the expected self-study effort. When scheduled during a full semester, the course will typically take two lectures per week, while an additional 3–4 hours will be required weekly for self-study. Including exam preparation (2 days), the total work load for the student will amount to about 170 hours.

Chapters 1–8 are considered the backbone of the course. Related topics are dealt with in Chapters 9–11. The latter are often part of other, more dedicated, courses, in which case they can be omitted from the course plan. This will reduce the total working load to about 120 hours. The reduced course would also fit nicely into a half-semester schedule consisting of three lectures a week and 5–6 hours of self-study.

For self-study rehearsal every chapter concludes with a series of Problems for which a solutions manual is available separately. General solution strategies are explained in the worked examples preceding each Problems section. A series of digital exercises is available for the reader to become familiar with the concepts of the corresponding topic and to practise, progressively, the required skills. They also include so-called diagnostic tests, to assess one's preparedness for the exam.

Literature

The mathematical models for the considered long-wave categories presented in this book are classical (with a few exceptions for an extension of the theory). Thus, no references are made to individual contributors. Instead, we refer to the textbooks on the subject listed below, to which we are much indebted. Only in cases of very specific results, and in cases where results bear the name of the originator, have individual references been given.

Chaudry, M. H. 1993. *Open-Channel Flow.* Prentice Hall.

French, R. H. 1985. *Open-Channel Hydraulics.* McGraw-Hill.

Henderson, F. M. 1966. *Open Channel Flow.* Macmillan.

Sturm, T. W. 2010. *Open Channel Hydraulics.* McGraw-Hill.

Ven Te Chow 1959. *Open-Channel Hydraulics.* McGraw-Hill.

Supplementary Materials

The book comes with a collection of supplementary materials, which is available online. Included are a solutions manual for the Problems at the end of each chapter, a series of digital tests (for use in a MapleTA environment), and some Python scripts for carrying out the computations and Problems of Chapter 11.

Acknowledgements

The authors thank the reviewers of the proposal for this book, as well as the editors at Cambridge University Press, for their helpful suggestions. Professor Ad Reniers and Dr. Marcel Zijlema are gratefully acknowledged for critically reading parts of the manuscript and for their constructive criticisms.

Jurjen Battjes
Robert Jan Labeur

1 Basic Equations for Long Waves

This chapter presents the derivation of the basic equations that we will use in analyses and calculations of unsteady free-surface flows in natural or man-made channels, e.g. tidal or fluvial channels, shipping canals and irrigation canals. We deal with a mass balance and a momentum balance integrated across the entire flow cross section, assuming a hydrostatic pressure distribution.

1.1 Approach

The principal subject of this book is the class of unsteady free-surface flows of water with a characteristic length scale that is far greater than the depth, the so-called long waves. Tides, storm surges and flood waves in rivers provide good examples of this category (contrary to ship waves or wind-generated waves, whose lengths are usually not large or even small compared with the depth).

We restrict ourselves to flows in relatively narrow, weakly curved conduits such as tidal channels and rivers, in which the main flow direction is determined by the geometry of the boundary, which is assumed to be given beforehand (excluding morphological changes such as meandering of rivers). In these cases, the bulk flow direction is known so that only the flow intensity (the discharge, say) is to be determined, in addition to the water surface elevation. A typical area of application, a long, slowly winding river reach with lateral side basins, is shown in Figure 1.1.

As expressed by their name, long waves are characterized by length dimensions that far exceed the depths. This implies that the curvature of the streamlines in the vertical plane is negligible, for which reason we will assume a *hydrostatic pressure distribution* in the vertical. This greatly simplifies the schematization and the calculations.

In bends, the flow is forced to change direction through a lateral variation of the water level, being higher at the outer bank and lower at the inner bank. This is essential in detailed computations of the spiral flow in bends, but it is irrelevant for the large-scale computations of longitudinal variations with which we are concerned. So we will ignore lateral variations in surface elevation. The height of this level above the adopted reference plane $z = 0$ is designated as h. This quantity is a function of the downstream coordinate s (measured along the axis of the conduit) and the time t, or $h = h(s, t)$.

Because the water level is assumed not to vary within the cross section, the same applies to the downstream pressure gradient driving the flow. Therefore, instead of working with the point values of the velocities within each cross section, it is feasible to work with

Fig. 1.1 Reach of the river Rhine in The Netherlands (Pannerdens Kanaal) with groyne fields and laterally connected side basins; from, https://beeldbank.rws.nl, Rijkswaterstaat / Bart van Eyck

cross-sectionally integrated flow velocities (i.e. the total volume flux or flow rate, hereafter referred to as the discharge Q, also for purely oscillatory flows).

Summarizing, we have two dependent variables (h and Q) that have to be determined as functions of the longitudinal coordinate and time:

$$h = h(s,t) \quad \text{and} \quad Q = Q(s,t)$$

This requires a so-called *one-dimensional flow model*, typified by the dependence on only one space coordinate.

This chapter presents the derivation of the basic equations that we will use in the analyses and calculations in the following chapters. We deal with a mass balance and a momentum balance integrated across the entire flow cross section.

1.2 Schematization of the Cross Section

As its name implies, a one-dimensional mathematical model for flow in open channels contains only one space coordinate, the streamwise coordinate s. As a consequence, the geometric description of the channel cross sections is possible only in terms of bulk parameters applicable to the entire cross section. The variation of the bed elevation and the bed roughness within the cross sections cannot be resolved because that would require a lateral coordinate.

The characterization of the cross section requires a distinction between the *transport* or *conveyance* of water, on the one hand, and its *storage*, on the other. There are situations

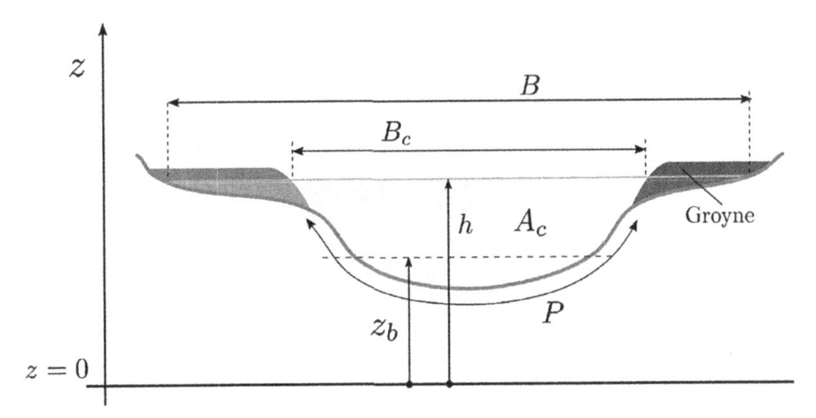

Fig. 1.2 Open conduit: cross section with groynes

where only a part of the wetted cross section contributes significantly to the conveyance. A typical example is provided by a river with a sequence of groynes normal to the flow, where the spaces between the groynes do contribute to the storage capacity but – in the case of low or moderate water levels – not to the conveyance capacity. In those cases it is necessary to distinguish between these two functions.

We designate the area, the width of the free surface and the mean depth of the conveyance cross section as A_c, B_c and d, respectively, where $d = A_c/B_c$. The wetted perimeter is P. See Figure 1.2. Storage takes place through a rise of the free surface, also between the groynes, so that in this respect the total width of the free surface (B) is the relevant parameter. It will be clearly indicated where we use the distinction between the total cross section and that of the conveyance part.

The bed elevation above the horizontal reference plane $z = 0$, averaged over the conveyance cross section, is denoted as z_b. The elevation of the free surface above the reference plane is designated as h. As stated above, and justified in Section 1.4.1, it is assumed to be laterally uniform at all times.

We use a length coordinate s along the streamwise axis that may be weakly curved and gently sloping. The longitudinal slope of the bed ($\tan\beta$), if nonzero, is assumed to be very small, allowing the approximations $\tan\beta \approx \beta$, $\sin\beta \approx \beta$ and $\cos\beta \approx 1$.

1.3 Mass Balance

Consider the mass in a control volume consisting of a slice of a water course with length Δs, containing the entire wet area of the cross section, from bed to free surface. See Figure 1.3.

Because of the free surface, pressure variations in environmental water systems are very limited. Therefore, we can neglect pressure-induced density variations. The water can then be considered as *incompressible*. In that case, the mass balance reduces to a *volume*

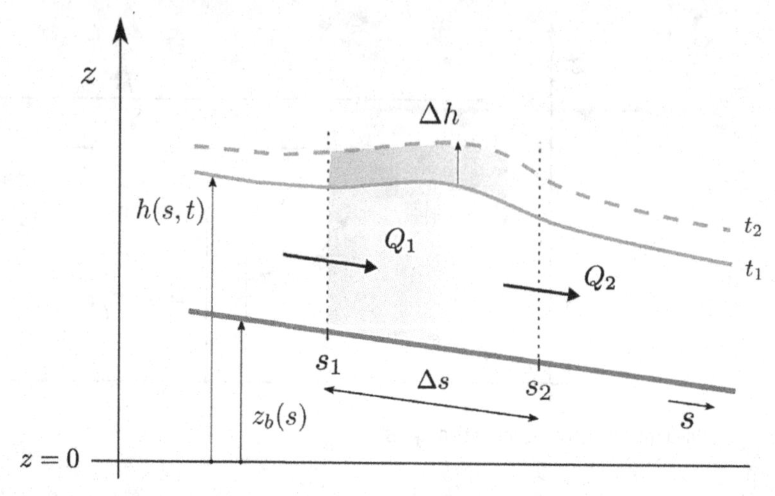

Fig. 1.3 Open conduit: longitudinal transect

balance, also called the *continuity equation*. To derive it, we consider the change in volume of the water in the control volume in a short time interval from $t = t_1$ to $t = t_2 = t_1 + \Delta t$.

The *volume flux* or *discharge* Q in a channel is defined as the volume of water passing a given cross section in a unit of time:

$$Q = \int\int u_s \, \mathrm{d}A = UA_c \tag{1.1}$$

in which u_s is the streamwise velocity at a point and U its value averaged over the conveyance cross section. The net influx of volume of water into the control volume, in the considered short time interval with duration Δt, is

$$(Q_1 - Q_2)\ \Delta t = -\Delta Q\, \Delta t \tag{1.2}$$

Suppose this is positive, i.e. there is more inflow than outflow. The difference is stored in the control volume, giving rise to an increase of the stored volume equal to $\Delta V = \Delta A \Delta s$ (see Figure 1.3).

Equating this storage to the net inflow yields $\Delta A\, \Delta s = -\Delta Q\, \Delta t$. Dividing by Δt and Δs, and taking the limit for $\Delta t \to 0$ and $\Delta s \to 0$, yields

$$\frac{\partial A}{\partial t} + \frac{\partial Q}{\partial s} = 0 \tag{1.3}$$

The storage is effected through a rise of the free surface by an amount Δh. Using the total width B of the free surface (not only that of the conveyance area) gives $\Delta A = B\Delta h$ (Figure 1.3), with which Eq. (1.3) can be written as

$$B\frac{\partial h}{\partial t} + \frac{\partial Q}{\partial s} = 0 \tag{1.4}$$

For given geometry of the cross section, which may vary with the downstream location s, the free-surface width B varies with time in a known manner through the time variation of h: $B = B(s, h(s,t))$. Therefore, Eq. (1.4), expressing mass conservation for the water (considered incompressible), is our first equation linking variations of the two unknowns Q and h. The second one, to be derived below, expresses momentum conservation.

The continuity equation (1.4) has been derived on the basis of a storage capacity that is continuously distributed along the length of the water course. However, it is not uncommon that there are lateral basins in communication with the main channel, such as harbours, remnants of previous meanders of a river, or dredged sand pits. These provide a discrete storage capacity that has to be taken into account in the overall mass balance. Some such basins can be seen in Figure 1.1.

Assuming that these basins are small compared with the length over which the exterior water level varies, the free surface inside can be assumed to be horizontal at all times, so that the surface elevation is a function of time only, written as $h_b(t)$. Moreover, if the entrance is sufficiently short and wide to allow an unobstructed in- and outflow, the water level in the basin will equal the exterior water level. Chapter 6 elaborates on these approximations in detail in the context of tidal basins.

The storage that can take place in a side basin can be incorporated in the continuity equation (1.4) by a local enlargement of the storage width B over a short longitudinal interval of the channel with length Δs, containing the connection with the basin, as sketched in Figure 1.4. In order to obtain a correct representation of the total storage through the continuity equation on this interval, B must locally be increased with $A_b/\Delta s$.

1.4 Equations of Motion

The formulation of Newton's second law for a flowing mass of water leads to a so-called *equation of motion*, which expresses a *balance between inertia, forcing and resistance*, where each of these in turn can consist of a number of contributions. It is important to be aware of this and to check the meaning of the various terms when writing or reading

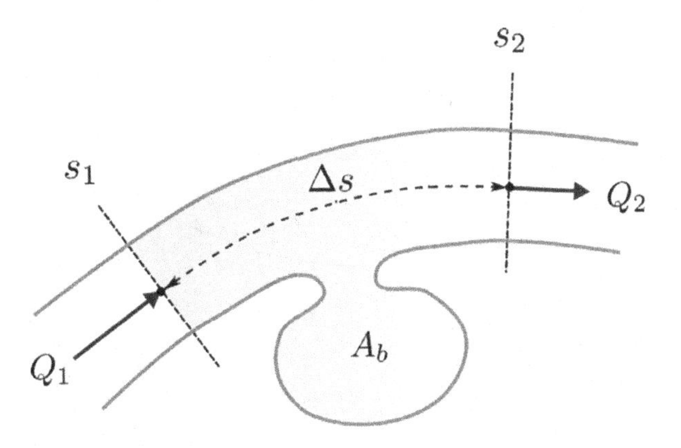

Fig. 1.4 Storage basin connected to a river reach

an equation of motion. It is also important for a good understanding of the phenomena involved to check whether one or more terms is negligible compared with another (in an equation consisting of three or more terms). We discuss this extensively in Chapter 2.

We will initially ignore flow resistance and start from the Euler equations for the acceleration of a fluid particle of an ideal (inviscid) fluid of constant density (ρ) under the action of gravity with gravitational acceleration g. The derivation of the Euler equations can be found in elementary textbooks on fluid mechanics and is not repeated here.

1.4.1 Euler Equations

We present the Euler equations in so-called natural coordinates, defined as follows. The streamwise coordinate is s; the normal coordinate is n, which lies in the local plane of curvature of the flow, pointing to the center of curvature; and the bi-normal coordinate is b, which is perpendicular to the plane of the local curvature (the so-called osculation plane); see Figure 1.5. For the nearly horizontal flows considered here, as in rivers and tidal channels, the bi-normal is nearly vertical; in fact, we will use the vertical coordinate z as the bi-normal coordinate.

Designating the respective particle velocity components as u_s, u_n and u_z, and the radius of curvature as r, and treating g and ρ as constants, the Euler equations can be written as

$$\frac{\mathrm{D}u_s}{\mathrm{D}t} = \frac{\partial u_s}{\partial t} + u_s \frac{\partial u_s}{\partial s} = -g \frac{\partial (z + p/\rho g)}{\partial s} \tag{1.5}$$

$$\frac{\mathrm{D}u_n}{\mathrm{D}t} = \frac{\partial u_n}{\partial t} + \frac{u_s^2}{r} = -g \frac{\partial (z + p/\rho g)}{\partial n} \tag{1.6}$$

$$\frac{\mathrm{D}u_z}{\mathrm{D}t} = \frac{\partial u_z}{\partial t} = -g \frac{\partial (z + p/\rho g)}{\partial z} \tag{1.7}$$

The total derivatives $\mathrm{D}u_s/\mathrm{D}t$ etc. signify the acceleration of a fluid particle following its motion, and p is the fluid pressure. The right-hand sides represent the forcing per unit mass due to gravity and pressure gradients, often expressed as the gradient of the so-called

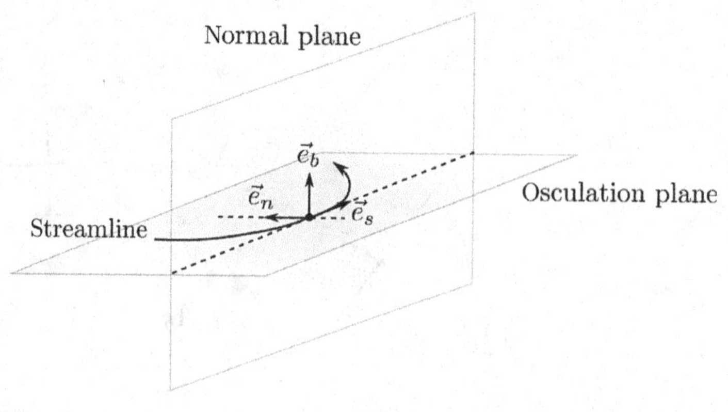

Fig. 1.5 Natural coordinate system

piezometric head h_p, defined as $h_p \equiv z + p/\rho g$. The Euler equations in three dimensions form the basis for the computation of rapidly varying flows.

The assumption of long waves implies that the vertical accelerations are neglected. It then follows from Eq. (1.7) that in this approximation the piezometric head is considered to be uniform in the vertical, or that *the vertical pressure distribution is hydrostatic*. The validity of this approximation is investigated quantitatively in Section 2.2.

The value of h_p at the free surface, and therefore at all points in the vertical, is $(h + p_{atm}/\rho g)$, in which h is the height of the free surface above the reference level $z = 0$, and p_{atm} is the atmospheric pressure at the air–water interface. Using this, and neglecting horizontal variations of the atmospheric pressure for the time being, the Euler equations in the longitudinal and lateral directions can be written as

$$\frac{\partial u_s}{\partial t} + u_s \frac{\partial u_s}{\partial s} = -g \frac{\partial h}{\partial s} \tag{1.8}$$

$$\frac{\partial u_n}{\partial t} + \frac{u_s^2}{r} = -g \frac{\partial h}{\partial n} \tag{1.9}$$

Note that the combined forcing by gravity and pressure gradients is now expressed in terms of the slope of the free surface, and is constant over the vertical. This is illustrated in Figure 1.6, showing a slice of water and the pressure forces acting on both of its sides. At a given elevation, the slope of the water surface gives rise to different pressures at both sides of the slice, but in the case of hydrostatic pressure, the difference δp is constant over the vertical.

Regarding Eq. (1.9), for the lateral motion, it is relevant to note that throughout this book we consider flows in relatively narrow conduits, so that the flow direction at each point is constant, from which it follows that $\partial u_n/\partial t = 0$. Bends force the flow to change direction. The associated centripetal acceleration u_s^2/r is forced by a lateral slope of the free surface, the latter being higher at the outside of the bend and lower at the inside. These lateral variations in the height of the free surface are crucial in detailed considerations of the spiral flow in bends, but their net effect on the large-scale streamwise motion, with which we are

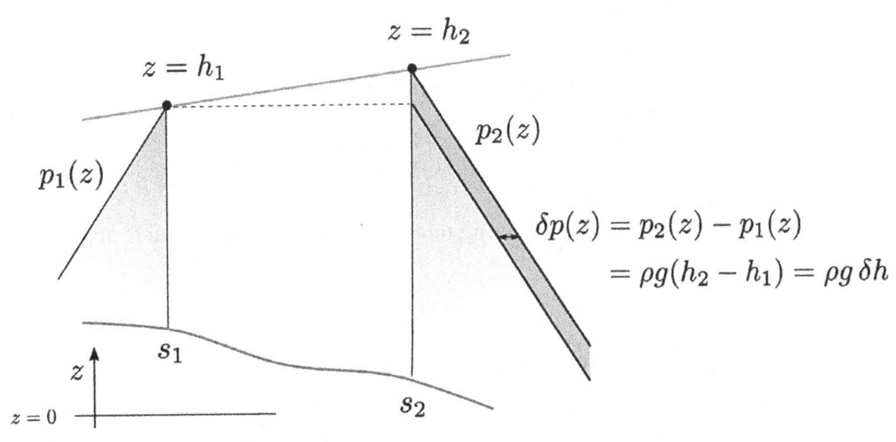

Fig. 1.6 Hydrostatic pressure and net horizontal forcing

concerned, is ignored. This implies that the forcing is considered uniform in the entire cross section and that only streamwise variations in surface elevation are taken into account. Therefore, in the remainder we ignore Eq. (1.9) and continue with Eq. (1.8), considering h to vary with t and s only.

So far we have considered the particle acceleration at a point of the cross section. In order to arrive at a one-dimensional model, we need cross-sectionally integrated or averaged values, as we did for the mass balance. The forcing is uniform in the cross section and, still neglecting resistance and effects of flow curvature in bends, so are the local particle accelerations and velocities. Therefore, in this approximation, Eq. (1.8) also applies to the cross-sectionally averaged flow velocity U, simply by replacing u_s with U:

$$\frac{\mathrm{D}U}{\mathrm{D}t} = \frac{\partial U}{\partial t} + U\frac{\partial U}{\partial s} = -g\frac{\partial h}{\partial s} \tag{1.10}$$

1.1.2 Flow Resistance

Our next task is to introduce flow resistance, which was left out of consideration in the above. In the context of a one-dimensional model, we consider boundary resistance to the bulk of the flow through the (conveyance) cross section, rather than internal flow resistance.

Let ΔW be the boundary resistance experienced by the water in the slice considered in Figure 1.3, with length Δs. Its value per unit area, averaged over the wetted boundary of the conveyance cross section, has the nature of a shear stress, written as τ_b (although, as we will see, processes other than local shear stress can contribute to the resistance): $\tau_b = \Delta W/(P\Delta s)$, in which P is the length of the wetted perimeter of the conveyance cross section.

The resistance per unit mass, $\Delta W/\,(\rho A_c \Delta s)$, can be written as $\tau_b/\rho R$, in which $R = A_c/P$, the so-called *hydraulic radius* of the conveyance cross section. For shallow, wide cross sections, the hydraulic radius is approximately equal to the laterally averaged depth of flow: $R \approx d$.

Adding the resistance force per unit mass (i.e. $-\tau_b/\rho R$) to the right-hand side of Eq. (1.10) yields the following *balance between inertia, forcing and resistance*:

$$\frac{\mathrm{D}U}{\mathrm{D}t} + g\frac{\partial h}{\partial s} + \frac{\tau_b}{\rho R} = 0 \tag{1.11}$$

For future reference, we write this equation in an alternative form as

$$\frac{\mathrm{D}U}{\mathrm{D}t} = g(i_s - i_f) \tag{1.12}$$

in which i_s is the free-surface slope, defined as

$$i_s \equiv -\frac{\partial h}{\partial s} \tag{1.13}$$

and i_f is the so-called friction slope, defined as

$$i_f \equiv \frac{\tau_b}{\rho g R} \tag{1.14}$$

Eq. (1.12) captures the flow dynamics in a nutshell, showing at a glance that the fluid acceleration results from an imbalance between the driving force and the resistance. For steady, uniform flow, these are in balance.

Reverting to Eq. (1.11), with the expanded form of the acceleration, we obtain

$$\frac{\partial U}{\partial t} + U\frac{\partial U}{\partial s} + g\frac{\partial h}{\partial s} + \frac{\tau_b}{\rho R} = 0 \tag{1.15}$$

This equation seems to have been first derived by De Saint-Venant (1871), although his name is usually associated with a similar equation in terms of the discharge Q; see Eq. (1.19) below.

In order to complete the formulation, we need a constitutive relationship between the resistance and the flow velocity. It is usually assumed that the resistance varies with the average velocity U as in uniform, steady turbulent flows. The flows of interest have high Reynolds numbers, typically of the order of 10^6, so that viscous effects can be ignored in the modelling, and the resistance varies in proportion to the square of the flow velocity, as in

$$\tau_b = c_f \rho |U| U \tag{1.16}$$

in which c_f is a dimensionless resistance coefficient. Methods of its estimation are dealt with in Section 9.3, on uniform flow. Typical values are in the range of 0.002–0.006. We will often use 0.004 in numerical examples.

The friction slope corresponding to Eq. (1.16) is

$$i_f = c_f \frac{|U|U}{gR} \tag{1.17}$$

Substituting Eq. (1.16) in to Eq. (1.15) yields

$$\frac{\partial U}{\partial t} + U\frac{\partial U}{\partial s} + g\frac{\partial h}{\partial s} + c_f\frac{|U|U}{R} = 0 \tag{1.18}$$

which completes the derivation of the *acceleration equation* for the averaged flow velocity U including boundary resistance.

In the derivation of Eq. (1.18) it has been assumed that the resistance is evenly distributed over the wetted perimeter of the cross section. This is a good approximation if the variations of depth, bed roughness and flow velocities within a cross section are moderate, but less so if they vary significantly.

In periods of high water in rivers, for example, the flood plains are covered and contribute to storage as well as conveyance, but the depth of flow over the flood plains is much less than it is in the main channel, and the resistance is usually much greater (due to vegetation, buildings etc.). Similar situations occur in estuaries with their main channels and shoals and in tidal channels bordered by tidal flats. These lateral variations in a cross section can be accounted for in the schematization by dividing the cross section into two or more subsections, each of them with its own characteristic width, depth and roughness. (An application of this for tidal propagation is presented in Section 7.4.2.)

1.4.3 Momentum Balance

Since the discharge Q is one of our two primary variables (h being the other one), we will now derive a momentum balance in terms of Q. For algebraic simplicity we will temporarily make no distinction between the total cross section and the conveyance part.

We could derive the momentum balance from Eq. (1.18) by substituting $U = Q/A_c$, but that requires several intermediate steps. It is simpler to start afresh by considering cross-sectionally integrated quantities from the outset, which at the same time allows us to express the effect of variations of the local particle velocity in a cross section. To this end, we consider (again) a control volume covering the entire wet cross section, as in Figure 1.3.

The streamwise momentum per unit length is the cross-sectional integral of ρu_s, which equals ρQ. The streamwise advection of streamwise momentum through a cross section is the cross-sectional integral of $\rho(u_s)^2$, which is written as $\alpha\rho U^2 A_c$, or $\alpha\rho Q^2/A_c$, in which α is a coefficient expressing the effect of the variations of the particle velocity within the cross section, defined as the cross-sectional average of $(u_s/U)^2$. Its value is always more than unity, but the deviations are usually rather small, and α is mostly ignored in practice. The driving force and the boundary resistance, both per unit mass, are given by the last two terms of Eq. (1.18).

Collecting the preceding contributions, we finally obtain the balance of streamwise momentum as follows (dropping the constant mass density ρ):

$$\frac{\partial Q}{\partial t} + \frac{\partial}{\partial s}\left(\frac{\alpha Q^2}{A_c}\right) + gA_c\frac{\partial h}{\partial s} + c_f\frac{|Q|Q}{A_c R} = 0 \tag{1.19}$$

In some applications, it is necessary to account for effects of the atmosphere on the flow. This requires additional terms in the momentum balance, shown in Box 1.1.

In the derivation of Eq. (1.19), it was assumed that the entire cross section contributes to the conveyance, such that $A = A_c$ and $B = B_c$. If this is not the case, a lateral exchange flow occurs between the main channel and the flood plains (in the case of river flow), requiring an additional term in Eq. (1.19), which, if inserted in the left-hand side, is given by $\rho U(B - B_c)(\partial h/\partial t)$; see Box 1.2.

Box 1.1 Inclusion of atmospheric forcings

In some applications, effects of a variable atmospheric pressure and of wind-induced shear at the free surface must be taken into account. These influences can be included in Eq. (1.19) by adding $p_{atm}/\rho g$ to h and by adding the streamwise component of the wind-induced shear force per unit length at the free surface (to be divided by ρ) given by $\tau_s B_c \cos\psi$, in which B_c is the width of the conveyance cross section, ψ is the angle enclosed between the wind direction and the direction of the flow, and τ_s is the wind-induced shear stress at the free surface: $\tau_s = c_D\,\rho_a\,W_{10}^2$ in which c_D is the wind drag coefficient (of order 0.001–0.002), ρ_a is the air density, and W_{10} is the 10-min averaged windspeed at a height of 10 m above the mean water level.

Box 1.2 Lateral exchange of streamwise momentum

If the storage width B is different from the conveyance width B_c, an additional term must be included in Eq. (1.19). To see this, consider a river with a main channel bordered by shallow flood plains, which contribute a negligibly small amount to the conveyance. The total width of the flood plain (summed over both river banks) is $B - B_c$. When the water rises, at the rate $\partial h/\partial t$, water is stored on the flood plain at the rate $(\partial h/\partial t)(B - B_c)\Delta s$, which is the result of a lateral volume flow from the main channel to the flood plain. This lateral flow carries streamwise momentum. Let us say that this amounts to ρU per unit volume. This implies a net lateral outflow of streamwise momentum from the conveyance cross section at a rate $\rho U(B - B_c)(\partial h/\partial t)$ per unit length. This comes in addition to the net outflow of streamwise momentum that results from the streamwise motion, given as $\partial\left(\alpha\rho Q^2/A_c\right)/\partial s$ in Eq. (1.19), and so explains the nature of the additional term noted above.

In practice, this additional term is often ignored because the streamwise velocity at the transition between the main channel and the shallow flood plain is usually (much) less than U, the flow velocity averaged over the relatively deep main channel. This is supposed to hold in particular in the case of flow from the flood plains back into the main channel ($\partial h/\partial t < 0$), it being assumed that this outflowing water carries no streamwise momentum. In some numerical models, the additional term is taken into account only for rising water and ignored when the water falls.

It depends on the circumstances whether all contributions to the momentum balance are important or whether one or two are negligible. In the latter case, the equation is reduced to a simpler form, allowing approximate solutions for a rough calculation. In general, one can say that resistance is less and less important as the motions vary more and more rapidly. On the other hand, inertia becomes negligible when the variations are very slow. These notions are elaborated in Chapter 2, and they form the basis for the subdivision of our total subject matter in parts that are treated in separate chapters in the remainder of this book.

1.5 Summary of the Long-Wave Equations

The continuity equation (Eq. (1.4)) and the momentum balance (Eq. (1.19)) together form the basis for analyses and computations of one-dimensional long-wave phenomena. They are known as the *equations of De Saint-Venant* or as the (one-dimensional) *shallow-water equations* (because the depth has been assumed to be very small compared with typical length dimensions). They are repeated here for convenience (dropping α from the momentum balance):

$$B\frac{\partial h}{\partial t} + \frac{\partial Q}{\partial s} = 0 \tag{1.20}$$

$$\frac{\partial Q}{\partial t} + \frac{\partial}{\partial s}\left(\frac{Q^2}{A_c}\right) + gA_c\frac{\partial h}{\partial s} + c_f\frac{|Q|Q}{A_c R} = 0 \tag{1.21}$$

These equations form a coupled set of hyperbolic partial differential equations for the two unknowns h (water level) and Q (discharge) as functions of location (s) and time (t). The geometric parameters A_c, R and B as well as the resistance coefficient c_f are supposed to be known functions of location (s), water level (h) and discharge (Q), so that mathematically speaking they are known, variable coefficients. The set of partial differential equations can be integrated if proper initial conditions and boundary conditions are provided.

Problems

1.1 What is the discharge through a conduit? What is its dimension?

1.2 Where is storage in free-surface flows taking place (mainly)?

1.3 Derive the volume balance equation for the flow in a river stretch provided with a sequence of groynes; cast it in differential form.

1.4 Derive the volume balance equation for the flow in a river stretch that is laterally connected to a basin (i.e. the situation of Figure 1.4).

1.5 Derive the volume balance equation (in differential form) for the flow in a river for which seepage of water into the subsoil (at a rate of q volume units per unit area of bottom and per unit time) has to be taken into account.

1.6 Derive an expression for the downstream force per unit volume in a flow with a sloping free surface.

1.7 Which contributions to the force in Problem 1.6 can be distinguished?

1.8 Check the dimensions of the individual terms in Eq. (1.19).

2 Classification and Analysis of Long Waves

The category of long waves encompasses different wave types, each with a different origin and with different dynamics, in the sense that the relative importance of the various physical processes, as expressed by the different terms in the equation of motion, can vary. In this chapter we will first give an overview of the various types of long wave and their generation, and we will verify formally that they can indeed be classified as 'long'. Next, we will estimate the relative magnitudes of terms in the equation of motion, which provides an ordering of wave types based on the importance of inertia relative to the resistance.

2.1 Types of Long Waves

In general, one can say that the faster the flow varies, the more important will be the inertia relative to the resistance, and the more it will be in balance with the net driving force. Before dealing with these dynamics we give short descriptions of the origin and typical characteristics of the different types of long waves:

- translatory waves (transient variations in discharge and water level, usually caused by operation of controls)
- tsunamis (sea waves generated by subsea earthquakes, volcanic eruptions etc.)
- seiches (standing oscillations in lakes, bays, harbours etc.)
- tides in oceans, shelf seas, estuaries and lowland rivers
- flood waves in rivers

2.1.1 Translatory Waves

As a result of manipulation (or breakdown!) of pumps or valves in the operation of locks, weirs, evacuation sluices, hydropower plants, etc., variations in discharge (δQ) can occur. These are accompanied by variations in water surface elevation (δh). Such disturbances travel as so-called *translatory waves* into the adjacent reaches of the conduit. The passage of such wave induces a rise in elevation in the case of an increase in discharge, and a lowering in the case of a decrease in discharge; see Figure 2.1. The particle velocities in a translatory wave are in one direction only (either forward or backward), which explains the name 'translatory waves', as opposed to 'oscillatory waves', in which the particles move back and forth.

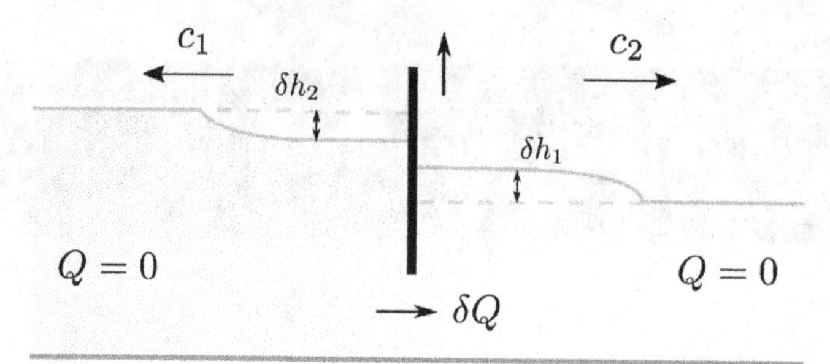

Fig. 2.1 Translatory waves after partial opening of a gate

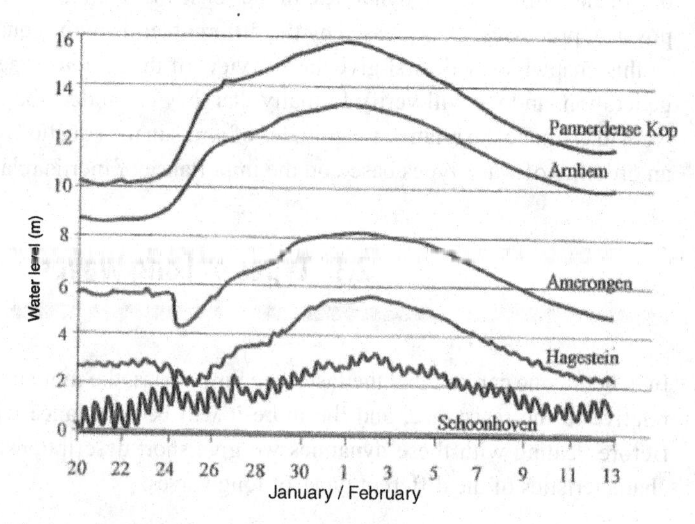

Fig. 2.2 Water level records along the river Rhine in The Netherlands during high water of January/February 1995; from the Department of Public Works, The Netherlands

Figure 2.2 shows water level records at five stations along the river Rhine during the high-water period of January/February 1995. The record near the city of Amerongen shows an abrupt lowering of the water level on January 24, which was the result of a negative translatory wave on the river Rhine that resulted from the raising of a movable weir located a small distance downstream, in anticipation of the approaching flood waters.

Translatory waves in navigation locks and navigation canals can cause hindrance to shipping as well as large forces in the mooring lines of moored ships, in some cases causing breaking. Such effects can be reduced through a more gradual operation of pumps or valves. We return to this matter in Chapter 5.

2.1.2 Tsunamis

Tsunamis are impulsively generated waves in oceans, shelf seas or (large) lakes, most commonly due to subsea earthquakes. Subsea volcanic eruptions or landslides are other possible causes of tsunamis.

One can distinguish four stages in the life of a tsunami: the generation, the propagation in relatively deep water from the source region to a coastal region, the enhancement and deformation in shoaling water, possibly up to breaking, and finally the run-up onto land.

Earthquakes causing tsunamis often occur in the subduction zones along the rims of tectonic plates. The accompanying sea bed motion has a plus–minus signature, generating a positive wave (elevation) travelling seaward and a negative wave (a depression) travelling landward. This is why the leading wave at the ocean side is usually one of elevation. This was e.g. the case for the Indian Ocean tsunami of December 26, 2004. Locations to the west of the source region off the Sumatra coast experienced an initial rise of the sea surface, as in the record shown in Figure 2.3 for a location near Male in the Maldives. In contrast, at the land side the leading wave is usually one of depression, resulting in an initial withdrawal of the seawater from the coast, exposing vast stretches of what is normally a subsea bottom. This should be a warning sign to humans who happen to be present at the site to seek high ground as fast as they can, but unfortunately it also triggers the curiosity of some, who venture seaward, unknowingly towards their almost certain death.

Depending on the distance involved, the propagation from the source region to a coastal area may last several hours, giving time to issue warnings to coastal populations. However, this is not the case when the source region is relatively close to land, as was the case for the Indian Ocean tsunami, and for the Tohoku (Japan) tsunami of March 11, 2011.

In deep water, the individual waves in a tsunami wave train are typically a hundred kilometres or more in length. This fact, combined with the relatively moderate wave heights offshore, often less than a meter, makes the steepness of tsunami waves in the deep ocean very low, causing them to pass unnoticed by ships in the deep ocean.

As the tsunami enters water of decreasing depth, the wavelengths shorten, the wave heights increase and the wave fronts steepen, possibly up to the point of breaking. This gives the tsunamis their often massive destructive power.

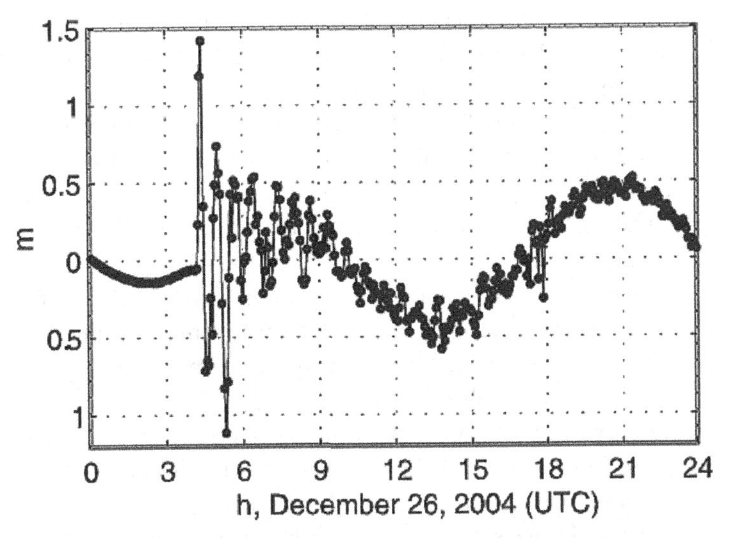

Fig. 2.3 Record of surface elevation measured offshore of Male, Maldives, on December 26, 2004; horizontal scale indicates time (h), vertical scale indicates surface elevation (m); from Merrifield et al. (2005)

In the fourth and final stage, the tsunami waves may overflow low-lying coastal areas, causing material damage and deaths, or they may run up against coastal mountain slopes, locally up to heights in excess of 35 m in the two major tsunamis mentioned above.

2.1.3 Seiches

Oscillations of water or other liquids in a drinking glass, a bathtub or other 'closed basins' are easily observable by everyone. They are called 'standing oscillations', as opposed to progressive, oscillatory waves. Similar standing oscillations occur in various natural water systems and at widely different scales, in closed basins as well as basins that are closed at one end and open at the other, comparable to an open organ pipe. Such oscillations have been first systematically observed in Lake Geneva, by Forel (1875), who called them 'seiches', a name whose origin is obscure but that has nevertheless been rooted in the scientific literature ever since, referring to standing oscillations in lakes, harbour basins and the like.

Seiches are *free* or *natural oscillations*; i.e. their periods are for given restoring force (gravity) determined by the geometry of the system, rather than by external forcing. A weak periodic forcing may be sufficient to generate a significant response, provided the excitation contains energy at the natural frequencies of the system. In such cases the system response is *resonant*. In this manner, the water in coastal harbour basins can oscillate with significant amplitude in response to low waves at sea, hardly discernable offshore, which in turn can be the result of oscillations in wind speed and atmospheric pressure during the passage of cold fronts (De Jong and Battjes, 2004).

Figure 2.4 gives an example for the port of IJmuiden, located at the North Sea coast in The Netherlands. The record for the offshore surface elevation in panel (a) shows minor disturbances, no more than about 0.1 m in height, superimposed on the astronomic tide. Figure 2.4b shows a similar record in the outer harbour with resonantly enhanced oscillations, with crest-to-trough heights up to 1.2 m and a period of about 35 min, which is the longest natural period of the semi-closed outer harbour basin.

As will be seen, friction is usually unimportant for harbour seiches because of their relatively short period and the large depths in which they occur. The largest contribution to the damping of harbour seiches is the seaward radiation of energy through the harbour mouth.

Seiches can have undesirable or even harmful effects. They can be a nuisance to shipping, and occasionally cause breaking of mooring lines of moored ships. They also affect the tidal window of passage of deep-draught ships through their influence on the instantaneous available depth.

In contrast to human-caused translatory waves, nothing can be changed in the cause of harbour seiches, i.e. the low, long waves incident from the sea (in the case of coastal harbours or bays) or the atmospheric forcing (in the case of lakes). Some alleviation of harmful effects can be achieved by reducing the resonance response factor, e.g. by optimizing the geometry of the harbour mouth, but navigational demands put a limit to

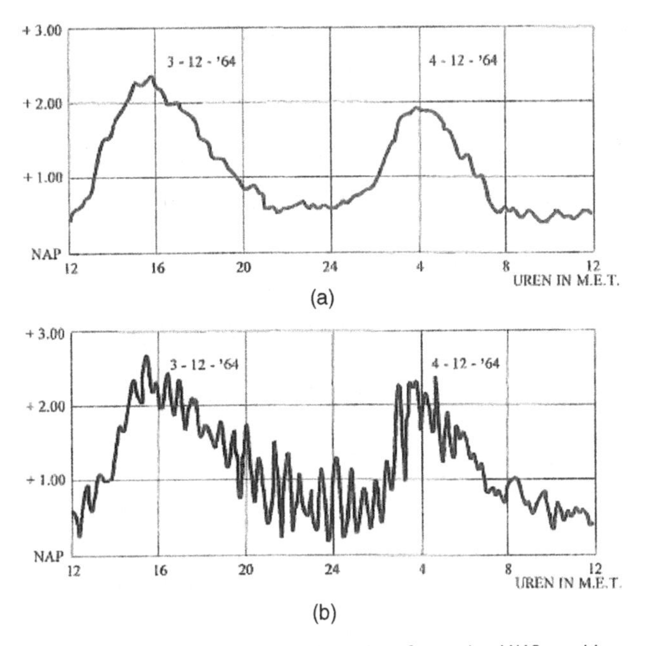

Fig. 2.4 Simultaneous records of the surface elevation (in metres above the reference level NAP, roughly equal to mean sea level); offshore of IJmuiden (a) and in the outer harbour of IJmuiden (b); from the Department of Public Works, The Netherlands

this. (It can even happen that narrowing the entrance results in higher seiches, because such narrowing may reduce the seaward radiation of seiche energy more than it hampers the excitation; this is the so-called harbour paradox (Miles and Munk, 1961).) Another option is adaptation of the geometry of the interior basins. This can shift the natural frequencies, but this is an improvement only in the case where the shift is away from the energetic frequencies in the excitation, which are usually not known *a priori*.

2.1.4 Tides

Tides are caused by the variations in time and space of the gravitational force exerted by the moon and the sun, in combination with the effects of the rotation of the earth. On a global scale, tides are oscillations in the ocean basins. On a smaller scale, the same is true for shelf seas such as the North Sea. From the oceans or shelf seas, tidal waves enter estuaries and bays where they may be damped or be resonantly enhanced, depending on the length and depth of the estuary or the bay. The tide in the Bay of Fundy (Nova Scotia, Canada) is an extreme example of strong enhancement, with a tidal range at the closed end up to some 16 m, world's highest.

In most locations, the semi-diurnal lunar tide (M_2) is dominant; its period equals the duration of half a moon-day, about 12 h 25 min. In some areas, e.g. South-east Asia or the Gulf of Mexico, the diurnal solar tide is dominant, due to resonance characteristics of adjacent ocean basins favouring these longer-period waves. See Figure 2.5.

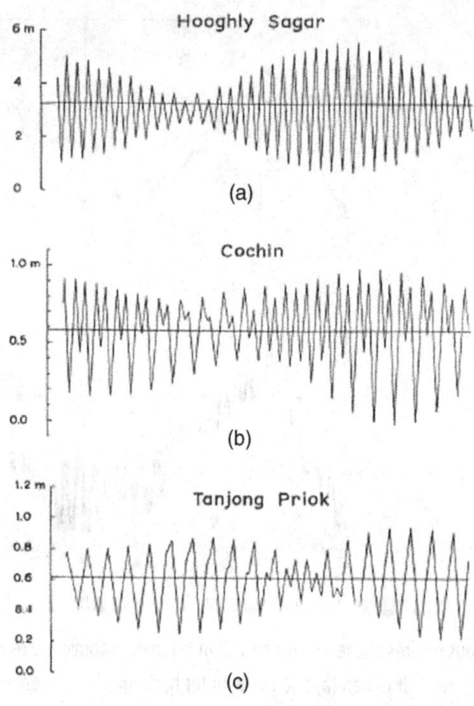

Fig. 2.5 Semi-diurnal tides (a), mixed tides (b) and diurnal tides (c) at three locations in South-east Asia; after Dronkers (1964)

Bed friction is relatively weak (compared with inertia) in ocean tides, because of the large depths, and can there be neglected in a first approximation, but in shelf seas and shallow tidal bays it is of equal significance as the inertia and cannot be neglected.

Tidal activity is visible in Figure 2.2, in the most downstream water level record in the lower reaches of the Rhine. Notice that the tide is significantly damped during high river stage, as a result of enhanced flow velocities and flow over the flood plains.

2.1.5 Flood Waves in Rivers

Enhanced precipitation and/or melting of snow in the catchment area of a river gives an enhanced discharge downstream, with some delay, and an associated rise in water level, a so-called *flood wave*. Figure 2.2 shows examples of the water surface elevation at several locations along a branch of the river Rhine in The Netherlands, separated by some 100 river km, during a flood wave of January/February 1995. The uppermost plot is for the most upstream location, and so on. All elevations are shown relative to the same reference level (NAP, roughly equal to mean sea level [MSL]).

Flood waves in lowland rivers can last for several days or even weeks (see the example of Figure 2.2). The corresponding variations in flow velocity are far slower than in translatory waves and even significantly slower than in tides. In a first approximation, inertia can be neglected relative to resistance.

2.2 A Condition for the Long-Wave Approximation

By definition, 'long waves' have a wavelength far greater than the depth in which they occur. As a consequence, the vertical particle accelerations are negligible, so that the pressure distribution can be considered to be hydrostatic. We shall derive a quantitative criterion for the validity of this basic assumption in the mathematical modelling of long waves.

It would not be sufficient to require that the vertical wave-induced acceleration be small relative to that of gravity (this requirement also applies to low-steepness short waves). An additional requirement is that the wave-induced pressure variation be almost uniform in the entire vertical.

Consider a situation where the local water level is raised by an amount ζ from an undisturbed situation. If the pressure were hydrostatic, the piezometric level (h_p) would be raised by the same amount at all points of the vertical (see Figure 2.6). Vertical accelerations ($a_z = \mathrm{D}u_z/\mathrm{D}t$, in which u_z is the vertical particle velocity) cause deviations from this hydrostatic pressure distribution.

To quantify this, we use Euler's equation for the vertical motion, with the forcing expressed as the gradient of the piezometric head:

$$a_z = -g\frac{\partial h_p}{\partial z} \tag{2.1}$$

It follows that the variation of h_p from the bottom to the free surface (Δh_p) can be expressed as

$$\Delta h_p = \int_0^d \frac{\partial h_p}{\partial z}\,\mathrm{d}z = -\frac{1}{g}\int_0^d a_z\,\mathrm{d}z \tag{2.2}$$

The approximation of hydrostatic pressure is valid if this difference is negligible compared to the wave-induced variation ζ, or $|\Delta h_p| \ll |\zeta|$.

To elaborate this further, we assume a sinusoidal motion of the vertical free-surface displacement with amplitude $\hat{\zeta}$ and (angular) frequency ω, as in $\zeta = \hat{\zeta}\cos\omega t$. Expressed in terms of amplitudes, the above inequality becomes

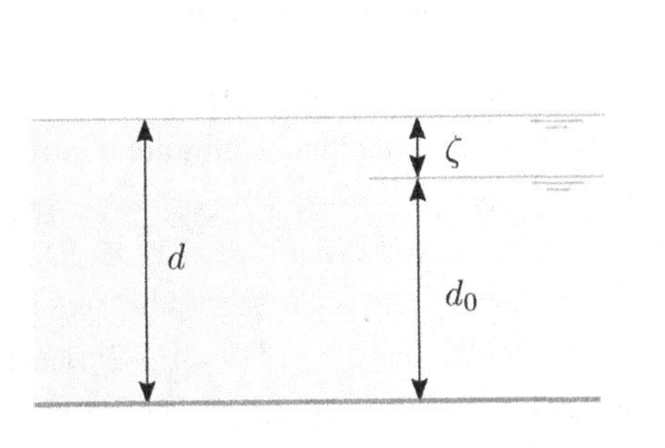

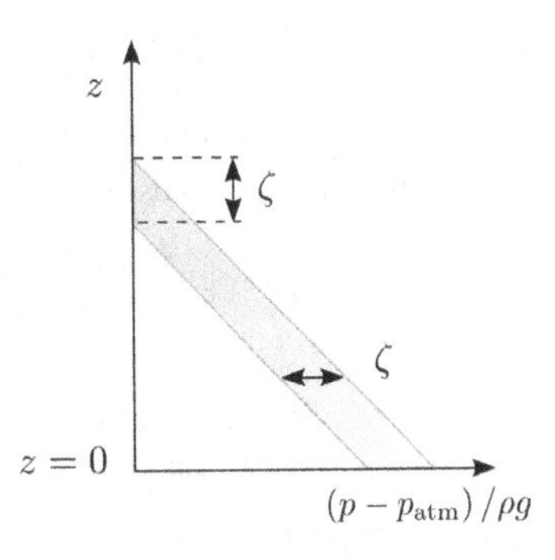

Fig. 2.6 Hydrostatic pressure distribution

$$\int_0^d \hat{a}_z \, dz \ll g\hat{\zeta} \tag{2.3}$$

The vertical motion dies out downward. Therefore, the left-hand side is smaller than $\hat{a}_{z,0}\, d$, in which $\hat{a}_{z,0}$ is the amplitude of the vertical acceleration at the surface, equal to $\omega^2\hat{\zeta}$. Thus, a sufficient condition for the validity of Eq. (2.3) is

$$\hat{a}_{z,0}\, d \ll g\,\hat{\zeta} \tag{2.4}$$

Because $\hat{a}_{z,0} = \omega^2\hat{\zeta}$, we finally obtain the following *condition for the validity of the approximation of hydrostatic pressure* for harmonic motion:

$$\frac{\omega^2 d}{g} \ll 1 \tag{2.5}$$

Box 2.1 shows that this condition can be expressed in terms of the wavelength/depth ratio.

Let us now investigate the implications of condition (2.5) for the most important categories of long waves.

For the **tides**, we consider the semi-diurnal tide M_2 in the deepest ocean trough on earth, with a depth of approximately 10^4 m. The period of the M_2 tides is 12 h 25 min, corresponding to an angular frequency $\omega = 1.405 \times 10^{-4}$ rad/s. For these conditions, $\omega^2 d/g \simeq 2 \times 10^{-5}$, so that even for these extreme depths the pressure in the tidal motion hardly deviates from being hydrostatic. Needless to say, this is even less the case in waters of more moderate depth. Therefore, the long-wave model is very well justified in tidal calculations. **Tsunami waves** have much shorter periods (higher frequencies) than the tides. Nevertheless, most of the energy contained in them corresponds to periods long enough to treat them as long waves. Taking a period of 10 min as an example, and a typical ocean depth of 4000 m, $\omega^2 d/g \simeq 0.04$, small enough to justify the long-wave approximation, even in the deep ocean. However, tsunamis may also contain some energy in higher frequencies, for which the long-wave approximation is not valid in ocean depths, requiring nonhydrostatic modelling.

Seiches have much shorter periods than tides, but these are important only in much shallower waters than the oceans. For a seiche with a typical period of about 10 min ($\omega \simeq 0.01$ rad/s) in water of about 20 m deep, $\omega^2 d/g \simeq 2 \times 10^{-4}$, so that also for these oscillations the long-wave approximation is very well justified.

Flood waves in lowland rivers can vary on the time scale of days, in any case more slowly than tides, at least in the lower river reaches, and the depths are relatively small, so that the

Box 2.1 The length of a 'long' wave

As we shall see below, long waves in water with a free surface and depth d have a speed of propagation c approximately equal to $\sqrt{gd}$. If we substitute this with $c = L/T = L\omega/2\pi$ into Eq. (2.5) we obtain the condition for hydrostatic pressure in the following form: $L \gg 2\pi\, d$. In other words: the wavelength should be much larger than the depth. This explains and justifies the name 'long waves'. A commonly used criterion is $L \gtrsim 20\, d$.

approximation of hydrostatic pressure is even better justified in these cases than it is for the tides.

Translatory waves usually result from manipulations of weirs, valves, pumps etc. These are assumed to be gradual to prevent hindrance or damage. In these cases the pressure can be assumed to be hydrostatic. However, in some cases, e.g. when they are the result of an accident, a sudden power outage, etc., they vary more rapidly, even as a shock wave. In such cases the long-wave assumption is locally invalid, but it can still be used in the reaches on either side. This is sufficient for the calculations if locally the shock conditions are used to connect the motions on either side of the shock wave.

Finally, it is obvious from Eq. (2.5) that this condition cannot be fulfilled by high-frequency (short-period) waves in waters of moderate or large depth. Examples are ship-generated waves and wind-generated waves, with periods of the order of seconds rather than minutes or hours. Take an example of wind-generated waves with a period of 6 s in water of 10 m depth. In that case, $\omega^2 d/g \simeq 1$, so that the pressure is not even approximately hydrostatic. Waves for which Eq. (2.5) is not (nearly) fulfilled are called *short waves*. These are the subject of a class of theory different from that for long waves and are not considered here.

2.3 Estimation of Terms

We are now going to investigate the relative importance of the different terms in the long-wave equations, in particular the equation of motion because that contains up to four terms that may bear ratios to each other of different orders of magnitude. This estimation does not have to be precise. What matters is to gain insight into the parameters determining which contributions (resistance, inertia, etc.) are important in any given situation and which ones are so unimportant that they can be neglected in a first approximation.

We start from the equation of motion in acceleration form as derived in the previous chapter:

$$\frac{\partial U}{\partial t} + U\frac{\partial U}{\partial s} + g\frac{\partial h}{\partial s} + c_f\frac{|U|U}{d} = 0 \qquad (2.6)$$

Here, we have replaced the hydraulic radius R by the cross-sectionally averaged flow depth d, which is justified for relatively flat cross sections.

2.3.1 Advective Acceleration Term

We first consider the *advective acceleration* ($U\partial U/\partial s$) in relation to the *local acceleration* ($\partial U/\partial t$), initially for a conduit whose cross section does not vary longitudinally (a prismatic conduit). Let $\mathcal{U}$ be a characteristic flow velocity and $\mathcal{L}$ a characteristic length scale, possibly a wavelength if it exists; see also Figure 2.7. In that case, $U\partial U/\partial s$ is of the order of magnitude $\mathcal{U}^2/\mathcal{L}$. Likewise, $\partial U/\partial t$ is of the order $\mathcal{U}/\mathcal{T}$, where $\mathcal{T}$ is a characteristic time scale of the motion, such as a wave period (if it exists). In that case, the ratio of the advective acceleration to the local acceleration is of the order $(\mathcal{U}^2/\mathcal{L})/(\mathcal{U}/\mathcal{T})$, or $\mathcal{U}\mathcal{T}/\mathcal{L}$.

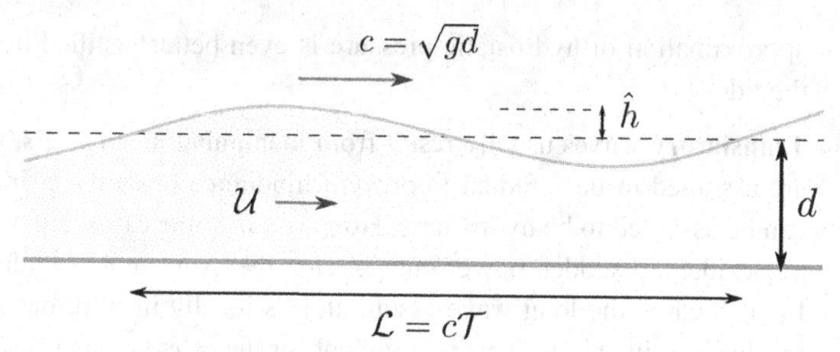

Fig. 2.7 Schematic of scaling parameters

The ratio $\mathcal{U}\mathcal{T}/\mathcal{L}$ can be interpreted geometrically as the ratio of the longitudinal particle displacement (of order $\mathcal{U}\mathcal{T}$) to the length scale of the flow ($\mathcal{L}$).

For waves in a prismatic conduit, the length scale $\mathcal{L}$ is determined by the flow and coupled to $\mathcal{T}$ through the relation $\mathcal{L} = c\mathcal{T}$, in which c is the wave propagation velocity, in which case $\mathcal{U}\mathcal{T}/\mathcal{L} = \mathcal{U}/c$. See Figure 2.7. As we will see further on, $c = \sqrt{gd}$ for long waves, in which case $\mathcal{U}/c$ equals $\mathcal{U}/\sqrt{gd}$, the Froude number Fr. (Similarly, in compressible flows for which c is the speed of sound, $\mathcal{U}/c$ is the Mach number.) Therefore, we can say that the lower the Froude number, the smaller is the relative magnitude of the advective acceleration. The preceding considerations apply in like manner to the momentum balance, Eq. (1.21), implying that the term arising from the advection of momentum, $\partial(Q^2/A_c)/\partial s$, is negligible if $Fr^2 \ll 1$.

The ratio $\mathcal{U}\mathcal{T}/\mathcal{L}$ also has an interesting link to the surface level amplitude ($\hat{h}$), which follows from scaling the continuity equation:

$$B\frac{\partial h}{\partial t} + \frac{\partial Q}{\partial s} = 0 \tag{2.7}$$

The terms $B\partial h/\partial t$ and $\partial Q/\partial s$ have orders of magnitude of $B\hat{h}/\mathcal{T}$ and $\mathcal{U}Bd/\mathcal{L}$, respectively, where B is the width of the channel. Since these terms equate it follows that $\hat{h}/\mathcal{T} = \mathcal{U}d/\mathcal{L}$ and, consequently, $\hat{h}/d = \mathcal{U}\mathcal{T}/\mathcal{L} = \mathcal{U}/c = Fr$.

For a given depth, therefore a given value of $\sqrt{gd}$, the Froude number (Fr) decreases with decreasing wave height. The ratio of the advective acceleration to the local acceleration decreases in like proportion. Therefore, *for relatively low waves ($\hat{h}/d \ll 1$) in a prismatic conduit, the advective acceleration is relatively small and can be neglected in a first approximation*. The same applies to the contribution of the advective momentum flux in the momentum balance.

In harbour oscillations and offshore tidal currents, the flow velocity is seldom more than 0.5 m/s. If the depth there exceeds 10 m, the Froude number is below 0.05, implying that the advective acceleration can be neglected in a good approximation. In tidal entrances, estuaries etc., flow velocities typically exceed 1 m/s, so that in these flows the advective acceleration is of some importance. The same holds for tides in shallow water, such as over tidal flats, where the flow velocities may be less than in the channels but the depths are smaller. In such flows, the advective acceleration is still not dominant but at the same time not negligible.

Box 2.2 **Varying geometries**

In nonprismatic conduits, local variations in the geometry can force small length scales, such that $\mathcal{U}^2/\mathcal{L} \gg \mathcal{U}/\mathcal{T}$, or $\mathcal{U}\mathcal{T}/\mathcal{L} \gg 1$. In such conditions, the balance tips to the other side: the advective acceleration is locally dominant. Flows through or over control structures are good examples of this. In such cases, the local acceleration ($\partial U/\partial t$) is unimportant. Neglecting it causes the time variation to vanish from the equation of motion: the flow is being modelled as quasi-steady, which means that at any instant it has fully adapted to the instantaneous boundary conditions, as if these were not changing in time.

The preceding estimates were restricted to waves in prismatic conduits, in which the length scale is determined by the time scale and the propagation velocity through $\mathcal{L} = c\mathcal{T}$. See Box 2.2 for scaling relations in the case of local variations in geometry, e.g. in control structures.

2.3.2 Resistance Term

We now turn to the importance of the flow *resistance* relative to the local acceleration. We first consider oscillatory motions, as in tides and seiches, in which the flow velocity varies in time with amplitude $\hat{U}$ and frequency ω. The local acceleration ($\partial U/\partial t$) is of the order of $\omega\hat{U}$, whereas the resistance per unit mass in Eq. (2.6) is of order $c_f\hat{U}^2/d$. This yields the following result for the ratio of the resistance to the local acceleration, here denoted as σ:

$$\sigma \equiv c_f \frac{\hat{U}}{\omega d} \tag{2.8}$$

The fraction $\hat{U}/(\omega d)$ allows for a simple physical interpretation: $\hat{U}/\omega$ is the amplitude of the horizontal displacement of the water particles, so that said fraction expresses how far the particles move back and forth relative to the flow depth.

Some typical, rounded values of σ for oscillatory motions have been collected in Table 2.1, using arbitrary but realistic values for the parameters, including $c_f = 0.004$. The entries have been placed in the order of ascending value of σ.

It is clear from the σ-values in Table 2.1 that resistance plays no role whatsoever in the instantaneous dynamics in tsunamis and seiches. The relative importance of resistance in tidal propagation depends mainly on the water depth. It is negligible in the oceans.

Table 2.1 Estimation of the relative resistance (σ) for different long-wave types

Category	T (min)	ω (rad/s)	d (m)	$\hat{U}$ (m/s)	σ
Tsunami	10	1.0×10^{-2}	4000	0.1	1×10^{-5}
Tide in the ocean	745	1.4×10^{-4}	4000	0.3	2×10^{-3}
Seiche	20	5.0×10^{-3}	20	0.5	2×10^{-2}
Tide in shelf sea	745	1.4×10^{-4}	50	0.5	3×10^{-1}
Tide in channel	745	1.4×10^{-4}	15	1.5	3×10^{0}
Tide over flats	745	1.4×10^{-4}	3	1.0	1×10^{1}

For tides in coastal waters, resistance is in general not negligible, and over tidal flats even dominant compared with the local acceleration.

Translatory waves are not included in the table. These typically have a time scale of the order of minutes or even less, so that here resistance is unimportant for the instantaneous dynamics. However, because its effect is always to resist motion, its influence can and will be important in the long run. Therefore, when dealing with such motions over long durations, compared with the time scale of the primary variations, resistance should be included in the modelling.

Flood waves in lowland rivers vary much more slowly than the tides, while their depths and flow velocities are comparable to those in tides in channels or over flats, so that in these waves inertia is negligible compared with the resistance.

Example 2.1 A tidal wave in an estuary has a length scale (wavelength) $\mathcal{L}$ of 500 km, a time scale (wave period) $\mathcal{T}$ of 12 h 25 min and a characteristic flow velocity $\mathcal{U}$ of 0.5 m/s. The depth d is 20 m and the resistance coefficient c_f equals 0.004. Give order of magnitude estimates for:

1. the local acceleration term
2. the advective acceleration term
3. the resistance term.

What can you conclude regarding the importance of the various terms?

Solution

Velocity variations are of order $\mathcal{U}$ over a time interval $\mathcal{T}$ and a spatial interval $\mathcal{L}$, respectively, yielding the following estimates:

1. $\partial U/\partial t \cong \mathcal{O}\left(\mathcal{U}/\mathcal{T}\right) = (0.50\,\text{m/s})/(45{,}000\,\text{s}) \approx (1.1 \times 10^{-5})\,\text{m/s}^2$.
2. $U\partial U/\partial s \cong \mathcal{O}\left(\mathcal{U}^2/\mathcal{L}\right) = (0.50\,\text{m/s}) \times (0.50\,\text{m/s})/(500{,}000\,\text{m}) \approx 5.0 \times 10^{-7}\,\text{m/s}^2$.
3. $c_f|U|U/d \cong \mathcal{O}\left(c_f\mathcal{U}^2/d\right) = 0.004 \times (0.50\,\text{m/s}) \times (0.50\,\text{m/s})/(20\,\text{m}) \approx 5.0 \times 10^{-5}\,\text{m/s}^2$.

The resistance term is most important in this case, but the local acceleration term cannot be neglected. The advective acceleration, however, is two orders of magnitude smaller than the resistance term and can be safely omitted at first.

2.4 Solution Methods

2.4.1 Complete Equations

The complete long-wave equations, Eqs. (1.20) and (1.21), are the basis of a large number of numerical models for the calculation of unsteady flows in open channels. Here, 'complete' refers to the fact that the equations have not been simplified by neglecting *a priori* terms that are expected to be small in a given application. Therefore, such models are suited, in principle, for all kinds of long-wave phenomena, provided these can be modelled as one-dimensional, which of course in itself is an approximation. Where this is not justified, one must resort to the long-wave equations in two horizontal dimensions.

In these numerical models, the long-wave equations are integrated in discretized form. Those techniques, the problems that may be encountered and the accuracy that can be achieved are dealt with in Chapter 11.

2.4.2 Simplified Equations

Based on the complete set of long-wave equations, various simplified forms have been developed and solved in order to obtain insight through simple calculations or analytical solutions. Each of those is tuned to a specific subset of long-wave problems for which certain terms in the equations of motion are estimated beforehand to be small, as was done in Section 2.3. By neglecting them, a simplified model results, which may allow analytical, preferably explicit solutions. This book focusses on these simplified models and their solutions because the primary purpose is to obtain insight into the dynamics of various long-wave phenomena that may be encountered in engineering practice. If and when more accurate quantitative answers are needed, one must resort to a validated numerical code.

The following subjects of wave propagation are dealt with in the remaining chapters:

- elementary wave equation, applicable to low, rapid waves: advective acceleration and flow resistance are neglected (Chapter 3)
- translatory waves: advective acceleration can be included; resistance neglected (Chapter 4)
- method of characteristics, showing fundamentals of wave propagation: particularly suited to high translatory waves; advective acceleration not neglected; resistance can be included but that is cumbersome (Chapter 5)
- harmonic method, suited for low-amplitude oscillatory progressive or standing waves such as seiches and tides: advective acceleration neglected, resistance included in linearized form (Chapter 7)
- flood waves in rivers: various approximations, inertia neglected (Chapter 8).

In addition, two chapters deal with highly reduced cases of unsteady motions that are not wave-like: the response of a short basin to harmonic excitation, in which inertia is negligible (Chapter 6), and flow through or over control structures in which storage is negligible (Chapter 9). Finally, a summary is provided of some relevant results for steady flow (Chapter 9).

Problems

2.1 What are 'long waves'?

2.2 Derive a condition for the validity of the long-wave approximation.

2.3 Mention a few categories of wave phenomena that belong to the class of long waves.

2.4 Check for each of the categories of wave phenomena in Problem 2.3 to which extent the long-wave approximation is justified.

2.5 For each of the categories of wave phenomena in Problem 2.3, choose some characteristic, realistic values of the most relevant parameters such as depth, flow velocity and time scale, and estimate the corresponding values of the ratios of various terms in the equation of motion.

2.6 What are so-called translatory waves?

2.7 Are translatory waves normally the result of natural processes or human intervention?

2.8 Argue why resistance is relatively unimportant in translatory waves.

2.9 What are seiches?

2.10 What is in general the cause of seiches?

2.11 Do seiches in essence belong to the class of progressive waves or to the class of standing waves?

2.12 Seiches in a given basin cannot have arbitrary frequencies. Why not?

2.13 Which mechanism is the major cause of energy loss of seiches?

3 Elementary Wave Equation

The considerations and analyses in this chapter are based on a strongly reduced set of equations, viz. those for the modelling of *low, long waves without resistance*. We assume a horizontal open channel without longitudinal variations in the channel cross section. The channel does not have to be straight, but for simplicity we refer to it as a prismatic channel. We account for storage and (local) inertia while neglecting advective accelerations (consistent with the restriction to low waves) and resistance (restricting the validity of the results to rapid variations). Based on these simplifying assumptions, a partial differential equation will be derived, the *wave equation*, so called because it has wave-like solutions. We present the wave equation in its most elementary form: linearized, with constant coefficients, no resistance, no forcing.

3.1 Simple Wave

3.1.1 Propagation

The notion of *wave propagation* is crucial in the present context. In order to develop a good understanding, we approach it in three steps of increasing complexity: a qualitative description of the propagation of a so-called simple wave, travelling into a region of rest; a quantitative description of the same in terms of algebraic relations derived from overall balance equations for a finite control volume; and a description in terms of a partial differential equation, the so-called wave equation.

We first consider a positive translatory wave, as sketched in Figure 3.1a (see also Figure 2.1). We recognize three regions: an undisturbed region, a region of established uniform flow and a travelling transient or wave front between them.

Thanks to the slope of the free surface in the wave front, a pressure gradient exists there that causes an acceleration of the water particles in the direction from the high-water side to the low-water side, indicated in the figure with a double arrow. At a given location, the water is initially at rest, while it accelerates during the passage of the wave front. Once the front has passed that location, the local free surface is again horizontal, the pressure gradient is zero, and the flow velocity is constant (we ignore flow resistance).

As a result of the difference in flow velocity across the transient, the passage of the transient causes a longitudinal compression of the water beneath it. Since water is almost incompressible, this longitudinal compression is compensated by a rise of the free surface. Because the wave considered here causes a rise in water level at a fixed point, we call it

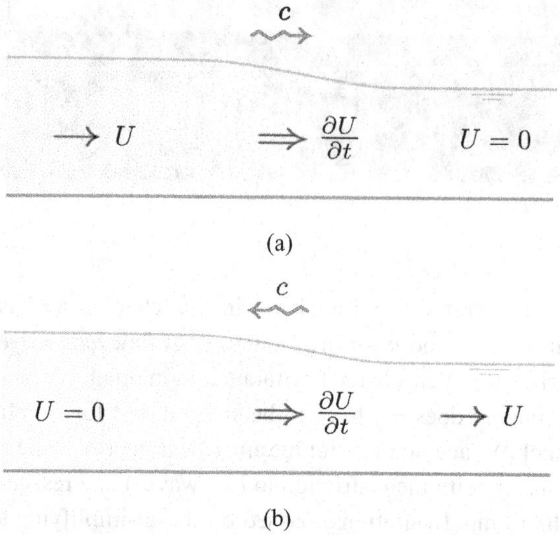

c

$\rightarrow U$ $\Rightarrow \frac{\partial U}{\partial t}$ $U = 0$

(a)

c

$U = 0$ $\Rightarrow \frac{\partial U}{\partial t}$ $\rightarrow U$

(b)

Fig. 3.1 Accelerations ($\Rightarrow$), flow velocities ($\rightarrow$) and propagation velocity ($\rightsquigarrow$) in a positive wave (a) and in a negative wave (b)

a positive wave; see Figure 3.1a. (In the analogous case in gas dynamics, we refer to a compression wave.)

Consider now the propagation of a negative wave, shown in Figure 3.1b. Notice that here the acceleration is again to the right, but now this is against the direction of wave propagation. Passage of this transient causes the particle velocities to go from zero to positive, opposite to the direction of wave propagation.

3.1.2 Balance Equations

Following the qualitative description given above, we will now quantify the arguments. We could start from the long-wave equations, which form a set of partial differential equations, but at this stage we prefer an algebraic formulation based on balance equations for a finite control volume between two fixed cross sections on either side of the transient. We restrict ourselves to a low disturbance entering quiescent water, which causes small variations in the elevation of the free surface (δh), the discharge (δQ) and the flow velocity (δU).

We consider the balances of mass and momentum during a time interval with duration Δt (Figure 3.2). We neglect compressibility of the water, so that the mass balance reduces to a volume balance. We further assume that the transient travels without change in shape (to be validated afterwards).

Volume Balance

This balance equates the net inflow of water, due to the difference in discharge δQ, to the storage at the free surface (the shaded area between the subsequent wave fronts in Figure 3.2) over a (storage) width B and with a rise in surface elevation δh, occurring in a finite time interval Δt:

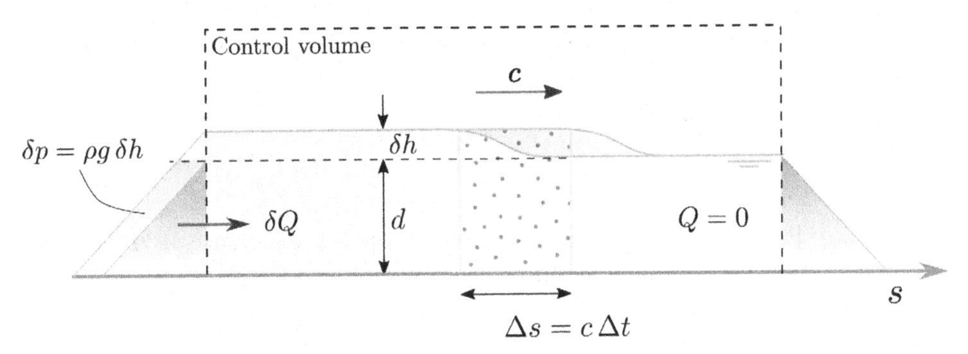

Fig. 3.2 Control volume containing the transient

$$\text{Net inflow} = \delta Q\, \Delta t = \text{Storage} = B \delta h\, \Delta s = B \delta h\, c \Delta t \tag{3.1}$$

or

$$\delta Q = Bc\, \delta h \tag{3.2}$$

Momentum Balance

The rise of the free surface (δh) gives rise to a pressure difference across the control volume given by $\rho g\, \delta h$, resulting in a net horizontal force on the flowing mass in the control volume equal to $\rho g A_c\, \delta h$. Here, we have neglected the small contribution at elevations above the undisturbed free surface (i.e. the small white triangle in the figure). Because of the restriction to a low disturbance, the advection of momentum, which is proportional to the square of the flow velocity, is negligible compared with the contribution by the wave-induced hydrostatic pressure.

During the time interval with duration Δt, the momentum inside the control volume (initially zero) increases with the amount $\rho\, \delta U A_c \Delta s = \rho\, \delta Q\, c\, \Delta t$ (see the dotted area in the figure, extending from the bottom to the free surface). The momentum balance becomes

$$\text{Momentum delivered} = \rho g A_c\, \delta h \Delta t = \text{Momentum gained} = \rho\, \delta Q\, c\, \Delta t \tag{3.3}$$

or

$$\delta Q = \frac{g A_c}{c} \delta h \tag{3.4}$$

Elimination of δQ and δh between Eqs. (3.2) and (3.4) yields the following important result for the long wave propagation speed:

$$c = \sqrt{\frac{g A_c}{B}} \tag{3.5}$$

Since $\delta Q = A_c\, \delta U$, Eq. (3.4) can be expressed in terms of the flow velocity as

$$\delta U = \frac{g}{c}\, \delta h \tag{3.6}$$

If the total cross section contributes fully to the conveyance, $A_c/B = A_c/B_c = d$, in which case the expression for the propagation speed reduces to

$$c = \sqrt{gd} \tag{3.7}$$

The expression for the flow velocity can then be written as

$$\delta U = \sqrt{\frac{g}{d}}\, \delta h = \frac{c}{d}\, \delta h \tag{3.8}$$

This also follows directly from Eq. (3.2).

The results derived here are valid only for (very) low waves ($Fr \ll 1$). They are used frequently in the following.

Example 3.1 As the result of a gate having been raised, a positive wave, with a rise in surface elevation (δh) of 0.15 m, propagates in a canal with an initial depth (d) of 5 m, a storage width (B) of 135 m and conveyance width (B_c) of 100 m, where the water is initially at rest.

1. Compute the wave speed c.
2. Compute the discharge Q through the gate.
3. Compute the flow velocity U behind the wave.
4. Compute the ratio of the advective momentum transfer ($\rho U^2 A_c$) relative to the pressure force ($\rho g A_c\, \delta h$).

Solution

Because the water is initially at rest, the total discharge following the passage of the transient (Q) equals δQ. For brevity, we will use the notation Q in the following.

1. Wave speed: $c = \sqrt{gA_c/B} = \sqrt{gdB_c/B} = 6.0$ m/s
2. Discharge: $Q = Bc\, \delta h = 122$ m^3/s
3. Flow velocity: $U = Q/A_c = 0.24$ m/s; this result also follows from $U = (g/c)\, \delta h = 0.24$ m/s.
4. Ratio of the advection of momentum relative to the pressure force: $U^2/g\, \delta h = 0.04$.

In the derivation given above, the advection of momentum was neglected relative to the pressure force. With the ratio of these respective terms being equal to 0.04 this assumption is indeed justified in this example.

3.2 Elementary Wave Equation

In this section, we derive and analyse a differential equation for long, low, frictionless waves in a prismatic conduit. It will appear that the simple wave of the preceding section is an important building block in the solution of this wave equation.

3.2.1 Derivation

The continuity equation is given by Eq. (1.20), which for easy reference is repeated here:

$$B\frac{\partial h}{\partial t} + \frac{\partial Q}{\partial s} = 0 \tag{3.9}$$

It is important to note that B is the full width of the free surface, available for storage, not the width of the conveyance cross section only.

On account of the simplifying assumptions mentioned above, the equation of motion, Eq. (1.21), reduces to

$$\frac{\partial Q}{\partial t} + gA_c\frac{\partial h}{\partial s} = 0 \tag{3.10}$$

In view of the assumption of low waves, the influence of a varying free-surface elevation on the parameters B and A_c will be neglected. Therefore, these do not vary in time. Furthermore, these parameters are supposed to be independent of the horizontal coordinate s as well. Mathematically speaking, we then deal with *a set of two first-order linear partial differential equations with constant coefficients*. These allow relatively simple solutions, compared with nonlinear equations with variable coefficients.

We now eliminate Q by differentiating Eq. (3.9) with respect to t and Eq. (3.10) with respect to s, with the result

$$\frac{\partial^2 h}{\partial t^2} - \frac{gA_c}{B}\frac{\partial^2 h}{\partial s^2} = 0 \tag{3.11}$$

Note that this equation is of second order as a result of including storage (through Eq. (3.9)) and inertia (through Eq. (3.10)). The interplay of these allows wave propagation, as we will see. Next, we ignore temporarily the findings of the preceding section, in particular the results for the propagation speed, and for brevity *define* a quantity c by

$$c \equiv \sqrt{\frac{gA_c}{B}} \tag{3.12}$$

Substitution of this into Eq. (3.11) yields

$$\frac{\partial^2 h}{\partial t^2} - c^2\frac{\partial^2 h}{\partial s^2} = 0 \tag{3.13}$$

Elimination of h instead of Q would have given a similar partial differential equation for the variable Q.

A partial differential equation with the form of (3.13) is known as 'the' wave equation because it is the most elementary form of all wave equations. It also applies to pressurized flow in pipes, which is treated in Appendix A, and to the vibration of the strings of a musical instrument. Notice that the wave equation (3.13) is linear, so that linear superposition applies: *a sum of solutions is also a solution.*

3.2.2 General Solution

Without derivation, we now state that the general solution $h(s, t)$ of Eq. (3.13) consists of the sum of an arbitrary function h^- of $(s + ct)$ and an arbitrary function h^+ of $(s - ct)$:

$$h(s,t) = h^+(s - ct) + h^-(s + ct) \tag{3.14}$$

This solution is due to d'Alembert, who derived it in 1746; see Strauss (1992). A similar expression applies to Q in terms of Q^+ and Q^-.

The wave equation (3.13) is linear. Therefore, in order to prove that the sum given by Eq. (3.14) is a solution, it is sufficient to prove this for h^+ and h^- separately. We begin with h^+. This function depends on s and t exclusively through the combination $s - ct$, which quantity we will represent as S^+, so $S^+ = s - ct$. Therefore, $h^+ = h^+(S^+)$, and because $\partial S^+/\partial s = 1$ and $\partial S^+/\partial t = -c$, it follows that

$$\frac{\partial h^+}{\partial s} = \frac{\mathrm{d}h^+}{\mathrm{d}S^+}\frac{\partial S^+}{\partial s} = \left(h^+\right)' \quad \text{and} \quad \frac{\partial h^+}{\partial t} = \frac{\mathrm{d}h^+}{\mathrm{d}S^+}\frac{\partial S^+}{\partial t} = -c\left(h^+\right)' \tag{3.15}$$

in which the prime on h^+ indicates an ordinary derivative of h^+ with respect to S^+. Continuing likewise, we obtain

$$\frac{\partial^2 h^+}{\partial s^2} = \left(h^+\right)'' \quad \text{and} \quad \frac{\partial^2 h^+}{\partial t^2} = c^2\left(h^+\right)'' \tag{3.16}$$

Substitution of these two expressions into the wave equation (3.13) shows that the latter is satisfied by any function $h^+ = h^+\left(S^+\right) = h^+(s - ct)$. The same applies to h^- and therefore also to their sum, which was to be proven.

We will now investigate the meaning of the solution, first for h^+ only. The latter depends on s and t solely through S^+, or through $s - ct$. Therefore, we observe no change in the local value of h^+ if we keep $s - ct$ constant in time, i.e. $\mathrm{d}s/\mathrm{d}t = c$, i.e. if we move in the positive s-direction with speed c. Stated another way: a point of constant h^+ moves with speed c in the positive s-direction. That is why the subscript + was chosen for this function. Because c is a constant in the present approximation, all points of the disturbance h^+ move with the same speed: the disturbance propagates without change of shape with velocity c, whose value is given by Eq. (3.12). Likewise, it can be shown that the function h^- represents a wave moving at speed c in the negative s-direction without change of shape.

It follows from the above that the quantity $h(s, t)$, defined in Eq. (3.14), obeys the wave equation (3.13). It can be shown also that Eq. (3.14) represents the *general solution* to this equation, which therefore consists of *two waves, propagating in opposite directions at a constant speed c without change in shape*. Needless to say, where both are present simultaneously, their sum does vary in shape. Whether such superposition actually occurs depends on the initial conditions and on the boundary conditions.

3.2.3 Total Derivative

The property of propagation without change in the local value of a disturbed quantity (for the + and − waves separately) can be formulated differently, in a manner that we shall use more in the following. To this end, the concept of the total derivative is important.

Suppose an observer travels with a velocity V in the s-direction, observing the local values of a quantity h. The change in h which is observed in a short time interval with duration Δt can be expressed as

$$\Delta h = \frac{\partial h}{\partial t}\Delta t + \frac{\partial h}{\partial s}\Delta s = \frac{\partial h}{\partial t}\Delta t + \frac{\partial h}{\partial s}V\Delta t \tag{3.17}$$

The observed change per unit time, in the limit as Δt goes to zero, is the so-called *total derivative* of h:

$$\frac{\mathrm{d}h}{\mathrm{d}t} = \frac{\partial h}{\partial t} + V\frac{\partial h}{\partial s} \tag{3.18}$$

Now consider h^+. It follows from Eq. (3.15) that it obeys

$$\frac{\partial h^+}{\partial t} + c\frac{\partial h^+}{\partial s} = 0 \tag{3.19}$$

Comparison with Eq. (3.18) shows that the left-hand side of Eq. (3.19) is the total derivative of h^+ for an observer moving with velocity c, which is zero according to Eq. (3.19). Therefore, this observer sees no change in the local value of h^+. Since this reasoning applies to all points of the disturbance, and the value of c is common to all of them (in the present approximation for low waves!), we conclude that Eq. (3.19) implies that the disturbance h^+ travels in the positive s-direction at speed c without change in shape. Therefore, another representation, equivalent to Eq. (3.19), is

$$\frac{\mathrm{d}h^+}{\mathrm{d}t} = 0 \quad \text{provided that} \quad \frac{\mathrm{d}s}{\mathrm{d}t} = c \tag{3.20}$$

in which $\mathrm{d}/\mathrm{d}t$ is the total derivative operator: $\mathrm{d}d/\mathrm{d}t = \partial/\partial t + c\,\partial/\partial s$. In this formulation, the *partial* differential equation (3.19), for a single dependent variable h^+ and two independent variables s and t, is replaced by a set of two *ordinary* differential equations for the two dependent variables h^+ and s and one independent variable t. Likewise, the following is valid for h^-:

$$\frac{\mathrm{d}h^-}{\mathrm{d}t} = 0 \quad \text{provided that} \quad \frac{\mathrm{d}s}{\mathrm{d}t} = -c \tag{3.21}$$

This kind of formulation will be used extensively in Chapter 5, dealing with the so-called method of characteristics.

3.3 Relation between Discharge and Free-Surface Elevation in a Progressive Wave

Variations in discharge and related variations in the free surface elevation have been considered in Section 3.1, using algebraic balance equations for mass and momentum in a finite control volume. Here, we rederive those results using the wave equation, applied to an arbitrary but low wave propagating in the positive s-direction into a region of uniform flow with water level h_0 and discharge Q_0. The corresponding water level h^+ obeys Eq. (3.19). Substitution of this equation into the continuity equation (3.9) yields

$$-Bc\frac{\partial h^+}{\partial s}+\frac{\partial Q^+}{\partial s}=0 \tag{3.22}$$

Integration of this with respect to s for constant Bc gives $Q^+ - Bc\,h^+ = \text{constant} = Q_0 - Bc\,h_0$. Expressed in terms of the changes with respect to the undisturbed situation, namely $\delta h^+ = h^+ - h_0$ and $\delta Q^+ = Q^+ - Q_0$, this becomes:

$$\delta Q^+ = Bc\,\delta h^+ \tag{3.23}$$

This result had already been seen in Section 3.1. A similar result applies to h^-:

$$\delta Q^- = -Bc\,\delta h^- \tag{3.24}$$

3.4 Solution for Arbitrary Initial Conditions

Because the wave equation is of second order in time, two initial conditions are needed for a well-posed problem. Here, we use the initial values of the disturbances in surface elevation (δh) and discharge (δQ) at all points of the channel considered, relative to an undisturbed state with uniform water level and discharge.

In Section 3.1, a disturbance travelling in the positive s-direction was considered. In the present section we consider the more general case of given arbitrary, small values of δh and δQ as functions of s for some initial instant $t = t_0$:

$$\delta h_0(s) = \delta h(s;t_0) \quad \text{and} \quad \delta Q_0(s) = \delta Q(s;t_0) \tag{3.25}$$

In order to determine the time evolution of this arbitrary disturbance, not necessarily travelling in one direction only, we separate it into two contributions that each do travel in one direction only:

$$\delta h = \delta h^+ + \delta h^- \quad \text{and} \quad \delta Q = \delta Q^+ + \delta Q^- = Bc\left(\delta h^+ - \delta h^-\right) \tag{3.26}$$

It follows from these equations that

$$\delta h^+ = \frac{1}{2}\left(\delta h + \frac{\delta Q}{Bc}\right) \quad \text{and} \quad \delta h^- = \frac{1}{2}\left(\delta h - \frac{\delta Q}{Bc}\right) \tag{3.27}$$

At the initial instant $t = t_0$, the values in the right-hand sides of Eq. (3.27) are known as functions of s. The same then holds for the initial values of δh^+ and δh^-. By translating the initial profile of δh^+ with speed c in the positive s-direction, and the initial profile of δh^- with speed c in the negative s-direction, and adding the results at each position for a chosen instant, the free-surface profile at that instant is found. The same method applies to the discharge.

If the water is at rest at the initial instant, the initial values of δh^+ and δh^- at each location are equal to half the local value of $\delta h_0(s)$, as follows from Eq. (3.27). Figure 3.3 gives an example for this case.

The preceding results are restricted to low waves, described by the linear wave equation, which allows linear superposition of elementary waves travelling in opposite directions to find the total solution. If the restriction to low waves is relaxed, linear superposition no longer applies, but we can still construct the total solution by considering component waves travelling in opposite directions, using the so-called method of characteristics. This is the subject of Chapter 5.

In the above, we have tacitly assumed that the disturbances can travel unimpeded in a prismatic channel, as if this were infinitely long. If and when the disturbance reaches a location where the channel geometry changes, reflection occurs. This is considered below.

3.5 Boundary Conditions

Because the wave equation is of second order in space (s), two boundary conditions are needed for a well-posed problem.

Where the canal has a *closed end*, at $s = s_c$ say, the discharge must be zero at all times. Mathematically, this is expressed as the boundary condition $Q(s_c, t) = 0$ for all times. In view of Eq. (3.26), this implies that $\delta h^- = \delta h^+$ at that location. In words: where an incident wave (δh^+), travelling in the positive s-direction, meets a closed end, a backward-propagating wave (δh^-) is generated whose height at the closed end equals that of the incident wave at that location at all times. At the closed end, the free-surface elevation is doubled. We say that *at a closed end, total positive reflection occurs*. (The qualification 'positive' refers to the free-surface elevation. The discharge is in fact negatively reflected because $\delta Q^- = -\delta Q^+$ at the closed end.)

The opposite situation occurs if the canal at some point, at $s = s_0$, say, is connected to a nontidal sea or a large lake or reservoir, in which the water level is not measurably affected by inflow into or outflow from the canal. This means that $\delta h(s_0, t) = 0$ at all times, implying that $\delta h^-(s_0, t) = -\delta h^+(s_0, t)$: *at an open end connected to a nontidal sea or a large reservoir, total negative reflection occurs*. It follows from Eq. (3.26) that at such open end, $\delta Q^- = \delta Q^+$, so that there the discharge is positively reflected. Thus, the discharge at an open end is doubled.

Numerical or physical models of flow in canals or rivers are often cut off at some distance from the study area, even when in reality the system extends further. Such cut-off creates an artificial, *open model boundary*. The condition to be imposed there is that

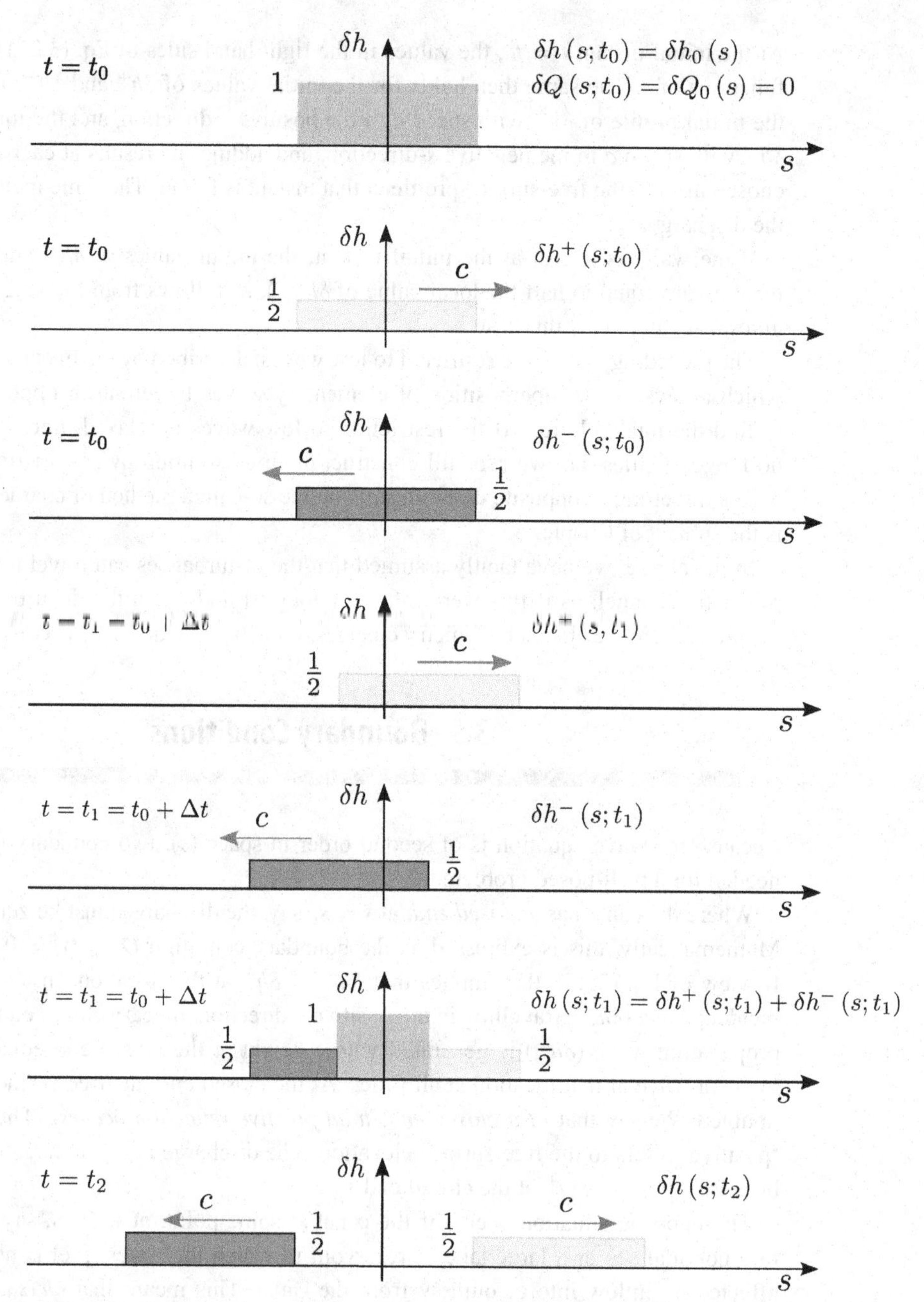

Fig. 3.3 Elementary wave solution for an initial discharge $Q_0 = 0$ and a uniform initial water level disturbance over a finite length

disturbances approaching that boundary from within the study area should not be reflected, so as to simulate reality in which the disturbances continue unimpeded. Thus, if the positive s-direction is from within towards the open boundary, δh^- should be zero at the open boundary at all times, implying the boundary condition $\delta Q - Bc\,\delta h = 0$; see Eq. (3.27).

3.6 Periodic Progressive and Standing Waves

A standard solution of the linear wave equation consists of harmonic motions. This section deals with such motions for the case of sinusoidal waves in a prismatic canal or basin. For brevity, we denote the elevation of the free surface above its mean value as ζ. Its amplitude is written as $\hat{\zeta}$, the wave period is T, and the angular frequency (i.e. the phase change per unit time) is $\omega = 2\pi/T$. The wavelength is L and the wave number (i.e. the phase change per unit propagation distance) is $k = 2\pi/L$.

If the wave is progressive, we have $L = cT$ in which c is the speed of propagation. Since we deal with free long waves, $c = \sqrt{gA_c/B}$, so that $L = \sqrt{gA_c/B}\,T$. Written in terms of frequency and wave number, this becomes

$$\frac{\omega}{k} = \sqrt{\frac{gA_c}{B}} \tag{3.28}$$

This is the so-called *dispersion equation*, linking frequency and wave number for the system considered. Note that it also applies to the superposition of two waves with the same frequency and wave number, as in the case of standing waves, even though the notion of a propagation speed does not apply to the latter category.

3.6.1 Infinitely Long Canal

In a prismatic canal of infinite length (in practice: in a long canal, away from the influence of boundaries), waves can propagate indefinitely in either direction without reflection, so-called *progressive waves*. We assume a sinusoidal wave to be propagating in the positive s-direction with speed c. The free-surface elevation above the mean water level can then be written as

$$\zeta = \hat{\zeta} \cos\left(2\pi \frac{s - ct}{L}\right) \tag{3.29}$$

Written in terms of frequency and wave number, this becomes

$$\zeta = \hat{\zeta} \cos(ks - \omega t) \tag{3.30}$$

The corresponding discharge is given by (see Eq. 3.23)

$$Q = Bc\hat{\zeta} \cos(ks - \omega t) \tag{3.31}$$

Note that the phase $ks - \omega t$ is constant for an observer moving at a speed $\mathrm{d}s/\mathrm{d}t = \omega/k$. That is why this speed ω/k is called the *phase speed.* An observer moving at this speed sees no variation in the local value of ζ and that of Q.

In a progressive wave as described above, the surface elevation and the discharge (reckoned positive in the propagation direction) are in phase; i.e. at each location they reach their maximum values at the same time.

For a wave progressing in the negative s-direction, we have

$$\zeta = \hat{\zeta} \cos (ks + \omega t) \tag{3.32}$$

and

$$Q = -Bc\,\hat{\zeta} \cos (ks + \omega t) \tag{3.33}$$

In this case, the surface elevation and the discharge are 180° out of phase (in opposite phase); i.e. at each location the maximum of one occurs at the same time as the minimum of the other. In other words, under the crests of the wave, where the surface elevation has its maximum, the discharge is minimal (maximal in an absolute sense, but in the negative s-direction).

When both waves, with equal frequency and amplitude, but propagating in opposite directions, are present simultaneously in the same canal, the resulting motion is given by superposition of the preceding expressions, with the result

$$\zeta = 2\hat{\zeta} \cos ks \cos \omega t = \hat{\zeta}_{st} \cos ks \cos \omega t \tag{3.34}$$

This expression represents a *standing wave* because as time goes on the resulting profile does not move forward or backward; it merely breathes up and down (see Figure 3.4b). The maximum amplitude of its surface elevation, $\hat{\zeta}_{st}$, is twice the amplitude of the two

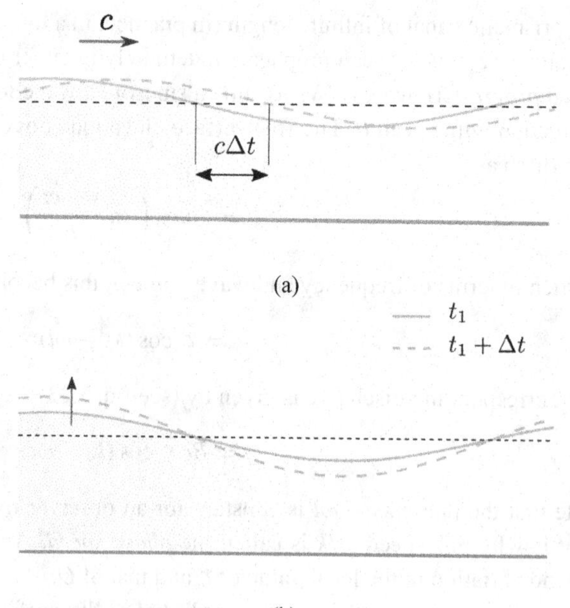

Fig. 3.4 Progressive (a) and standing (b) periodic waves

opposing component progressive waves of which it consists. The corresponding discharge is given by

$$Q = 2Bc\,\hat{\zeta}\,\sin ks\,\sin\omega t = Bc\,\hat{\zeta}_{st}\,\sin ks\,\sin\omega t \tag{3.35}$$

We see that in a standing wave, the surface elevation and the discharge are 90° ($\pi/2$ radians) out of phase, both in s and in t (they are in quadrature). At the locations where $\cos ks = 0$, the surface elevation is zero at all times. Those points are called *nodes*. At these locations, the local amplitude of the discharge is maximal. The distance between adjacent nodes is one-half wavelength. At the locations where $\cos ks = \pm 1$, the local amplitude of the surface elevation is maximal, but the discharge is zero at all times. These points are called *antinodes*. The distance between adjacent antinodes is one-half wavelength.

In an infinitely long canal, progressive waves and standing waves can exist without constraints on the frequency and wave number and (in the case of standing waves) on the locations of the nodes and antinodes. This changes when the canal (or basin) is semi-infinitely long or has a finite length, as considered in the following.

3.6.2 Semi-Infinitely Long Canal with a Closed End

In a canal with a closed end, progressive waves cannot exist for an unlimited duration because sooner or later they either vanish towards infinity or are reflected at the closed end. In the latter case, a standing wave develops with zero discharge, thus an antinode, at the closed end and at all points at a distance of an integer number of half wavelengths away from it. If we choose $s = 0$ at the closed end, the motion is described by Eqs. (3.34) and (3.35). There are no restrictions on the allowable values of the frequency or the wave number.

3.6.3 Closed Basin

In a finite-length basin, extending from $s = 0$ and $s = \ell$, say, closed at both ends, the boundary conditions $Q = 0$ at $s = 0$ and $Q = 0$ at $s = \ell$ have to be fulfilled at all times. Thus, we have a *standing wave* with an antinode at each end, and possibly one or more antinodes between them. Therefore, the basin contains a whole number of half wavelengths. This implies that only a discrete set of wave numbers and wavelengths is permitted: $L_n = 2\ell/n$ or $k_n = n\pi/\ell$ for $n = 1, 2, \ldots$. Figure 3.5a shows the surface profile in the standing wave for $n = 3$. The value of n equals the number of nodes in the closed basin.

Because we deal with free oscillations, $\omega_n/k_n = \sqrt{gA_c/B}$ for all n. Thus, the discrete set of wave numbers k_n corresponds to a discrete set of frequencies given by $\omega_n = \sqrt{gA_c/B}\,k_n, n = 1, 2, \ldots$, referred to as the set of *natural frequencies*, corresponding to the natural oscillations of the water mass in the closed basin. Of these, ω_1 is the fundamental frequency, corresponding to an oscillation with only one node, halfway the length of the canal (just one-half wavelength in the basin). The frequencies corresponding to $n = 2, 3, \ldots$ are higher harmonics.

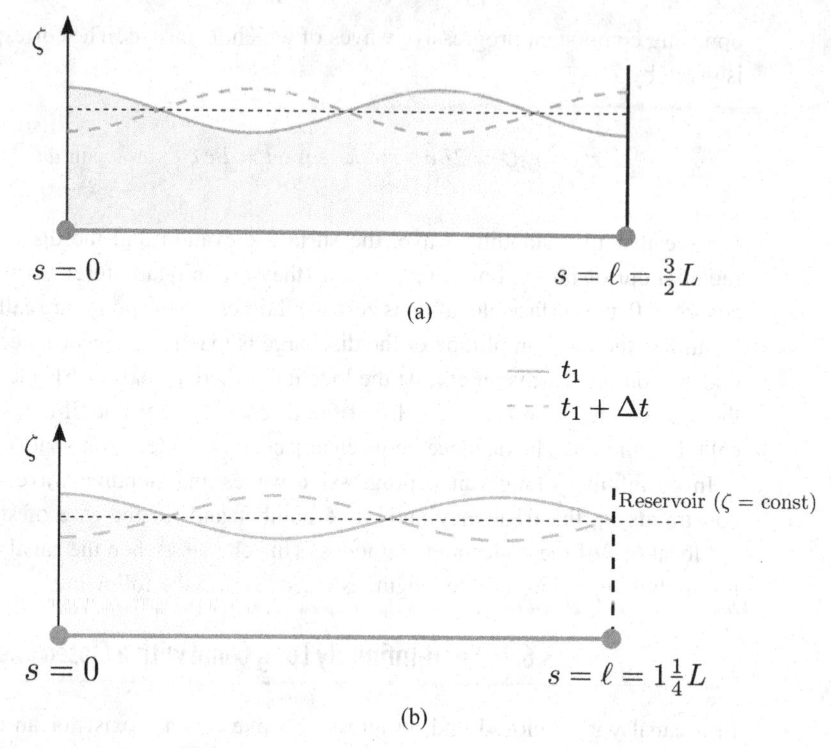

Fig. 3.5 Free standing wave: closed basin, $n = 3$ (a); semi-closed basin connected to a reservoir, $n = 2$ (b)

3.6.4 Semi-Closed Basin Connected to a Reservoir or Tideless Sea

Consider now a basin closed at one end, where $s = 0$, say, and connected to a large reservoir of constant elevation at the other end (the open end), where $s = \ell$, say. This corresponds to the boundary conditions $Q = 0$ at $s = 0$ and $\zeta = 0$ at $s = \ell$ for all times. These conditions are fulfilled in a standing wave with an antinode at the closed end and a node at the open end, requiring $\cos k\ell = 0$ or $k_n\ell = \frac{1}{2}\pi + n\pi$ for $n = 0, 1, 2, \ldots$. Written another way: $\ell = (2n + 1)\frac{1}{4}L$: the basin contains an odd number of quarter wavelengths. The value of n equals the number of nodes *inside* the basin, not counting the node at the open end. See Figure 3.5b.

As above, there is a set of natural frequencies, here given by $\omega_n = (\frac{1}{2}\pi + n\pi)\sqrt{gA_c/B}/\ell$ for $n = 0, 1, 2, \ldots$. The fundamental frequency ω_0 applies to the case where there is just one-quarter wave in the basin.

For a natural oscillation in a semi-closed basin connected to a reservoir, the discharge amplitude is maximum in the entrance and the surface elevation amplitude is maximum at the closed end.

Example 3.2 A prismatic basin, closed at one end, is connected at its open end to a reservoir. The basin dimensions are $B = 600$ m, $B_c = 300$ m (the width of the conveyance cross section), $d = 6$ m (the depth of the conveyance cross section) and $\ell = 6$ km. There is a natural oscillation in the basin with one node in the interior. The amplitude of the water surface elevation at the closed end ($\hat{\zeta}_{st}$) is 0.5 m. Calculate:

1. the period (T) of the oscillation
2. the amplitude of the discharge at the mouth ($\hat{Q}(\ell)$)
3. the amplitude of the flow velocity at the mouth ($\hat{U}(\ell)$).

Solution

It follows from the given number of nodes that the basin length equals $\frac{3}{4}$ wavelength, or $L = cT = \frac{4}{3} \times 6000\,\text{m} = 8000\,\text{m}$.

1. The wave speed $c = \sqrt{gA_c/B} = \sqrt{gdB_c/B} = 5.42$ m/s, from which the wave period $T = L/c = 8000$ m / 5.42 m/s = 1475 s.
2. $\hat{Q}(\ell) = Bc\,\hat{\zeta}_{st}\,|\sin k\ell|$. Since $\ell = \frac{3}{4}L$, we have $k\ell = \frac{3}{2}\pi$ and $\sin k\ell = -1$, it follows that $\hat{Q}(\ell) = (600\text{ m}) \times (5.42\text{ m/s}) \times (0.5\text{ m}) = 1627\,\text{m}^3/\text{s}$.
3. $\hat{U}(\ell) = \hat{Q}(\ell)/A_c = \hat{Q}(\ell)/(dB_c) = (1627\,\text{m}^3/\text{s})/(1800\,\text{m}^2) = 0.90$ m/s.

3.6.5 Semi-Closed Basin Connected to a Tidal Sea

Instead of a semi-closed basin connected to a reservoir or tideless sea, with a constant water level, we now consider such a basin connected at its open end to a tidal sea. The tide-induced up-and-down motion at the basin mouth generates oscillations of the water mass inside the basin at the tidal frequency, not necessarily related to any of the natural frequencies of the basin. In such cases we speak of a *forced oscillation* of the water mass inside the basin.

As before, Q must be zero at the closed end. A standing wave with an antinode at the closed end can fulfill this condition. Its amplitude is such that ζ at the mouth equals the value being forced by the tide (Figure 3.6). The ratio of the amplitude of ζ at the closed end to that at the open end equals $1/|\cos k\ell|$. This goes to infinity as $|\cos k\ell|$ goes to zero; i.e. the basin contains a whole number of quarter waves. In such cases we speak of *resonance*, occurring if the frequency of the forcing equals one of the natural frequencies of the basin.

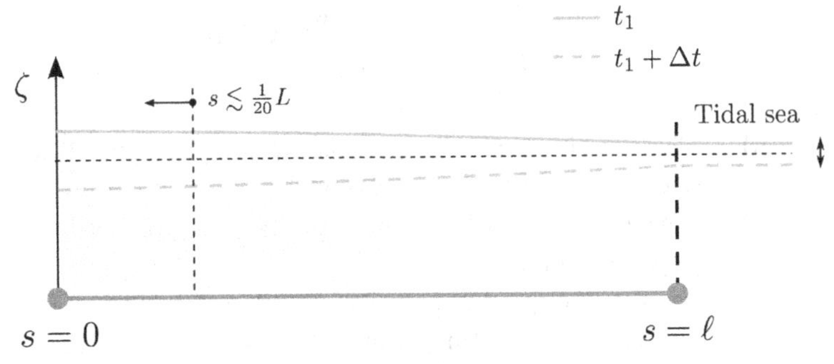

Fig. 3.6 Forced standing wave: semi-closed basin connected to a tidal sea

As opposed to the case of (near) resonance, with strongly amplified motions in the interior of the basin, we now consider a *short basin*, i.e. one whose length is small compared with the wavelength: $\ell \ll L$, or $k\ell \ll 2\pi$, so that $\cos k\ell \simeq 1$. This implies a very small phase difference from the mouth to the closed end (if there were a progressive wave in the basin), so that the water level responds almost in unison to the tidal forcing at the mouth, rising and falling with the tide but being virtually horizontal at all times. The wave character of the motion in the basin can then be disregarded. This response is called the *pumping mode* or *Helmholtz mode*.

Example 3.3 Consider the same (prismatic) basin as in Example 3.2, now connected at its open end to a tidal sea with an M_2-tide with a surface elevation amplitude of 1.5 m. The period of the M_2-tide is T = 12 h 25 min, or 44,700 s, so the tidal frequency $\omega = 1.4 \times 10^{-4}$ rad/s.

1. Determine the response of the basin in terms of the surface elevation amplitude.
2. Calculate the discharge amplitude in the open end of the basin.
3. Calculate the discharge amplitude halfway between the open end and the closed end

Solution

The value of c is the same as in Example 3.2, or 5.42 m/s, and the wavelength $L = cT =$ 242 km, which is more than 20 times the basin length (ℓ): the pumping mode approximation applies.

1. In view of the open connection between the basin and the sea, we can equate $\hat{\zeta}_{st}$ to the amplitude of the offshore tide (1.5 m).
2. The discharge amplitude in the entrance $\hat{Q}(\ell) = B\ell\omega\hat{\zeta}_b = (600\,\text{m})(6000\,\text{m})(1.4 \times 10^{-4}\,\text{rad/s})(1.5\,\text{m}) = 756\,\text{m}^3/\text{s}$.
3. In a short, prismatic basin the discharge varies linearly with the distance of the cross section to the closed end. Therefore, the discharge halfway the length of the basin is half of that in the mouth.

The pumping mode approximation neglects the influence of resistance, which usually needs to be accounted for in tidal calculations (Chapter 6).

Problems

3.1 What does it mean to linearize the equation of motion? How can this be achieved? What is a condition for its validity?

3.2 Explain why a small storage capacity in a river leads to an enhanced speed of propagation of flood waves.

3.3 Calculate the speed of propagation of low, long waves without resistance in water with a depth of 4000 m (ocean), 50 m (shelf sea) and 5 m (estuary).

3.4 Calculate the corresponding wavelength for an M_2-tide and for a tsunami with a period of 10 min. First verify whether the latter can be regarded as a long wave in the given depths.

3.5 Derive a relation between the variations in water level and discharge for a long wave propagating without change in shape in open water.

3.6 A pumping station begins discharging water at a rate of $40\,m^3/s$ onto an evacuation canal with a depth of 4 m and a width of 50 m. Calculate the speed of the resulting translatory wave propagating into the canal, the associated rise of the free surface and the particle velocity.

3.7 The following values of the surface elevation and flow velocity are given for an initial disturbance (at $t = 0$) in an infinitely long, 5 m deep canal:

$$(\zeta, U) = \begin{cases} (0.50\ \mathrm{m}, 0) & \text{for} \quad s < 0 \\ (0, 0.50\ \mathrm{m/s}) & \text{for} \quad s \geq 0 \end{cases}$$

Calculate and plot the values of ζ and U as functions of s at $t = 10$ s.

3.8 Same as in Problem 3.7, now with the following initial values:

$$(\zeta, U) = \begin{cases} (0.50\ \mathrm{m}, 0) & \text{for} \quad -100\ \mathrm{m} < s < 0 \\ (0, 0.50\ \mathrm{m/s}) & \text{for} \quad 0 \leq s < 100\ \mathrm{m} \\ (0, 0) & \text{elsewhere} \end{cases}$$

Calculate and plot the values of ζ and U as functions of s for the instants $t = 10$ s and $t = 20$ s.

3.9 Sketch the two functions $\cos(\omega t - ks)$ and $\cos \omega t \cos ks$ for ks between -2π and 2π and $\omega t = 0, \frac{1}{6}\pi, \frac{1}{2}\pi, \pi$ and 2π. Verify that the first of these two functions represents a progressive wave advancing at a speed ω/k.

3.10 Can a periodic, progressive wave exist continually in a basin closed at one end or at both ends?

3.11 Argue why a free periodic oscillation in a closed basin is possible for a countable set of frequencies only. Same for a semi-closed basin.

3.12 Verify how the natural frequencies of a prismatic basin are affected by the width of the free surface, the width and depth of the conveyance cross section, and the length. Which of these has the most influence?

3.13 Calculate the three longest natural periods of a semi-closed prismatic basin with a length of 30 km, a conveyance cross-sectional area of $6 \times 10^3\ \mathrm{m}^2$ and a width of the free surface of 600 m.

3.14 Continuing with the second of the three modes of the previous question: sketch a longitudinal profile of the surface elevation and the discharge at the instant $\omega t = \frac{1}{4}\pi$ (using the phases as in Eqs. (3.34) and (3.35)). Check the relation between surface elevation and discharge in a qualitative sense, including their signs.

3.15 Same as in the two preceding questions, now for a basin closed at both ends.

3.16 A semi-closed basin with the dimensions as in Problem 3.13 is subjected to a forced oscillation. Sketch the longitudinal profiles of the surface elevation and calculate the ratio (r) of the amplitude of the surface elevation at the closed end to that at the open end, for the periods $T = 3\,\mathrm{h}$ and $T = 6\,\mathrm{h}$, respectively. Interpret these results considering the periods of the forced motion in relation to the natural periods of the basin.

3.17 Consider the situation of the preceding question for $T = 6\,\mathrm{h}$, with the additional information that the amplitude of the surface elevation at the closed end is 1 m. Calculate and plot the amplitudes of the discharge at the following distances from the closed end: $s = 1\,\mathrm{km}$, 2 km, 3 km, 5 km and 10 km, and interpret the results.

4 Translatory Waves

This chapter deals with the modelling of translatory waves. The archetype of a translatory wave is a transient disturbance of discharge and surface elevation, travelling between two regions of uniform flow (including the state of rest as a special case). We will discuss the generation and propagation of these waves, considering uniform as well as non-uniform canals. The effect of bed resistance on translatory waves is marginal initially, but it can become important over long travelling distances. For this reason we will quantify the associated damping mechanism. A special type of translatory wave is the tidal bore, whose spectacular appearance alone, besides its intricate physics, would already motivate its treatment here.

4.1 Introduction

Translatory waves are usually the result of operation of engineering structures, such as

- locks
- control structures
- pumping stations
- hydropower plants.

Manipulation of these structures gives rise to varying discharges onto the adjacent canal reaches. For example, the levelling of the water in a lock chamber is accompanied by a discharge into or from the adjacent canal reach, which gradually increases from zero to maximum and then back to zero. This results in a transient disturbance in the canal in which the variation of the surface elevation mirrors that of the discharge, going from zero to an extreme (either a maximum or a minimum) and back to zero. The other three cases usually involve a gradual shift to a new setting, which is then left unchanged for quite some time; in such cases, the variation in discharge and surface elevation is monotonic (instead of going through a maximum or a minimum), and the resulting translatory wave forms a monotonic, moving transition between two different states of uniform flow.

The propagation of low translatory waves, and the relation between the changes in discharge and surface elevation, have already been considered in the preceding chapter. Here, we return to the subject by presenting some extensions dealing with propagation in nonprismatic channels and with the process of gradual damping due to boundary resistance, which is relevant in the case of long travel times and long propagation distances. All of this is based on the assumption of low waves, allowing linearization. Following this, the constraint to low waves is relaxed, and the most important nonlinear effects of a finite wave height are considered. The final section of this chapter presents some field data and a comparison thereof with the theory.

4.2 Low Translatory Waves in Uniform Channels

For easy reference, we reproduce here the principal results from the preceding chapter, valid for low, frictionless waves in a prismatic conduit. The propagation speed is given by

$$c = \sqrt{\frac{gA_c}{B}} \tag{4.1}$$

and the variations in discharge, surface elevation and flow velocity are related by

$$\delta Q = Bc\,\delta h \qquad \text{and} \qquad \delta U = \frac{g}{c}\,\delta h \tag{4.2}$$

These equations apply to the wave as it propagates. In order to determine the state of motion, the discharge must be specified at the location where the wave is generated.

Consider for instance the case of discharge through a gradually varying opening. In order to calculate the time-varying discharge, the flow through the opening is modelled as quasi-steady. The instantaneous discharge then depends on the instantaneous opening and on the instantaneous head difference as for steady flow, for which a Bernoulli-type relation applies (see Chapter 9):

$$Q(t) = \mu A(t)\sqrt{2g\,(h_1\,(t) - h_2\,(t))} = \mu A(t)\sqrt{2g\Delta h\,(t)} \tag{4.3}$$

$A\,(t)$ is the instantaneous area of the opening, μ an empirical discharge coefficient of order 1, dependent on the specific geometry, and h_1 and h_2 are the instantaneous water levels on the upstream side and the downstream side. Operation of the gate changes the value of the initial head difference across it, which must be taken into account in the calculations.

Example 4.1 A movable gate in a barrier between a canal (A_c = 80 m^2, B = 30 m) and a large reservoir is partially opened to an effective flow cross section μA of 4 m^2. Initially, the water in the canal is at rest, and the surface level is 3 m below that in the reservoir. Calculate:

1. an estimate of the discharge through the gate opening (Q_0), neglecting the influence of the wave height on the discharge
2. the discharge (Q) through the gate opening, taking into account the influence of the wave height on the discharge
3. the resulting wave height (δh) in the canal.

Solution

The discharge $Q = \mu A\sqrt{2g\,\Delta h}$. Initially, $\Delta h = \Delta h_0 = 3$ m, changing to $\Delta h = \Delta h_0 - \delta h$, once the gate is opened, where $\delta h = Q/Bc$ is the height of the wave propagating into the canal (the water level in the reservoir is not affected).

1. Neglecting the influence of the wave height on the discharge gives the estimate $Q_0 = \mu A\sqrt{2g\,\Delta h_0} = 30.7\ \mathrm{m^3/s}$.
2. Substitution of $\Delta h = \Delta h_0 - Q/Bc$ into the discharge relation gives a second-degree algebraic equation for the discharge whose solution (rounded) is $Q = 29.7\ \mathrm{m^3/s}$.
3. The resulting wave height $\delta h = Q/Bc = 0.20$ m (using $c = \sqrt{gA_c/B} = 5.11$ m/s).

The wave height δh is about 7% of Δh_0. The initially estimated discharge Q_0 is therefore approximately 3.5% too high, since the discharge varies in proportion to the square root of the head difference.

It follows from the above that the height and the wave-induced particle velocities in a translatory wave that results from the opening of a gate, a valve, or a similar operational device can be controlled through the discharge (Eq. (4.2)), which in turn can be controlled through the maximum opening (Eq. (4.3)). In the case of lock-generated waves, the levelling is controlled such that only acceptably low and slow variations in water level and weak currents are produced in the lock chamber and in the approach canal, in order to avoid hindrance to shipping.[1]

Let us now investigate how the time-varying conditions at a control structure, resulting from its operation, manifest themselves in the adjacent canal reaches. We take the example of a pumping station that delivers a time-varying discharge, from a time $t = 0$ until $t = T$, with a maximum value Q_{max}, as shown in Figure 4.1a (a hypothetical, nonrealistic variation). As soon as the discharge starts, a disturbance propagates into the canal reach, in the positive s-direction, say. At a time $t = t_1$, after the pumps have been switched off ($t_1 > T$), its front and its end have advanced through a distance of ct_1 and $c(t_1 - T)$, respectively. This is shown in the longitudinal profile of the discharge at time $t = t_1$, in

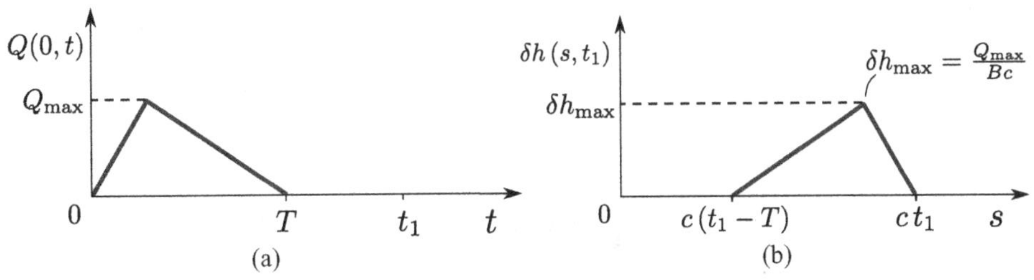

Fig. 4.1 Translatory waves in a canal; discharge variation in time at a fixed point (a) and wave height variation in space at a given instant (b)

[1] As a typical example, the following upper limits were used in the Seine–Nord Europe Canal project (PIANC, 2015): 30 cm for the height of the wave and 35 cm/s for the flow velocity (controllable through a limit on Q), and 1 in 1000 for the surface slope (controllable through a limit on $\partial Q/\partial t$).

Figure 4.1b. It can be seen that this spatial profile is in a sense the mirror image of the time variation in panel (a). This corresponds to the fact that for propagation in the positive *s*-direction, the variations depend on $s - ct$, in which s and t have opposite signs.

4.3 Propagation in Non-Uniform Canals

4.3.1 Rapidly Varying Cross Section

As was noted in Section 4.1, translatory waves are often the result of manipulation of man-made hydraulic structures in man-made canals. Although these canals are for the main part prismatic, they do have variations in cross section, e.g. in the vicinity of navigation locks. Furthermore, there may be a lateral harbour, a natural basin connected to the canal, or a junction between three or more canal reaches.

We will investigate how these geometric variations affect the propagation of translatory waves. We do this on the assumption that the extent of the transition is much shorter than the wavelength, allowing us to treat the transitions as abrupt. Moreover, as elsewhere in this section, the wave height is assumed to be small so that linear approximations can be used.

Consider a transition in canal geometry between two prismatic reaches 1 and 2, such that the conveyance cross-sectional area and the width of the free surface change from $(A_{c,1}, B_1)$ to $(A_{c,2}, B_2)$ (see Figure 4.2). We expect an effect on the ongoing wave as well as reflection to a certain extent. Therefore, we suppose that we have to deal with three waves:

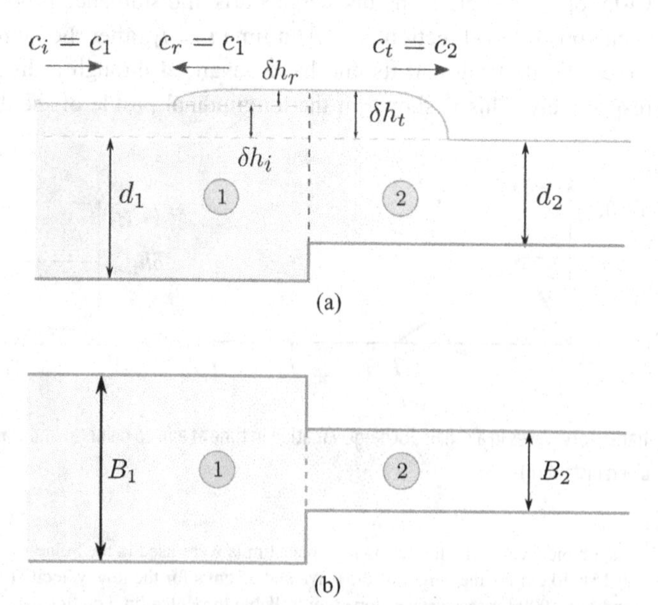

Fig. 4.2 Translatory waves at a transition in canal geometry; longitudinal transect (a) and top view (b)

incident (from reach 1), transmitted (into reach 2) and reflected wave (back into reach 1), with the following parameters:

- incident: $\delta Q_i, \delta h_i$ in reach 1 with $A_{c,1}, B_1$
- reflected: $\delta Q_r, \delta h_r$ in reach 1 with $A_{c,1}, B_1$
- transmitted: $\delta Q_t, \delta h_t$ in reach 2 with $A_{c,2}, B_2$

We assume low waves and neglect wave–wave interactions and their effect on the propagation speed, which in this approximation has the constant values $c_1 = \sqrt{gA_{c,1}/B_1}$ in reach 1, common to the incident wave and the reflected wave, and $c_2 = \sqrt{gA_{c,2}/B_2}$ in reach 2. Because of continuity of discharge and elevation at the transition (neglecting the respective velocity heads, in accordance with the linear approximation), we have

$$\delta h_i + \delta h_r = \delta h_t \tag{4.4}$$

and

$$\delta Q_i + \delta Q_r = \delta Q_t \qquad \text{or} \qquad B_1 c_1 (\delta h_i - \delta h_r) = B_2 c_2\, \delta h_t \tag{4.5}$$

(Q is reckoned positive in the direction of wave incidence.) We now define the following dimensionless ratios:

$$r_t \equiv \frac{\delta h_t}{\delta h_i}, \qquad r_r \equiv \frac{\delta h_r}{\delta h_i} \qquad \text{and} \qquad \gamma \equiv \frac{B_2 c_2}{B_1 c_1} = \sqrt{\frac{A_{c,2} B_2}{A_{c,1} B_1}} \tag{4.6}$$

With these definitions, Eqs. (4.4) and (4.5) are transformed into

$$1 + r_r = r_t \qquad \text{and} \qquad 1 - r_r = \gamma r_t \tag{4.7}$$

which yield the following expressions for the ratios of the heights of the reflected wave and the transmitted wave to the height of the incident wave:

$$r_r = \frac{1-\gamma}{1+\gamma} \qquad \text{and} \qquad r_t = \frac{2}{1+\gamma} = 1 + r_r \tag{4.8}$$

The following points are noted with respect to this result:

- There is no reflection at transitions for which the product Bc remains constant ($\gamma = 1$), even though B and c change.
- Waves experiencing a reduction in the value of Bc are positively reflected and their transmission is enhanced.
- If $B_2 c_2 \ll B_1 c_1$, the reflection is almost 100%, and the transmitted wave height is almost twice the incident wave height (though the discharge in the transmitted wave is relatively small because it is proportional to $B_2 c_2$).
- In the case of a strong enlargement of the cross section ($B_2 c_2 \gg B_1 c_1$), as for waves in a canal approaching a lake or reservoir, the reflection is negative and almost 100% in absolute value, whereas the transmitted wave height is quite small, going to zero in the limit as γ goes to ∞. (This property was already used in the preceding chapter.)

Notice that the designations of positive and negative reflection refer to the surface elevation. A positive reflection of the surface elevation implies a negative reflection for the particle velocity.

If there is a connection of three or more canal reaches, or between the canal and a side basin, the reflection and the transmission of a wave approaching the transition through one of these can be calculated by lumping the others into one equivalent canal, having a Bc-value equal to the sum of the Bc-values of the constituent canals that are being approached by the incident wave. Because the waves considered are long compared with the extent of the junction, they are insensitive to an abrupt change of propagation direction. This has been confirmed empirically by Thijsse (1935).

The height of transient translatory waves can be significantly reduced by a local widening or by a basin connected to the canal. When a positive wave reaches the mouth, a certain volume of water flows into the basin, thereby reducing the volume in the wave travelling down the canal, while at the same time negative reflection occurs. Soon, a mixture of partially reflected waves occurs, resulting in a confused pattern. This is illustrated in Figure 4.3 for the situation of a lock (point A) with a downstream approach canal (AB) and a lateral harbour (BD) at some distance from the lock. The translatory waves generated by the levelling of the lock are transmitted along the canal as well as into the harbour.

Let the Bc-value of the harbour mouth be twice that of the canal. The total Bc-value of these is three times that of the canal itself, so the waves approaching the harbour mouth experience a γ-value of 3. This gives $r_r = -0.5$, or 50% negative reflection back to the lock, and $r_t = 0.5$, or 50% transmission down the canal and into the harbour. Since $\gamma = 1$ for reflected waves approaching the harbour mouth from the closed end of the harbour, these waves are not reflected at the mouth but fully transmitted into both canal reaches.

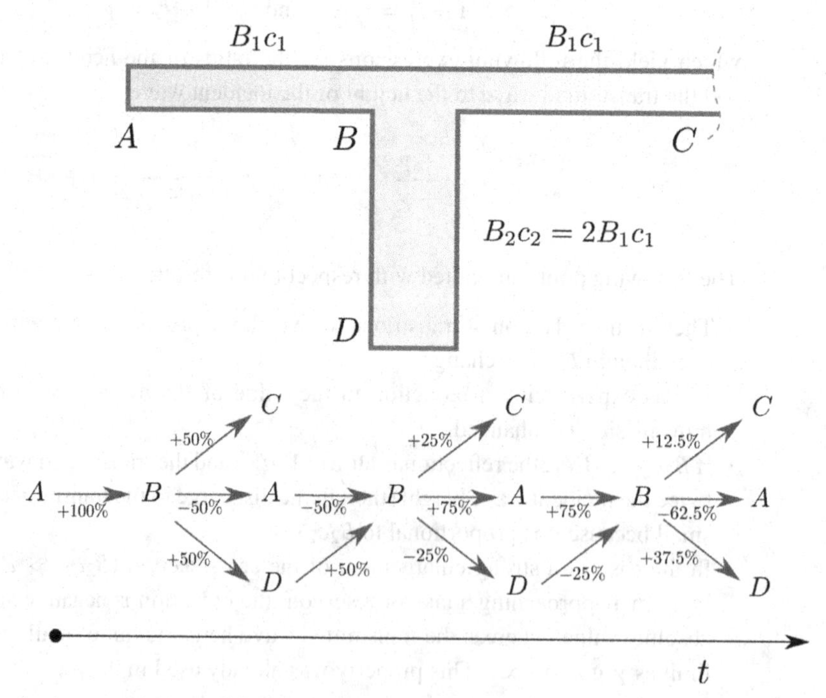

Fig. 4.3 Wave reflection near the shipping lock in Born (The Netherlands); wave travel time in channel AB approximately equals that in harbour BD; wave height is expressed as fractions of the initial wave height

We see that the presence of the harbour causes the height of the initial wave transmitted down the canal beyond the harbour mouth to be only 50% of the original height in the upstream canal reach. Of course, in the long run the total volume of water discharged from the lock must flow down the canal. The remaining 50% is temporarily stored in the harbour and in the upstream canal reach, between the harbour and the lock, undergoing multiple reflections and being released onto the canal going downstream in smaller portions, spread out over time. (See the sequence of transmitted waves travelling downstream, towards point C in Figure 4.3.) Note that this process of wave height reduction is effective only if the basin is of sufficient length compared with the length of the translatory wave. If it is relatively short, the wave reflected off the closed end of the harbour basin joins the primary wave shortly after the latter first reached the mouth, thereby hardly changing its height or profile.

Example 4.2 At a transition between two prismatic canal reaches, the values of A_c and B change from $150\,\text{m}^2$ and 50 m to $250\,\text{m}^2$ and 80 m, respectively. A translatory wave with a height $\delta h_i = 0.35$ m approaches the transition from the side of the narrower canal reach. Calculate:

1. the height of the reflected wave (δh_r)
2. the height of the transmitted wave (δh_t)
3. the discharge (Q) at the transition shortly after reflection of the incoming wave.

Solution

The wave speeds ($c = \sqrt{gA_c/B}$) in the narrow and wide reaches are 5.43 m/s and 5.54 m/s, respectively. This gives a Bc ratio $\gamma = 1.63$ (from narrow towards wider reach).

1. Reflected wave: $r_r = (1-\gamma)/(1+\gamma) = -0.24$ giving $\delta h_r = -0.24 \times 0.35\,\text{m} = -0.08\,\text{m}$
2. Transmitted wave: $r_t = 2/(1+\gamma) = 0.76$ giving $\delta h_t = 0.76 \times 0.35\,\text{m} = 0.27\,\text{m}$
3. Discharge: $Q = Q_i(1 - r_r)$, where Q_i is the discharge of the incoming wave ($95\,\text{m}^3/\text{s}$). So $Q = 95\,\text{m}^3/\text{s} \times 1.24 = 118\,\text{m}^3/\text{s}$

Since the discharge is continuous at the transition it equals the discharge of the transmitted wave $Q_t = Bc\,\delta h_t$ (using Bc of the wide canal reach).

4.3.2 Gradually Varying Cross Section

In the above, we have dealt with abrupt transitions, at which partial reflection occurs, possibly with a change in sign, and a change in the height of the transmitted wave. Let us now consider the wave propagation in the case of gradual profile variations, such that the changes of width and depth within a wavelength are relatively small.

A simple model for the resulting wave height variations is due to Green (1837). His result can be derived on the basis of energy conservation. To see this, we first consider a low translatory wave of finite length, progressing without change of form and height in a prismatic canal in which initially the water is at rest. Its energy moves along with it.

Let us consider the energy contained in a long wave. Disturbing the surface of a water mass in equilibrium in the field of gravity requires work to be done against gravity. A rise in water level over a distance ζ (for brevity, this symbol is used here instead of δh) corresponds to a raised mass $\rho\zeta$ per unit area, with its center of mass at a height $\frac{1}{2}\zeta$ above the equilibrium level, corresponding to an amount of work against gravity given by $\frac{1}{2}\rho g\zeta^2$ per unit area. (Alternatively, this result can be obtained by integrating the gravity potential per unit mass, gz, from $z = 0$ to $z = \zeta$.)

The kinetic energy per unit area is $\frac{1}{2}\rho U^2 d$. For a low progressive wave we have $U = c\zeta/d$; see Eq. (3.8). Substituting this into the expression for the kinetic energy shows that the potential and kinetic energy have the same value (this is true in lowest order for all free waves) and that the total energy per unit area is $\rho g\zeta^2$. It follows that the total energy of a wave with length L in a canal of width B can be written as $\rho gBL\zeta_{rms}^2$, in which ζ_{rms} is the root mean square value of ζ, where the mean is taken over the entire wave.

For waves propagating in a channel of slowly varying cross section, such that the changes in depth and width within the length of the wave are relatively small, the reflection is negligible, and the total energy of the forward-progressing wave may be assumed to be conserved during the propagation. This implies that the wave height ζ_{rms} varies in proportion to $(BL)^{-1/2}$.

If the propagation speed c varies in space, the front and the rear of the wave move at different speeds, causing a change in wavelength. However, the propagation speed is assumed to depend on the location only, independent of time or the wave properties, so that all points of the wave go through the same sequence of time intervals to traverse given space intervals. This implies that the time it takes for a wave of finite length to pass a fixed point is the same for all points. Since this duration is L/c, it follows that the wavelength L varies in proportion to $c = \sqrt{gd}$, which leads to *Green's law*:

$$\zeta_{rms} \propto B^{-1/2} d^{-1/4} \tag{4.9}$$

It can be shown that Green's law and the result for the transmission at an abrupt transition, presented in the preceding section, are mutually consistent (see Box 4.1).

Box 4.1 Green's law and abrupt transitions

Consider the continuous profile variation as the limit of a sequence of small but abrupt variations. For a weak but abrupt transition, we write $\gamma = B_2c_2/B_1c_1 = 1 + \delta(Bc)/B_1c_1 = 1 + \epsilon$ in which $\epsilon = \delta(Bc)/B_1c_1$ is small compared with unity, going to zero in the limit. According to Eq. (4.8), the corresponding value of the transmission coefficient is $r_t = 2/(1+\gamma) = 2/(1+1+\epsilon) = 1 - \frac{1}{2}\epsilon + \mathcal{O}(\epsilon^2)$. On the other hand, from Green's law we have $r_t = \zeta_2/\zeta_1 = (B_2c_2/B_1c_1)^{-1/2} = (1+\delta(Bc)/B_1c_1)^{-1/2} = (1+\epsilon)^{-1/2} = 1 - \frac{1}{2}\epsilon + \mathcal{O}(\epsilon^2)$, which for small ϵ is the same as for the abrupt transition.

4.4 Damping of Translatory Waves

As stated in Section 4.1, the flow in a translatory wave is assumed to vary so rapidly that resistance can be neglected. This is a reasonable assumption on the time scale of these variations, but at longer times the effect of the continuously working resistance cannot be ignored, since it causes a gradual damping of the wave.

In the following, we will make an estimate of the rate of damping of a transient translatory wave propagating through a region of rest, as for a translatory wave in a canal caused by the levelling of the water in a navigation lock, in which case the discharge goes from zero to a maximum and then back to zero. We base the calculation on a bulk energy balance for the entire wave.

It was seen above that the total energy per unit area is $\rho g \zeta^2$. This decreases as a result of the work done to overcome boundary resistance, at the rate $\tau_b U$ or $c_f \rho |U|^3$ per unit area, where we have used the quasi-steady approximation for the boundary resistance given by Eq. (1.16). This is somewhat questionable for rapidly varying flow as in translatory waves, as was seen in Chapter 2, but it should suffice for an indication. Integrating the above-mentioned quantities over an interval of the canal encompassing the translatory wave at all times, and equating the results, we obtain

$$\frac{\mathrm{d}}{\mathrm{d}t} \int \rho g \zeta^2 \, \mathrm{d}s = - \int c_f \rho |U|^3 \, \mathrm{d}s \tag{4.10}$$

We substitute $U = c\,\zeta/d$ and introduce the dimensionless surface elevation profile, written as $\eta = \zeta/\hat{\zeta}$, in which $\hat{\zeta}$ is the maximum of ζ, the variable whose evolution in time we seek. Eq. (4.10) then becomes

$$\frac{\mathrm{d}}{\mathrm{d}t} \left(\rho g \hat{\zeta}^2 \int \eta^2 \, \mathrm{d}s \right) = -c_f \rho \frac{c^3}{d^3} \hat{\zeta}^3 \int |\eta|^3 \, \mathrm{d}s \tag{4.11}$$

Neglecting the change in shape (not the height) of the free-surface profile allows us to take the integral in the left-hand side outside the differential operator and to define the following time-independent dimensionless profile shape parameter α:

$$\alpha \equiv \frac{\int \eta^2 \, \mathrm{d}s}{\int |\eta|^3 \, \mathrm{d}s} \tag{4.12}$$

It may be argued that a quadratic damping deforms the surface elevation profile, because the higher parts are damped more strongly than the lower ones as a result of the quadratic resistance, but that is a secondary effect. Suppose that the initial profile is a half-sine function; the corresponding value of α is $\frac{3}{8}\pi$, or about 1.2. When the wave flattens, the value of α reduces, with a minimum of 1 in the case of a completely flat profile of uniform height. This justifies the approximation of a constant α in the present order-of-magnitude calculation.

Substituting Eq. (4.12) into Eq. (4.11), we obtain the following ordinary differential equation for the time-varying maximum surface elevation:

$$\frac{\mathrm{d}\hat{\zeta}^2}{\mathrm{d}t} = -\frac{c_f}{\alpha} \frac{c^3}{g d^3} \hat{\zeta}^3 \tag{4.13}$$

which can be integrated to yield

$$\frac{1}{\hat{\zeta}(t)} = \frac{1}{\hat{\zeta}_0} + \frac{c_f}{2\alpha} \frac{c^3}{g d^3} t \tag{4.14}$$

in which $\hat{\zeta}_0$ is the initial value of $\hat{\zeta}$. This is a hyperbolic type of damping, as is common for systems with quadratic resistance. Recasting this relation into an expression proportional to $\hat{\zeta}(t)$, and substituting $t = s/c$ with $c = \sqrt{gd}$, to transform the time dependence into a space dependence, the following final result is obtained:

$$\hat{\zeta}(s) = \frac{\hat{\zeta}_0}{1 + s/S} \tag{4.15}$$

in which S is a relaxation length defined by

$$S = \frac{2\alpha}{c_f} \frac{d^2}{\hat{\zeta}_0} \tag{4.16}$$

It can be seen from Eq. (4.15) that S represents the propagation distance over which the initial amplitude is halved (ca. 10–100 km for realistic canal parameters and wave heights). The above result seems to have first been published by Thijsse (1935), although without derivation[2] and with a different definition of the present factor α. A comparison with field data is presented in Section 4.6.

Thijsse also showed that due to the gradual damping, a positive transient translatory wave leaves a slight rise of the free-surface level ($\delta\zeta$) behind, equal to the integral of the friction slope over the length of the wave. This compensates for the loss of height of the wave (whose length does not increase since the front and rear end travel at the same speed) so that the total volume is conserved. Since this integral itself diminishes in time, so does $\delta\zeta$, setting up a weak slope behind the wave and a corresponding weak flow, until in the end a state of rest is reached (in absence of new disturbances) in which the total volume released into the canal is spread evenly over the entire canal length.

4.5 High Translatory Waves

We will now abandon the assumption of (very) low waves, and do not make the associated simplifying approximations of neglecting the advective acceleration and the effects of a variable surface elevation on the channel cross-section geometry (the instantaneous values of A and B). Because the resulting equations are nonlinear, their mathematical treatment is much more complicated than it is for low waves described by linear equations. In the

[2] Thijsse did give a derivation in his hydraulics lectures at the Delft University of Technology (written lecture notes of the author JB, 1960), although this was based not on an energy balance but on the Euler equation with quadratic resistance.

present chapter, we will mention only a few characteristic consequences of the inclusion of nonlinear terms. A more complete treatment is presented in Chapter 5.

4.5.1 Wave Deformation

Within the framework of the linear theory in the preceding chapters, the effects of the presence of a (low) disturbance on the velocity of propagation were neglected. As a consequence, propagation in a prismatic channel without change in shape is possible (in the case of absence of resistance). In the case of higher waves, their effect on the instantaneous depth and thereby on the velocity of propagation is no longer negligible. Various theories have been developed to account for this, all having in common that *the waves deform as they propagate*.

The occurrence of wave deformation can be made plausible, without a formal mathematical derivation, using the relation $c = \sqrt{gA_c/B} = \sqrt{gd_B}$, in which d_B is the average depth of the entire cross section (not of the conveyance cross section only). This relation was derived for low waves propagating without change in shape, but it should by approximation also be valid for slowly varying waves, using the instantaneous local depth. This implies that higher portions of a wave propagate at greater speed than lower portions.

Furthermore, the given expression for c applies to the wave speed relative to the water mass ahead of it. The effect of the wave-induced (as opposed to pre-existing) flow velocity on the wave propagation speed was neglected in the previous section. By taking it into account, the wave speed relative to the bottom becomes $U \pm c$, with c the wave speed relative to the water as given above, the + and − signs being applicable to waves propagating in the positive and the negative s-direction, respectively. For a positive wave propagating in the positive s-direction, the wave-induced flow velocity U is positive, enhancing the velocity of propagation relative to the bottom for the higher parts of the wave even further. It follows that the wave deforms: positive waves steepen (see Figure 4.4), negative waves flatten out. Because a positive wave steepens, it may eventually develop into a shock wave, a so-called bore. An exact expression for the speed of a point of constant d and U, including the effects of finite wave height and wave-induced flow velocity, is derived in Chapter 5, using the method of characteristics (Eq. (5.24)). The result for a wave advancing into quiescent water with depth $d = d_0$ is given by

$$\frac{\mathrm{d}s}{\mathrm{d}t} = U + c = 3\sqrt{gd} - 2\sqrt{gd_0} \tag{4.17}$$

in which $\mathrm{d}/\mathrm{d}t$ denotes the total derivative, following a point of constant depth d. This shows explicitly how the propagation speed increases with increasing depth. For a finite but small relative wave height ζ/d_0, in which $\zeta = d - d_0$, this can be approximated as

$$\frac{\mathrm{d}s}{\mathrm{d}t} = 3\sqrt{gd_0\left(1 + \zeta/d_0\right)} - 2\sqrt{gd_0} = \sqrt{gd_0}\left(1 + \tfrac{3}{2}\zeta/d_0\right) + \mathcal{O}\left((\zeta/d_0)^2\right) \tag{4.18}$$

In the derivation of Eq. (4.17), friction was not considered. Its effect is a gradual damping of the wave. If the rate of friction-induced damping outweighs the rate of inherent steepening, it may even prevent a shock wave from developing. Another reason why a

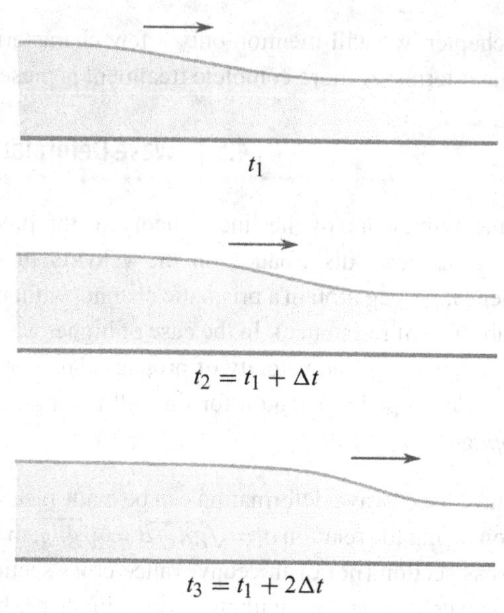

Fig. 4.4 Steepening of a positive wave

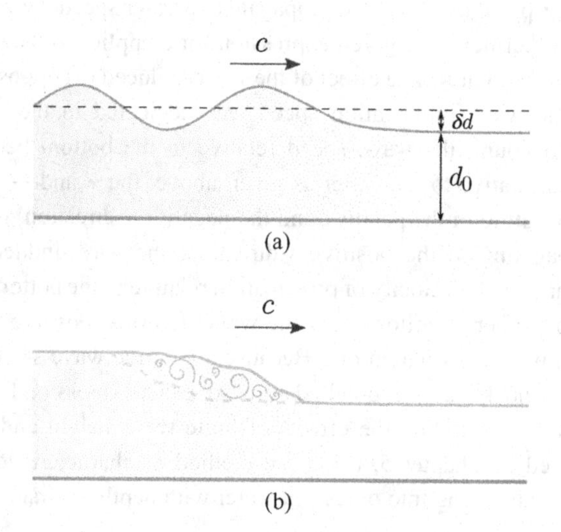

Fig. 4.5 Undular bore (a) and turbulent bore (b)

steepening wave does not necessarily develop into a bore is the fact that as the waves become steeper, vertical accelerations increase, so that the pressure deviates more and more from being hydrostatic. This effect, not included in the long-wave approximation, delays the formation of a shock wave. (In fact, it allows the presence of certain nonlinear waves that do not deform at all.)

There are two types of bores. For sufficiently low bores (relative bore height $\delta d/d_0$ less than about 0.28), the bore is undular; its surface is smooth and wavy (Figure 4.5a). Higher bores are turbulent; in fact, they are travelling hydraulic jumps (Figure 4.5b).

4.5.2 Tidal Bores

Shock waves in free-surface flows can develop as a result of human manipulation of control structures, pumping stations etc., but also as the result of natural processes. Within the latter category, the so-called *tidal bores* are the most common and best known. They occur at a number of coastal bays and estuaries on earth at rising tide, particularly spring tide, appearing as an undular or a turbulent front moving inland, causing a significant rise in the local surface elevation in a short time. An overview of tidal bores occurring around the world is given in Chanson (2012).

As an illustration, we present some details of the bores occurring in the estuary of the River Severn in Great Britain (Rowbotham, 1983). The Severn discharges into the Bristol Channel. The tides, approaching from the Atlantic Ocean, where they already have a significant amplitude, are enhanced in this channel with a factor of about 3 as a result of its funnel shape, narrowing in the inland direction. As a result, the tides in the Severn estuary are quite strong, reaching 15 m tidal range at equinoctial spring tides.

The incoming, rising tide front is steepened as it propagates inland. For sufficiently high tides, a tidal bore develops somewhere in the estuary. This happens about 250 times a year. Most of the time, the bore occupies the entire width of the estuary, advancing at a speed of a few meters per second. Depending on the height of the tide and the local depths, the bore may be undular or turbulent; both types can be present simultaneously: undular in the deeper parts of a cross section and turbulent near the shallow banks, as can be seen in Figure 4.6. An individual bore in the Severn estuary exists for about 2 h and traverses a distance of about 30 km, after which it vanishes.

The highest tidal bores occur in the Qiantang River in Hangzhou, in the south-east of China, which can reach a height of 4 m. They travel at a speed of some 40 km/h, spanning

Fig. 4.6 Tidal bore in the River Severn (UK); from Rowbotham (1983)

the entire width of the estuary of several kilometres. Their approach, accompanied by a strong roar, can be heard from a great distance.[3]

Whether or not a tidal bore develops depends on the strength of the tide and on the longitudinal profile of the width and depth of the estuary. A funnel shape and decreasing depths are conducive to bore formation. Due to dredging, several tidal bores have disappeared from the scene. An example is the bore in the Seine river in France, which was a regular phenomenon prior to major dredging works, even reaching Paris where its rushing between the river banks, in the heart of this metropole, was a popular sight for residents and tourists.

4.5.3 Bore Propagation

In this section, we derive an expression for the velocity of propagation of a bore, or shock wave in open water, with height Δd, which forms a moving transition between two regions of uniform flow with depths d_0 and $d_1 = d_0 + \Delta d$, respectively. Ahead of the wave, the water is at rest ($U_0 = 0$); behind it the flow velocity is U_1. We ignore whether the shock is undular or turbulent. Contrary to the derivation presented in Section 3.1, using low-wave approximations, we will here include the advection of momentum and use a more exact expression for the total pressure force.

The long-wave theory does not apply to the flow in the shock region. Instead, we use balance equations for the mass and momentum inside a control volume encompassing the shock, indicated in Figure 4.7 by the chain line. The volume balance per unit width reads

$$U_1 d_1 = c\,\Delta d = c\,(d_1 - d_0) \tag{4.19}$$

and the momentum balance

$$\frac{1}{2}\rho g {d_1}^2 + \rho {U_1}^2 d_1 - \frac{1}{2}\rho g {d_0}^2 = \rho U_1 d_1 c \tag{4.20}$$

Elimination of U_1 from these two equations yields after some algebraic manipulation

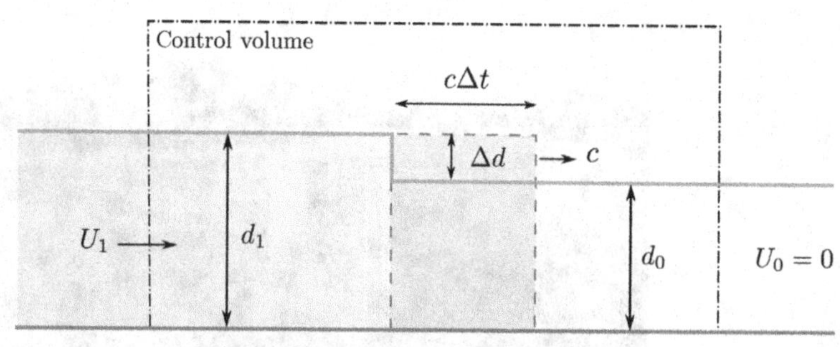

Fig. 4.7 Control volume containing a shock wave propagating into quiescent water

[3] A number of spectacular videos and photographs can be seen on YouTube where there is also some erroneous information, such as that the bore height can reach 9 m and that it is an annual phenomenon; the latter misinformation is probably due to a mix-up with the fact that it is common for Chinese people to gather at the site of the bore on the occasion of one of the annual Chinese festivals.

$$c = \sqrt{g \frac{d_0 + d_1}{2} \frac{d_1}{d_0}} \tag{4.21}$$

This is the propagation speed of a shock wave entering quiescent water. More generally, we can say that this is the propagation speed relative to the water ahead of the wave. If the initial velocity U_0 is nonzero, the wave speed relative to the fixed bed is $U_0 + c$. Because $d_1 > d_0$, it follows that $c > \sqrt{gd_0}$ (and even $c > \sqrt{gd_1}$). We see that the finite wave height causes the wave propagation speed to exceed the linear-theory value.

4.6 Field Observations

4.6.1 Observations in the Twenthekanaal

Thijsse (1935) has presented the results of an extensive measurement campaign in a canal in The Netherlands (the Twenthekanaal), aimed at a verification of the theory of translatory waves. Although the data are old, they are unmatched in scope, both with respect to the propagation distance (75 km) and propagation time (more than 4 h) and with respect to the number of aspects that have been covered. For this reason, we present some of these data and results here.

Layout

The measurements were taken in a canal under construction, situated between the river IJssel (a branch of the Rhine) and a location 28 km inland (see Figure 4.8). A lock and a pumping station are situated side by side in the canal at a distance of about 3 km from the junction with the river, for which reason the canal is locally split into two short, parallel branches.

At a distance of some 25 km from the lock, the canal (under construction) is closed by a temporary barrier. At km 17, a small stream ('Bolksbeek') issues into the canal.

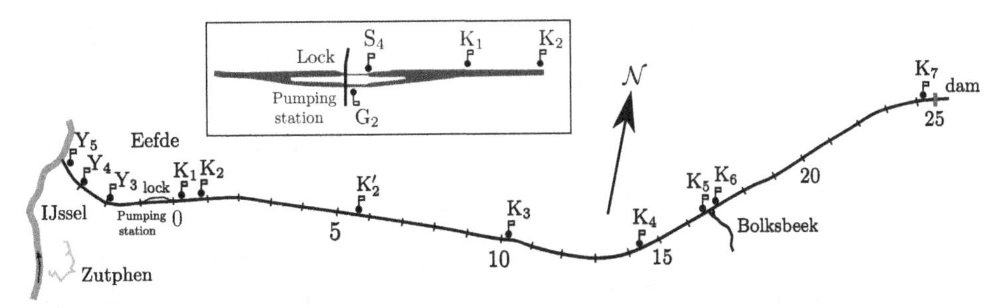

Fig. 4.8 Plan of the Twenthekanaal: lock and pumping station situated between stations Y_3 and K_1 (see inset for details); upstream gauging stations K_1 through K_7; numbers indicate distance (km) from lock; after Thijsse (1935)

Downstream of this point, i.e. towards the lock, the canal dimensions are larger than those upstream to accommodate the occasional discharge from this stream.

At the day of the observations, the water level in the river was about 7 m below that in the canal. Levelling of the lock took about 7 min, with a maximum discharge of some 50 m^3/s. Simultaneous with the levelling of the lock, and for more than 4 h after that, surface elevations were observed in a series of points in the lock chamber and along the canal. Here, we use only the data for the 25 km-long canal reach upstream from the lock (stations K_1 through K_7; see Figure 4.8).

Primary Wave

Levelling of the lock results in a negative translatory wave travelling upstream into the canal. It is negatively reflected at the junction with the canal branch connecting to the adjacent pumping station (see Figure 4.9, inset), and partially transmitted into the upstream canal. As time goes on, the reflected waves are re-reflected at the lock, the pumping station and the junction, resulting in a confused pattern, as in Figure 4.3.

Figure 4.9 shows a negative wave as observed and calculated at station K_1, the first station in the prismatic canal, taking account of the multiple reflections and partial transmissions. Overall, the agreement is good, particularly during the first 12 min or so, in which the major changes occur, lending support to the validity of the theoretical relations involved, i.e. $c = \sqrt{gA_c/B}$, $\zeta = Q/Bc$, and the expressions for the reflection and the transmission at abrupt transitions.

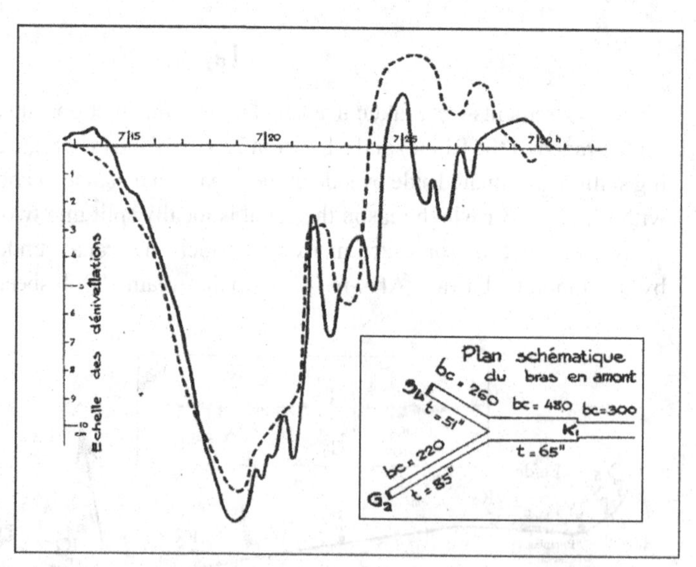

Fig. 4.9 Observed (solid) and calculated (dashed) surface elevation of a negative translatory wave in the Twenthekanaal at station K_1 due to levelling of the lock; the inset shows the schematization of the short canal reaches between the lock (S_4), the pumping station (G_2) and the beginning of the prismatic canal reach (station K_1), with corresponding travel times (in s) and Bc-values (in m^2/s); after Thijsse (1935)

Thijsse did not present details of his calculation, but we can retrace some of the steps. The discharge into the lock chamber was calculated from the time variation of the water level in the lock. From this, the surface elevation in the resulting negative translatory wave and its multiple reflections was computed using the Bc-values of the short, schematized canal reaches between the lock and the prismatic canal, shown in Figure 4.9, inset. At the site of the lock, and in absence of reflections, the surface elevation at the moment of maximum discharge (51 m^3/s) is estimated at $-Q/Bc = -19.6$ cm. This primary wave passes the junction ($r_t = 0.54$) and the transition to the prismatic canal reach at station K_1 ($r_t = 1.23$), so that its height at station K_1 is estimated at $-0.54 \times 1.23 \times 19.6$ cm $= -13.0$ cm. The observed minimum is -13.4 cm, but this includes the effects of multiply reflected lower disturbances of earlier origin, arriving at K_1 at the same time as the minimum elevation of the primary wave. We cannot reproduce Thijsse's calculation of this effect with precision because the time variation of Q is insufficiently known.

Propagation and Reflection

Figure 4.10 presents the observed surface elevations at the stations K_1 through K_7 as a function of time. It shows at a glance the path of the leading edge of the main wave (thick black lines) from the lock towards the barrier at 25 km from the lock, back to the lock and then again towards the barrier, a total distance of 75 km, which is traversed in about 4 h.

The theoretical speed of propagation of the leading edge is $\sqrt{gd_0}$, because at that point the wave height is still zero. Thijsse noted that the value determined from the observed wave path in s, t-space agreed with this theoretical value in some stretches, but fell short by up to 7% in others. He made an estimate of the reduction in the theoretical propagation speed due to resistance, and found that that could not explain the discrepancy (and taking this into account would of course harm the agreement where that was found in absence of resistance).

Close inspection of Figure 4.10 shows that the trough of the wave is lagging behind the leading edge. This can be ascribed to the decrease in depth, and the counteracting flow velocity, which result from the negative wave. Thijsse states that the ratio between the two observed speeds was well represented by the factor $\left(1 + \frac{3}{2}\zeta/d_0\right)$. This agrees with the first-order approximation to the exact value, given in Eq. (4.18).

Where the primary wave reaches the Bolksbeek, some 17 km upstream from the lock, it experiences a reduction in Bc-value from 250 m^2/s to 185 m^2/s approximately, with a weak positive reflection as a result, observed and theoretically about 15%. (Note that 'positive reflection' means reflection without change of sign, so that a positively reflected negative wave is in fact negative.) Though weak, this reflected wave is clearly identifiable up to the lock where it is again positively reflected, from where it can be traced all the way back to the Bolksbeek (thick grey lines). These observations are consistent with the theory.

Damping

To investigate the rate of damping, we must use data on purely progressive waves, without interference with possible reflections. Keeping this in mind, we can start the comparison

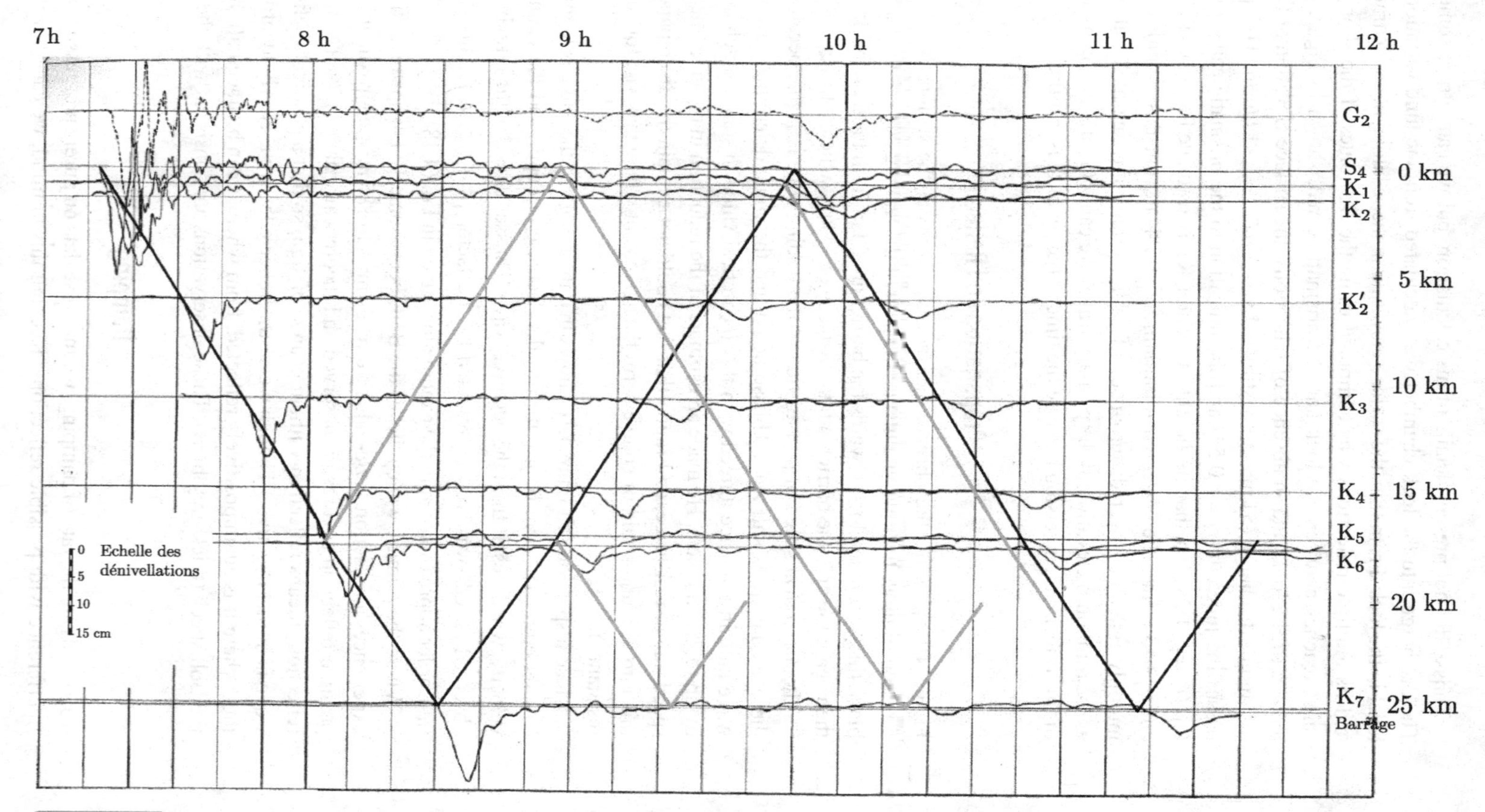

Fig. 4.10 Measured surface elevations in the Twenthekanaal showing multiple reflections and gradual damping; lock situated at station S_4; after Thijsse (1935)

Table 4.1 Computed and observed wave heights along various sections of the Twenthekanaal

Path	Length	$\hat{\zeta}_0$ observed	S	$\hat{\zeta}_e$ observed	$\hat{\zeta}_e$ computed
K_2–K_4	14 km	−13.1 cm	55 km	−10.4 cm	−10.5 cm
K_4–K'_2	9 km	−5.3 cm	135 km	−3.7 cm	−4.9 cm
K'_2–K_4	9 km	−3.3 cm	230 km	−2.9 cm	−2.8 cm

between observed and calculated values at station K_2, the first station in the prismatic reach of the canal, and follow this primary wave to station K_4 and hence back to K'_2 and from there again to K_4.

Table 4.1 lists the values of the observed minimum elevation at the start (subscript 0) and at the end (subscript *e*) of each of these three paths, the values of the respective damping length scale *S* defined in Eq. (4.16), in which we have used a representative canal depth of 3.5 m, $\alpha = 3\pi/8$ and $c_f = 0.004$, and finally the values of the minimum elevations computed with Eq. (4.15). The agreement is better than might be expected in view of the approximations involved in the modelling of the resistance.

4.6.2 Observations in the Approach Canal to the Lanaye Lock

A remarkable effect of lock-induced translatory waves occurs in the approach canal between the Lanaye lock in Belgium and the river Meuse at the border between Belgium and The Netherlands, namely the occurrence of a free oscillation of the water mass in the canal (Figure 4.11), generated by the levelling of the lock (Bertrand and Hiver, 1998; see also PIANC, 2015).

The origin of the oscillation can be explained as follows. The positive surge generated in the approach canal by the levelling operation is almost fully but negatively reflected at the junction with the river, some 1700 m downstream from the lock, because the river has a much larger cross section than the canal, and its *Bc*-value counts twice because the transmitted wave can travel along the river in both the downstream and the upstream directions.

A particular circumstance conducive to the generation of a free oscillation of the water mass in the canal is that the travel time of a translatory wave from the lock to the river and back again happens to be very near to the duration of levelling of the lock, both about 9 min. Therefore, at the time when the levelling process is completed, the negative wave reflected off the river has just reached the lock. At that instant, the surface elevation of the outgoing primary wave is more or less annulled by the negatively reflected wave over the entire length of the canal, assuming that the time variation of the surface elevation of the primary, outgoing wave is symmetric with respect to the time of occurrence of its maximum value, 4.5 min after the start. The particle velocities, doubled by the negative reflected wave, are then at their maximum, directed towards the river.

The situation just described corresponds to that in a quarter-wavelength free oscillation of the canal (see Chapter 3) at the instant when this goes through its equilibrium configuration, i.e. a horizontal water surface and an outgoing flow, with the boundary

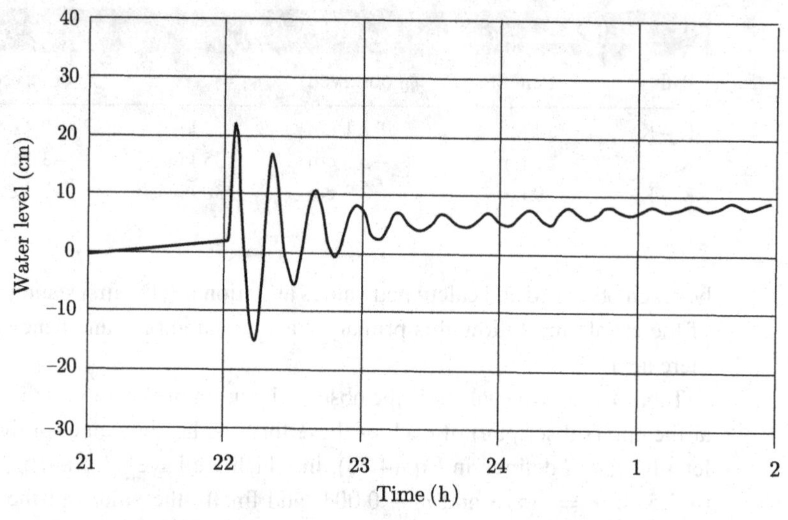

Fig. 4.11 Free oscillations in the approach canal to the Lanaye lock; from Bertrand and Hiver (1998)

conditions $Q = 0$ at the lock and $\zeta = 0$ (say) at the junction with the river. From that moment on the free oscillation sets in, as shown in Figure 4.11.

Data on the discharge and the geometry of the canal allow a rough check on the theoretical expressions for the propagation speed and the initial height of the wave. The comparison cannot be precise because a range of values has been given for the width and depth of the canal, rather than point values.

PIANC (2015) lists the following data: maximum discharge $Q = 108\,\text{m}^3/\text{s}$, approach canal length 1700 m, free-surface width B from 50 to 70 m, depth d from 4.0 to 4.5 m. Using the mid-range value of d, we obtain $c = \sqrt{gd} \approx 6.45$ m/s, giving a travel time between lock and river of 1700 m/(6.45 m/s), or about 4.4 min, close to the observed value of 4.5 min. Likewise, using the mid-range value $B = 60$ m, the initial amplitude of the surface elevation is estimated at $\zeta = Q/Bc \approx 28$ cm, compared with an observed value of 24 cm.

Problems

4.1 What are translatory waves?

4.2 A control structure connects an irrigation canal ($A_c = 100\text{ m}^2, B = 30$ m) with a reservoir. At an initial head difference of 4 m, a gate is abruptly opened to an effective flow area $\mu A = 5\text{ m}^2$. Calculate the discharge and the height of the resulting translatory wave in the canal for two cases: (a) neglecting the effect of the wave on the head difference and (b) taking that effect into account.

4.3 Verify the percentages in the scheme in Figure 4.3.

4.4 Answer the questions in Example 4.2 for waves approaching the transition from the other side.

4.5 Mention a few theoretical consequences of the distinction between so-called low translatory waves and high translatory waves.

4.6 Why do (high) translatory waves deform?

4.7 What is a characteristic difference between the deformation of positive translatory waves and of negative ones?

4.8 How does resistance affect the deformation of translatory waves?

4.9 Mention some factors favouring the formation of tidal bores.

4.10 Why are there fewer tidal bores at present than in older days?

4.11 Derive an expression for the velocity of propagation of a bore entering a region with depth d_0 and flow velocity U_0.

5 Method of Characteristics

A general feature of wave phenomena is the transmission of information and energy through a physical system at a finite speed. A disturbance brought about somewhere in the system, e.g. due to operation of a control structure in an irrigation system, reaches other locations after a finite time. Insight into this phenomenon is important both for the purpose of effective control of water levels and discharges in the system and for performing the required computations. The so-called method of characteristics lends itself particularly well to this purpose because it makes visible how disturbances travel through the system and it enables their computation. It was developed by Massau (1878).

5.1 Introduction

In this chapter, we use the mass balance and the momentum balance without the low-wave approximations. Flow resistance is not included except for a minor reference.

As before, we restrict ourselves to one-dimensional systems, schematically represented by the s-axis, and consider the varying position of a disturbance in the course of time. This can be represented as a curve in the *(s, t)-plane* whose slope $\mathrm{d}s/\mathrm{d}t$ equals the local propagation speed of the disturbance. Such curves are called *characteristics*. They portray how information travels through the system, as illustrated in Figure 4.10.

The balance equations for mass and momentum for one-dimensional wave phenomena form a set of two partial differential equations for two dependent variables, such as the depth (d) and the discharge (Q), as functions of two independent variables (s, t). The two dependent variables are called *state variables*. The instantaneous values of these can be represented as a point in the *state plane*, a plane with the two state variables as coordinates.

Given a set of sufficient initial and boundary conditions, the solution of the set of partial differential equations is determined. Expressed in terms of d and Q, this solution is a set of values $d\,(s, t)$ and $Q\,(s, t)$, which can be represented as a surface in the (d, s, t)-space and the (Q, s, t)-space, respectively, the so-called *integral surfaces*, depicted schematically in Figure 5.1. In finite-difference methods for the integration of the partial differential equations, the values of d and Q are determined for a set of points in the (s, t)-plane chosen beforehand, for instance according to a rectangular grid with finite differences Δs, Δt (see Chapter 11). In the method to be dealt with in this chapter, the computations proceed along specific paths in the (s, t)-plane, the characteristics. The advantage of this is that (as we will see) the set of two partial differential equations in s, t (the balances of mass and momentum)

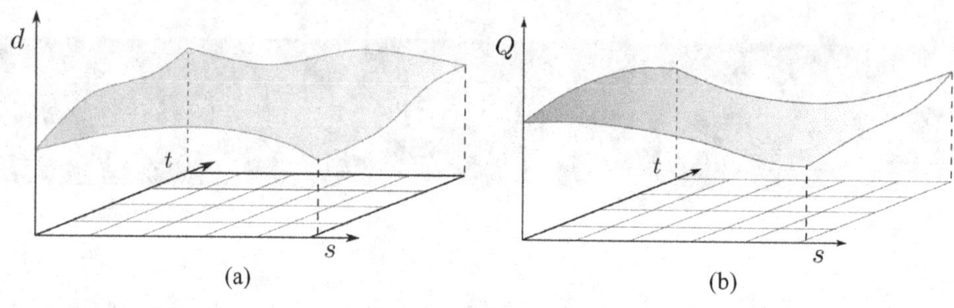

Fig. 5.1 Integral surfaces for d (a) and Q (b)

is replaced by a set of ordinary differential equations in t, simplifying the solution. Note that a similar operation was carried out on h^+ in Section 3.2.3, where the partial differential equation given by Eq. (3.19) was replaced by the ordinary differential equations given by Eq. (3.20).

As stated above, the slope of a characteristic (ds/dt) equals the local propagation speed, which depends on the depth and the flow velocity, which are not known beforehand. That is why the position of the characteristics has to be determined as part of the solution. For low waves, the propagation speed does not depend on the disturbance, in which case the slope of the characteristics is determined beforehand. Needless to say, this simplifies the solution greatly. This low-wave approximation has already been used extensively in preceding chapters.

The introductory remarks made above have given a broad indication of the essence of the method of characteristics. The formal mathematical formulation is presented in Section 5.2, which is followed by an outline of the principle of the application in Section 5.3. Examples are given in subsequent sections.

5.2 Mathematical Formulation

For simplicity, we restrict the formulation to two-dimensional motions (i.e. no lateral variations) over a horizontal bottom. We describe it in terms of the state variables U and d.

We recall the long-wave equations from Chapter 1. The continuity equation for one-dimensional motion is

$$\frac{\partial d}{\partial t} + \frac{\partial Ud}{\partial s} = 0 \tag{5.1}$$

whereas the equation of motion (without resistance) reads

$$\frac{\partial U}{\partial t} + U\frac{\partial U}{\partial s} + g\frac{\partial d}{\partial s} = 0 \tag{5.2}$$

(Because the bottom has been assumed to be horizontal, we can replace $\partial h/\partial s$ by $\partial d/\partial s$.) Instead of using d as one of the state variables, we use its equivalent, c, defined by $c \equiv \sqrt{gd}$. This simplifies the algebra without loss of information. Note that it is not necessary to assign a physical meaning to this quantity at this stage (as if we have no prior knowledge about it from preceding chapters).

By substitution of $d = c^2/g$ into Eq. (5.1) and division of the result by c/g, and substitution of $d = c^2/g$ into Eq. (5.2), we obtain

$$\frac{\partial 2c}{\partial t} + c\frac{\partial U}{\partial s} + U\frac{\partial 2c}{\partial s} = 0 \tag{5.3}$$

and

$$\frac{\partial U}{\partial t} + U\frac{\partial U}{\partial s} + c\frac{\partial 2c}{\partial s} = 0 \tag{5.4}$$

Adding these two equations yields

$$\frac{\partial (U + 2c)}{\partial t} + (U + c)\frac{\partial (U + 2c)}{\partial s} = 0 \tag{5.5}$$

For brevity, we define

$$R^+ = U + 2c \tag{5.6}$$

Using this shorthand notation, Eq. (5.5) becomes

$$\frac{\partial R^+}{\partial t} + (U + c)\frac{\partial R^+}{\partial s} = 0 \tag{5.7}$$

The left-hand side of this equation represents the total derivative of R^+ with respect to t for an observer moving at a speed $\mathrm{d}s/\mathrm{d}t = U + c$ in the positive s-direction (see Section 3.2.3). Because the right-hand side is zero, such observer sees no change in the value of R^+. Stated another way: a constant value of the state variable $U+2c$ is propagated at speed $U + c$ along the s-axis. Therefore, a formulation equivalent to Eq. (5.7) is

$$\frac{\mathrm{d}R^+}{\mathrm{d}t} = 0 \quad \text{provided that} \quad \frac{\mathrm{d}s}{\mathrm{d}t} = U + c \tag{5.8}$$

Next, we perform the same operations with change of sign. Subtracting Eq. (5.4) from Eq. (5.3), the result is

$$\frac{\partial (U - 2c)}{\partial t} + (U - c)\frac{\partial (U - 2c)}{\partial s} = 0 \tag{5.9}$$

Defining

$$R^- = U - 2c \tag{5.10}$$

we obtain the following equivalent of Eq. (5.8):

$$\frac{\mathrm{d}R^-}{\mathrm{d}t} = 0 \quad \text{provided that} \quad \frac{\mathrm{d}s}{\mathrm{d}t} = U - c \tag{5.11}$$

The velocities $U \pm c$ are called the *characteristic velocities*, and curves in the (s, t)-plane for which $\mathrm{d}s/\mathrm{d}t = U \pm c$ are called *positive characteristics* and *negative characteristics*, to be labeled as K^+ and K^-, respectively. (These are labels, not quantities.) The so-called *characteristic relations* (5.8) and (5.11) imply that R^+ is constant along K^+ and that R^- is constant along K^-.

The quantities R^+ and R^- (in some places taken together as $R^\pm$ for brevity) are called *Riemann invariants*. Notice that the condition that $R^\pm$ = constant, or $U \pm 2c$ = constant, corresponds to straight lines in the (U, c)-plane.

It follows from the preceding results that information is transmitted through the system at speeds $U \pm c$, or with speeds $\pm c$ relative to the fluid. This gives a physical meaning to the quantity c, which formerly was only a short-hand way of writing $\sqrt{gd}$ (ignoring knowledge from previous chapters).

5.3 Principle of Application

5.3.1 General Procedure

The essence of the manner in which the characteristic method is applied is as follows. Suppose that the state of motion is known for two closely neighbouring points in the (s, t)-plane, say points 1 and 2 in Figure 5.2a, meaning that the values of U and c are known at these points, shown as points S_1 and S_2 in the (U, c)-diagram in Figure 5.2b. Because the values of the state variables U and c are known for points 1 and 2, so are the characteristic velocities $ds/dt = U \pm c$ for these two points. These are velocities along the s-axis, but they are also directions or slopes in the (s, t)-plane.

Now draw a short straight line segment in the (s, t)-plane, starting at point 1 in the direction given by $ds/dt = U_1 + c_1$ (assumed positive, which will always be the case for subcritical flow, regardless of the sign of U_1). This is by definition a portion of a positive characteristic K^+. Likewise, we draw a line segment starting at point 2 with slope $ds/dt = U_2 - c_2$ (assumed negative). This is a segment of a negative characteristic K^-. Strictly speaking, the characteristics are curved in general, but this can be ignored over sufficiently short intervals, which can be achieved by starting at closely neighbouring points.

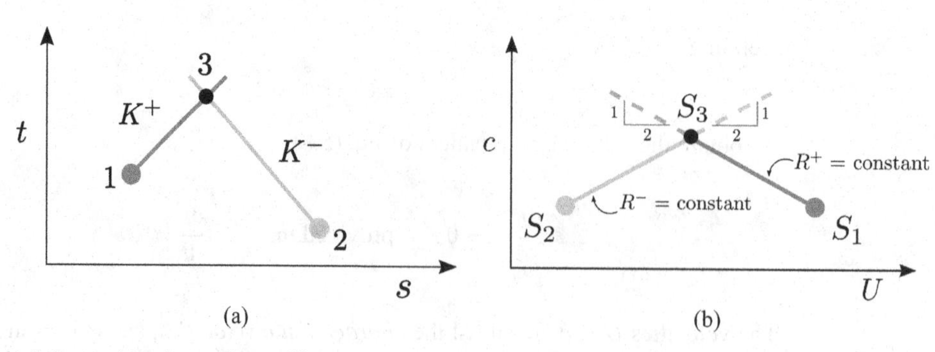

Fig. 5.2 (s, t)-diagram (a) and state diagram (b)

The Riemann invariants R^+ and R^- are constant along the characteristics K^+ and K^-, respectively. Physically, this means that information about the state of motion at points 1 and 2 is carried to point 3, which determines the state of motion at that point:

$$R_3{}^+ = R_1{}^+ \quad \text{and} \quad R_3{}^- = R_2{}^- \tag{5.12}$$

or, in terms of the state variables U and c:

$$U_3 + 2c_3 = U_1 + 2c_1 \quad \text{and} \quad U_3 - 2c_3 = U_2 - 2c_2 \tag{5.13}$$

The values of U_3 and c_3 are easily calculated from these two linear algebraic equations:

$$U_3 = \frac{1}{2}(U_1 + U_2) + (c_1 - c_2) \quad \text{and} \quad c_3 = \frac{1}{4}(U_1 - U_2) + \frac{1}{2}(c_1 + c_2) \tag{5.14}$$

The various states of motion and the connections between them can be portrayed graphically in the (U, c)-diagram: the point S_3, representing the state at point 3, can be found from the states S_1 and S_2 as the point of intersection of the straight line $R^+ = U + 2c =$ constant, starting at point S_1, and the straight line $R^- = U - 2c =$ constant, starting at point S_2.

The computation can proceed in like manner as for point 3, by starting at an additional point 4 with given state of motion, which yields a new point 5 where the state of motion follows from that at points 2 and 4. This can be continued for many more points, so covering the (s, t)-plane with a network of positive and negative characteristics, as shown in Figure 5.3. The states of motion at the nodes of this network are determined from those in the starting points. This is elaborated in following paragraphs.

5.3.2 Characteristics

As we have seen above, *characteristics are paths in the (s, t)-plane along which the information travels.* This is of great importance for a good understanding of wave propagation, as is needed for a proper operation of control structures, pumping stations etc.

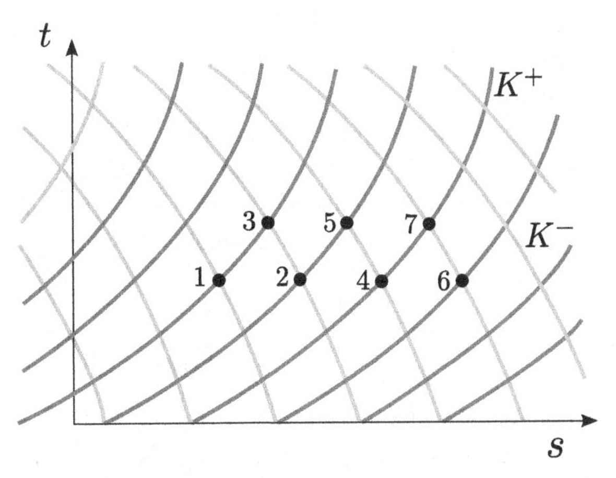

Fig. 5.3 The (s, t)-plane with several characteristics

Box 5.1 Curved characteristics

Since the characteristic velocities depend on the instantaneous, local depth and flow velocity, they are not known beforehand but must be determined as part of the solution, which therefore proceeds in consecutive small steps. Moreover, in regions where U and d vary, so does $U \pm c$. In those cases, the characteristics are curved. Using straight-line segments instead is an approximation valid for weak disturbances and/or small steps in s and t. This is in contrast to the (U, c) state diagram: the condition of constant Riemann invariants is graphically represented as straight lines in the (U, c)-plane, no matter how large the variations.

The slope of a characteristic ($\mathrm{d}s/\mathrm{d}t$) equals the characteristic velocity ($U \pm c$). If $|U| = c$ (*i.e.*, Fr $= 1$), the flow is called critical. In subcritical flow, for which $|U| < c$, the sign of $\mathrm{d}s/\mathrm{d}t = U \pm c$ equals that of $\pm c$, so the slopes of the positive and the negative characteristics have opposite signs. Stated another way: $\mathrm{d}s/\mathrm{d}t > 0$ for K^+, and $\mathrm{d}s/\mathrm{d}t < 0$ for K^-. This means that *in subcritical flow, information is transmitted downstream as well as upstream*. In supercritical flow, $|U| > c$, and the sign of $\mathrm{d}s/\mathrm{d}t = U \pm c$ equals that of U, for the positive as well as for the negative characteristics. Therefore, *in supercritical flow, information is transmitted in the downstream direction only.*

In general, the characteristics are curved, but in some circumstances they are straight or they can locally be approximated as such; see Box 5.1.

It is important to be aware that the *position* of the points in the (s, t)-plane (where? when?) is determined by the *characteristics*, whereas the *state of motion* (what?) is determined in the (U, c)-plane using the *Riemann invariants*. Different points in the (s, t)-plane can very well have the same state of motion (steady, uniform flow), represented by a single point in the (U, c)-plane.

5.3.3 Boundary Conditions

A boundary condition prescribes the values of a state variable, or a relation between two state variables (such as a dependence of discharge on water level) at one of the boundaries of the computational domain, as a function of time from the moment at which the initial conditions have been prescribed. Two boundary conditions are required. The flow regime (sub- or supercritical) determines at which boundary or boundaries these should be imposed, as will be made clear in the following.

Consider therefore a prismatic canal of finite length, from $s = 0$ to $s = \ell$, say, in which the values of U and d, so U and c, are known at time $t = t_0$ (the initial conditions), allowing the start of the construction of a network of characteristics and the determination of the associated values of the state variables. Boundary conditions are specified for $t > 0$ at $s = 0$ and/or at $s = \ell$.

Subcritical Flow

Assuming subcritical flow, in which at each point the two characteristic velocities ($\mathrm{d}s/\mathrm{d}t = U \pm c$) have opposite signs (or, the characteristics have opposite slopes), we obtain a network as shown in Figure 5.4, in which we have chosen four points (1, 2, 3, 4) in the

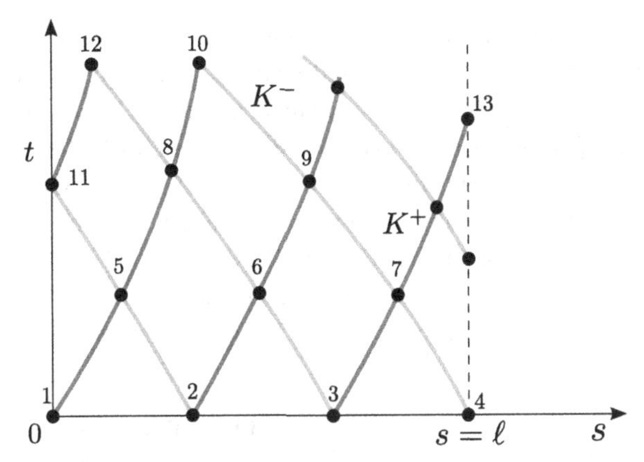

Fig. 5.4 Subcritical flow: characteristic paths for given initial conditions and boundary conditions

computational domain to start the computation. (This is just an illustration. In practical applications, it is likely that more points would be required.)

In the manner described in Section 5.3.1, the solution can be obtained successively at points 5 through 10, in a more or less triangular region bordered by the characteristics K^+ issuing at $t = 0$ from the left boundary $s = 0$ (point 1), and K^- issuing at $t = 0$ from the other boundary $s = \ell$ (point 4).

The values of the Riemann invariants at point 10 equal those at points 1 and 4, but the characteristics linking these points to point 10 depend also on information at points in between, in fact the entire interval from $s = 0$ to $s = \ell$. The latter interval is called the *domain of dependence* of point 10. Boundary conditions at $s = 0$ and at $s = \ell$ for $t > 0$ have no influence at point 10 (or in the triangular domain 1–4–10).

Conversely, considering an individual location such as $s = s_2$, the initial condition there can influence the state of motion only in the domain between the positive characteristic (2–6–9) and the negative characteristic (2–5–11) issuing from that point at $t = 0$, the so-called *domain of influence* of point s_2.

In order to extend the solution beyond the domain of dependence (the triangle 1–4–10), we continue the negative characteristic issuing from point 2 up to the left boundary, $s = 0$. This yields point 11 and the value of R^- in that point ($R^-_{11} = R_2{}^-$). In order to find the solution, a second relation is required. This cannot be delivered by a positive characteristic, because such characteristic cannot reach point 11 from the region where information is available. That is why a boundary condition is needed at the left boundary, and only one. (With two boundary conditions in $s = 0$, the problem would be overdetermined, with three relations for only two variables.)

Proceeding in this manner, we obtain two relations between the two state variables at the left boundary, viz. the values of R^- obtained from the initial condition, and the other one being provided by the boundary condition. With this information, positive characteristics issuing from the left boundary can be constructed, entering the domain of computation. See e.g. point 12, where $R^+_{12} = R_{11}{}^+$, and where R^-_{12} is also known, from the negative characteristic through that point ($R^-_{12} = R_3{}^-$). Thus, the solution at point 12 is known, etc.

Proceeding in like manner from the other boundary $s = \ell$, we see that there too a boundary condition is required, and only one. We can conclude that in subcritical flow, one boundary condition is required at each of the two boundaries. In other words, *in subcritical flow, one boundary condition is required at the upstream boundary and one at the downstream boundary*. This is consistent with Section 5.3.2, where it was noted that in subcritical flow, information is transmitted downstream as well as upstream.

The influence of the left boundary enters the computational domain along positive characteristics. When the positive characteristic issuing from point 1 reaches the right boundary, the influence of the left boundary condition is felt in the entire domain.

Supercritical Flow

Next, we consider supercritical flow, in which the two characteristic velocities at each point ($\mathrm{d}s/\mathrm{d}t = U \pm c$) have the same sign (or, the characteristics are slanting in the same direction), equal to that of U. Assuming $U > 0$, implying $\mathrm{d}s/\mathrm{d}t > 0$, we obtain a network of characteristics slanting to the right, as shown in Figure 5.5.

In contrast to the situation for subcritical flow, the left boundary (the upstream boundary) cannot be reached by negative characteristics issuing from the domain $s > 0$, where information is available from the initial conditions. If we want to know the solution at the left boundary for some time $t > 0$, as at point 8, the required information must be delivered entirely by the (two) boundary conditions there. Thus, *in supercritical flow, two boundary conditions are required at the upstream boundary* (e.g. depth and flow velocity). This is consistent with Section 5.3.2, where it was noted that in supercritical flow, information is transmitted downstream only.

Now consider the downstream boundary. The positive characteristic issuing from point 3, for example, reaches the downstream boundary at point 7. But that point can also be reached by a negative characteristic issuing closer to the downstream end because such characteristics also slope to the right (downstream). Thus, the values of both Riemann invariants at point 7 are known, and the state of motion there is fully determined. A downstream boundary condition is not necessary. In fact, it would make the problem overdetermined.

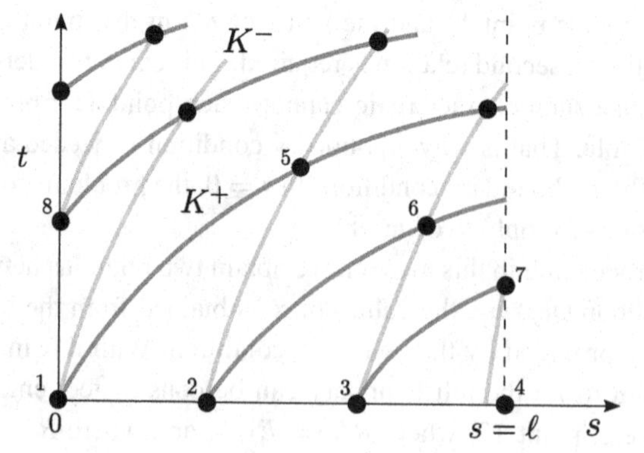

Fig. 5.5 Supercritical flow: characteristic paths for given initial conditions and boundary conditions

Transcritical Flow

Suppose now that, in supercritical flow conditions, after some time a gate at the downstream boundary is closed. Then what? Closing a gate effectively imposes a downstream boundary condition $Q = 0$ in $s = \ell$. In that case, the local flow becomes subcritical, and negative characteristics issuing from the closed downstream end enter the computational domain. At some point these reach the region of supercritical flow, which point is also reached by a positive characteristic and by a negative one issuing in the region of supercritical flow. The solution is overdetermined at that point. It becomes multivalued; i.e. two values of the surface elevation exist simultaneously at the same location, implying the existence of a shock wave or bore.

At the location of the shock, there are five unknowns: the two state variables on each side, and the velocity of the shock. An equal number of relations is available for their determination: three Riemann invariants obtained from the three characteristics intersecting at the location of the shock, and two equations for the balances of mass and of momentum across the shock (as derived in the preceding chapter). Together, the available information is just sufficient to obtain a unique solution.

5.3.4 External Forces

So far, external forces such as resistance or wind action have been ignored, as well as a possible bed slope, so as to present the method of characteristics in its most elementary form. This restriction is now relaxed.

Let F be the resultant of the external forces per unit mass acting in the positive s-direction. We add this to the right-hand side of Eq. (5.2). Repeating the procedure to go from the balance equations of mass and momentum to the characteristic form, we, obtain, instead of Eqs. (5.8) and (5.11),

$$\frac{\mathrm{d}R^{\pm}}{\mathrm{d}t} = F \quad \text{provided that} \quad \frac{\mathrm{d}s}{\mathrm{d}t} = U \pm c \tag{5.15}$$

Instead of being truly invariant, the Riemann variables will now change along the corresponding characteristics at a rate given by the forcing term F.

Referring to the situation sketched in Figure 5.2, the Riemann variable R^+, corresponding to the positive characteristic K^+, will change between points 1 and 3 according to

$$\Delta R^+_{1,3} \equiv R^+_3 - R^+_1 = \int_{K^+_{1,3}} F(s,t)\,\mathrm{d}t \tag{5.16}$$

where the domain of integration $K^+_{1,3}$ is the path from point 1 to point 3 along the characteristic K^+. Similarly, along the characteristic path $K^-_{2,3}$ from point 2 to point 3, the corresponding change of R^- is given by

$$\Delta R^-_{2,3} \equiv R^-_3 - R^-_2 = \int_{K^-_{2,3}} F(s,t)\,\mathrm{d}t \tag{5.17}$$

These variations in R^+ and R^-, respectively, must be added to the right-hand sides in Eq. (5.12), giving

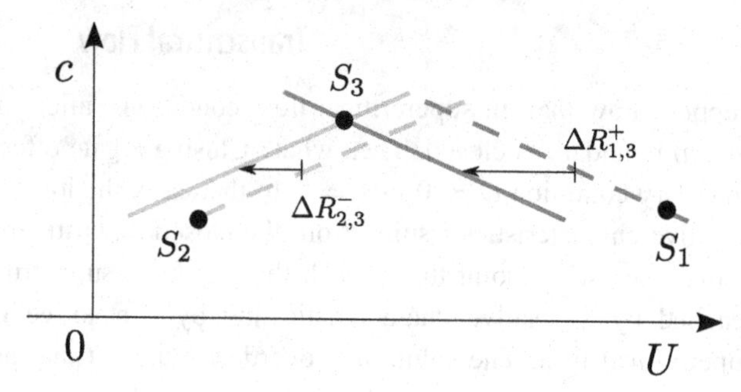

Fig. 5.6 Inclusion of forcing terms in the state diagram in the case of a resistance force F_r

$$R_3^+ = R_1^+ + \Delta R_{1,3}^+ \quad \text{and} \quad R_3^- = R_2^- + \Delta R_{2,3}^- \tag{5.18}$$

or, in terms of the state variables U and c:

$$U_3 + 2c_3 = U_1 + 2c_1 + \Delta R_{1,3}^+ \tag{5.19}$$

and

$$U_3 - 2c_3 = U_2 - 2c_2 + \Delta R_{2,3}^- \tag{5.20}$$

As a result, the point S_3 does not lie on the line $R^+ = R_1^+$ (dashed line in Figure 5.6), but on another (solid) line parallel to this. This line can be found by shifting the first one such that $\Delta R^+ = \Delta U + 2\Delta c = \Delta R_{1,3}^+$. If we measure this shift along lines of constant c (constant depth), or 'horizontally', then $\Delta c = 0$ and we have $\Delta U = \Delta R_{1,3}^+$. Similarly, the line $R^- = R_2^-$ has been shifted over a distance $\Delta U = \Delta R_{2,3}^-$.

If resistance is the only external force ($F = F_r = -c_f|U|U/d$, say), F opposes U, and the lines of constant $R^\pm$ must be shifted in the direction of $U = 0$, regardless of the direction of the flow; see Figure 5.6. This results in a gradual decrease in absolute value of the flow velocities, or a damping of the motion, as expected for the effect of resistance.

5.4 Graphical Solution Procedure

The method of characteristics lends itself to solving the one-dimensional shallow water equations graphically. This leads to an elegant solution procedure that provides insight into the behaviour of solutions, highlighting in particular the influence of the initial and boundary conditions. For practical applications the graphical method has since long been replaced by computer tools.[1] Practising this method remains useful, however, to become familiar with the essential ingredients of wave propagation problems, which is why we treat it here.

[1] The method of characteristics is frequently used in numerical models to formulate absorbing boundary conditions. Some numerical methods are intrinsically based on the method of characteristics (so-called characteristics-based methods). These are particularly suited to compute compression wave, bores, transcritical flows and the like; see also Section 11.4.

For ease of illustration we linearize the characteristic velocities. Although the ensuing solution procedure is valid for infinitesimal disturbances only (low waves), the essential features of the method are retained.

In the case of weak disturbances entering a region with uniform depth d_0 and flow velocity U_0, the characteristic velocity can be approximated as $\mathrm{d}s/\mathrm{d}t \simeq U_0 \pm c_0 = U_0 \pm \sqrt{gd_0}$. In this approximation, the characteristics are straight and independent of the solution. We will use this approximation in following graphical representations and elaborations

Also, we will draw the (s, t)-diagram 'upside down', compared with Figure 5.2; i.e. we plot the direction of increasing time downwards, as in Figure 5.7. The advantage of this is that the positive characteristics K^+ in the (s, t)-plane are slanting in more or less the same direction as the lines $R^+ =$ constant in the (U, c)-plane, and similarly for K^- and the lines of constant R^- (except for supercritical flow).

5.4.1 Initial Value Problem

The long-wave equations (5.1) and (5.2), or their equivalent characteristic relations (5.8) and (5.11), require a set of two conditions in t and two conditions in s for a well-posed problem.

In order to be able to march forward in time with the integration of the basic equations, we impose *two initial conditions* that specify the state of motion through the values of two independent state variables at some initial instant $t = t_0$ in the entire s-domain of calculation.

As an example, we consider the evolution of an initial disturbance of finite length in a long canal, such that the canal boundaries do not (yet) affect the solution. We start with the same initial conditions as in the example presented in Section 3.4, i.e. a state of rest everywhere with an undisturbed depth d_0 and a small rise of the free surface (to a depth d_1) over a finite interval from $s = s_1$ to $s = s_2$, say (Figure 5.8). The undisturbed state is labelled as I, the disturbed state as II. The intervals in which these states exist initially are indicated along the upper line in Figure 5.9. The states themselves are shown in Figure 5.10.

We introduce the initial conditions step by step, beginning with the construction of a small network of characteristics issuing at $t = 0$ from the undisturbed region where $s < s_1$

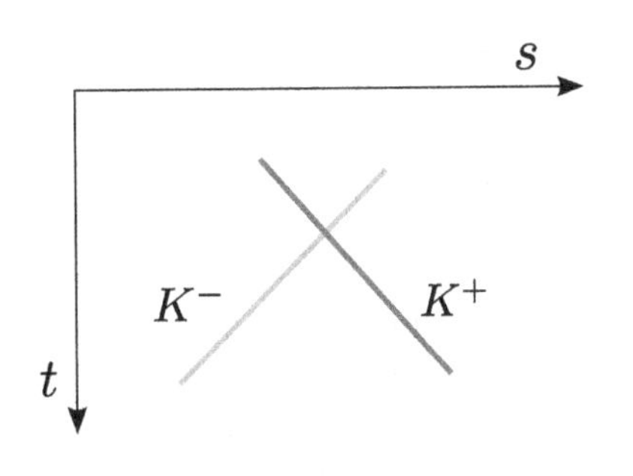

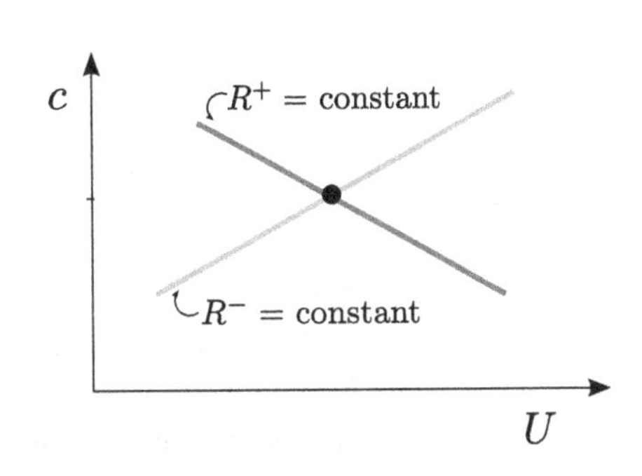

Fig. 5.7 Orientation of (s, t)-diagram and (U, c)-diagram

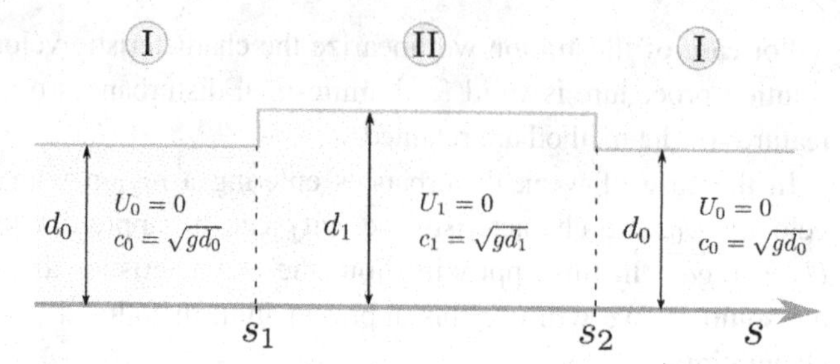

Fig. 5.8 Example: initial condition

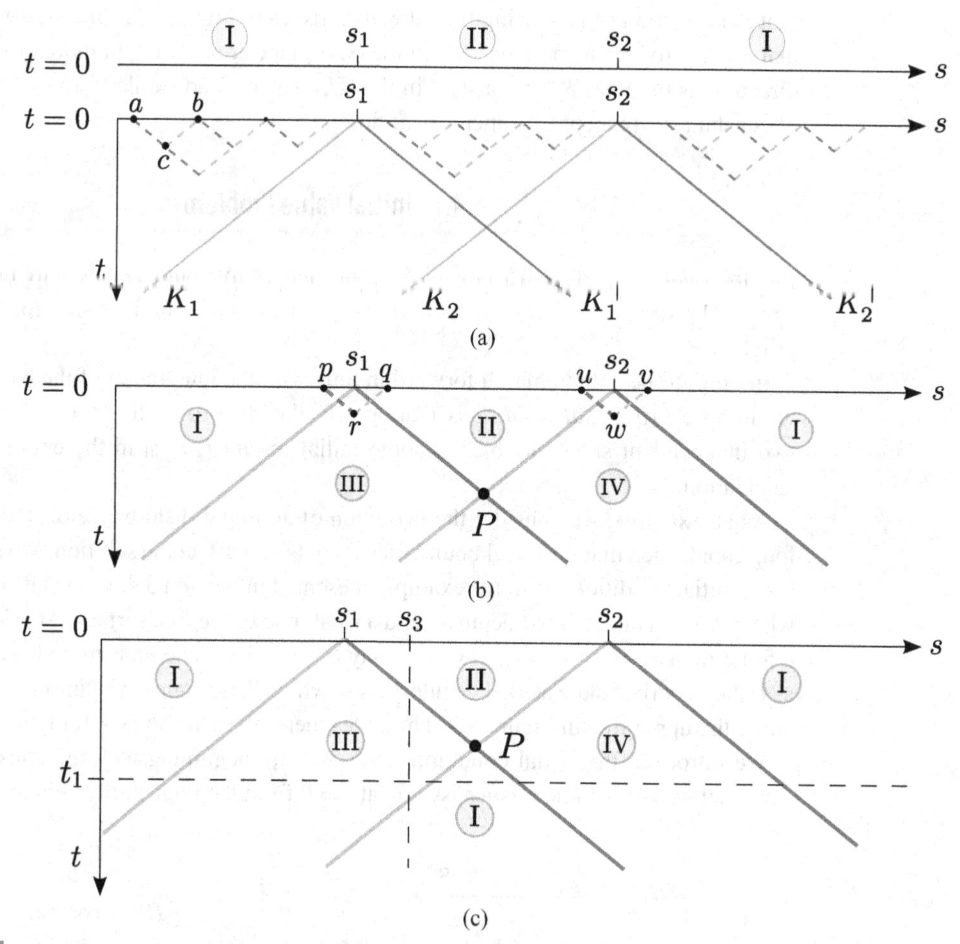

Fig. 5.9 Example: characteristics and flow states in the (s, t)-diagram

(Figure 5.9a). We draw a positive characteristic from a point a and a negative one from a point b. These intersect at a point c. Because the points a and b share the same undisturbed state of motion (I, actually a state of rest), the corresponding lines of constant $R^{\pm}$ in the (U, c)-diagram all pass through the point I, so that the same state (I) exists at point c. The same is true for neighbouring points. Therefore, this undisturbed state (I) exists in the entire

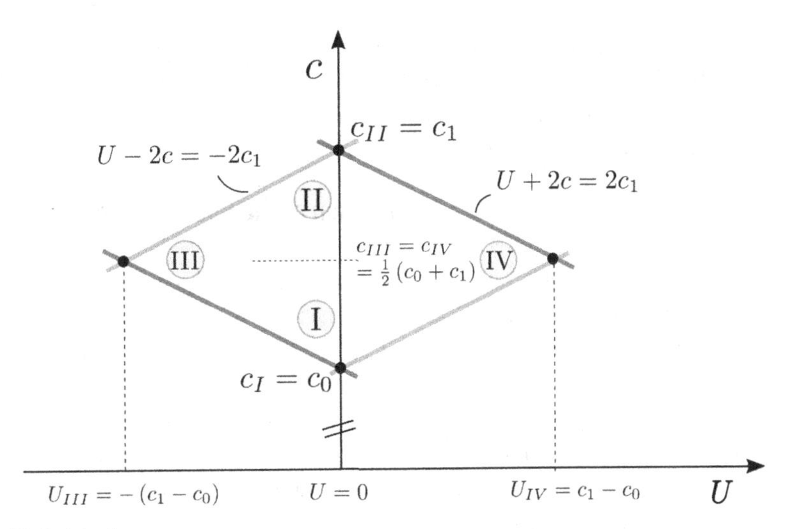

Fig. 5.10 Example: (U, c)-state diagram

domain to the left of the negative characteristic K_1^-. This is because the disturbance issuing from $s = s_1$ has not yet arrived at those faraway points in the restricted time. (We could have stated this from the outset, but here we have derived the result formally to illustrate the application of the method of characteristics.) The same applies to the region to the right of K_2^+.

A similar consideration, applied to the triangular domain between the s-axis and the two characteristics K_1^+ and K_2^-, will show that the state II (still) exists in that domain. The disturbances originating at the locations $s = s_1$ and $s = s_2$ have not yet reached the points inside this triangular domain.

Next, we draw a positive characteristic issuing from the left region with initial state I (point p) and a negative characteristic issuing from the region with initial state II (point q) (Figure 5.9b). They intersect at point r. The state at this point can be found in the state diagram of Figure 5.10 as the intersection of the line $R^+ =$ constant through point I and the line $R^- =$ constant through point II, yielding the state III valid for point r and other nearby points between K_1^- and K_1^+. Likewise, starting at points u and v, we find the state IV for point w and nearby points between K_2^- and K_2^+.

The characteristics K_1^+ and K_2^- intersect at point P. This point marks the corner of a domain, centered between s_1 and s_2, in which the undisturbed state I has returned, as shown in Figure 5.9c.

Figure 5.9c gives an overview of the results. It is clear that the initial disturbance is effectively split into two disturbances, one propagating to the left and the other to the right along the s-axis. At first, these two partial disturbances overlap, but after some time they are separated, after which the undisturbed state is restored in the center, extending over an increasingly long interval of the s-axis as the two disturbances propagate to the left and to the right.

Using the finalized diagrams, the variation of the flow state in time at a particular location can be found by plotting a *vertical* line $s =$ constant in the (s, t)-diagram. Intersection of the line $s = s_3$ in Figure 5.9c with the principal characteristics (those separating the different flow states) gives the time instances at which the state at point s_3 changes. The alternate flow states (subsequently II, III and I) are obtained from the corresponding state diagram, Figure 5.10. Similarly, a *horizontal* line $t = t_1$ in the (s, t)-diagram will give the spatial variation of the flow state at time t_1.

5.4.2 Inclusion of the Boundary Conditions

We will now illustrate how the boundary conditions are introduced, continuing with the preceding example.

Suppose that at some point $s = \ell$, somewhere to the right of point $s = s_2$ in Figures 5.8 and 5.9, the canal connects to a reservoir with a constant water level. This location is indicated in the extended (s, t)-diagram of Figure 5.11. Ignoring the velocity head, the connection with the reservoir implies the boundary condition $d =$ constant $= d_0$ at $s = \ell$, which is represented by a straight line $c =$ constant $= c_0 = \sqrt{gd_0}$ in the extended (U, c)-state diagram in Figure 5.12.

We have seen in the above that the initial conditions of the present example give rise to two disturbances, one travelling to the left and one travelling to the right. We focus on the latter. Its leading edge is represented by the characteristic K_2^+ in Figure 5.9. The initial state of rest (state I) exists in the domain to the right of this characteristic up to the instant when the disturbance arrives at the boundary, labelled as point Q in Figure 5.11. From that moment on, the influence of the boundary at $s = \ell$ comes into play.

To determine the resulting state of motion, we draw a positive characteristic issuing at $t = 0$ from a point d in the region of the initial disturbance, with state II. This characteristic intersects the boundary at an instant labelled as e. Along this characteristic, we have $R^+ = U + 2c =$ constant $= 2c_1 = U_e + 2c_e$. Together with the boundary condition $c_e = c_0 = \sqrt{gd_0}$, this yields $U_e = 2(c_1 - c_0)$. These relations have been plotted in the extended (U, c)-diagram of Figure 5.12. The corresponding state of motion, with $c = c_0$ and $U = 2(c_1 - c_0)$, is labelled as state V. The domain where this exists has been indicated in Figure 5.11.

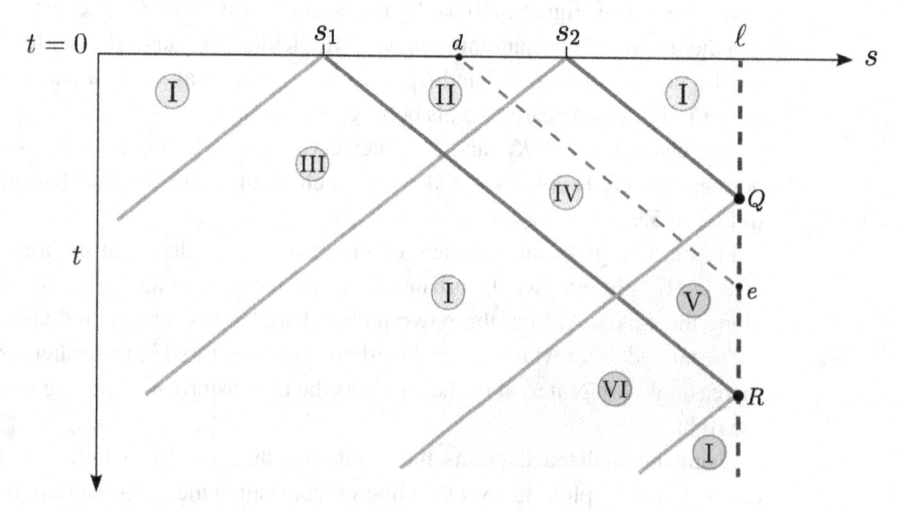

Fig. 5.11 Example: boundary conditions, extended (s, t)-diagram; effect of boundary condition noticeable in shaded area

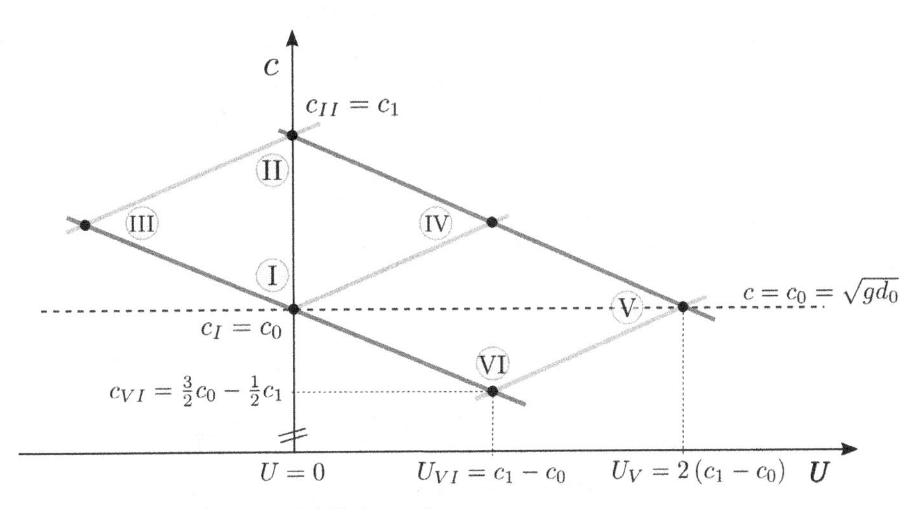

Fig. 5.12 Example: boundary conditions, extended (U, c)-state diagram

Notice that the flow velocity in the disturbance approaching the reservoir is $U_{IV} = c_1 - c_0$, and that it is doubled due to the negative reflection of the travelling disturbance against the reservoir, as we saw in Chapter 4 in the context of translatory waves.

After some time the trailing edge of the right-travelling disturbance arrives at the boundary, labelled R in Figure 5.11, inducing yet another state of motion. It is determined through a negative characteristic issuing from point e and a positive one issuing from the s-interval with the initial, undisturbed state of motion (I). The result is state VI, as indicated in the (s, t)-diagram of Figure 5.11 and in the state diagram of Figure 5.12.

Boundary Types

In the preceding example, the boundary condition of a constant depth was plotted as a horizontal line in the state diagram. Its intersection with an outgoing characteristic determines the new state of motion, including the effect of the boundary. Likewise, the condition of a closed boundary is represented by a vertical line $U = 0$ in the state diagram.

Instead of prescribing a constant value of U or d, it is also possible to prescribe a relation between them. This is for instance the case when the boundary represents a control structure (weir, orifice), for which a *discharge relation* has to be specified as the boundary condition. Given that the head loss over such structures is proportional to the local velocity head, the following relation between d and U will hold locally (see Chapter 9):

$$d = h_B - \xi \frac{|U|\,U}{2g} \tag{5.21}$$

where ξ is the head loss coefficient of the structure and h_B is the water level in the water system outside the structure (a lake or sea to which it is connected) with respect to the local bed level in the canal. The notation using the absolute value of the velocity warrants

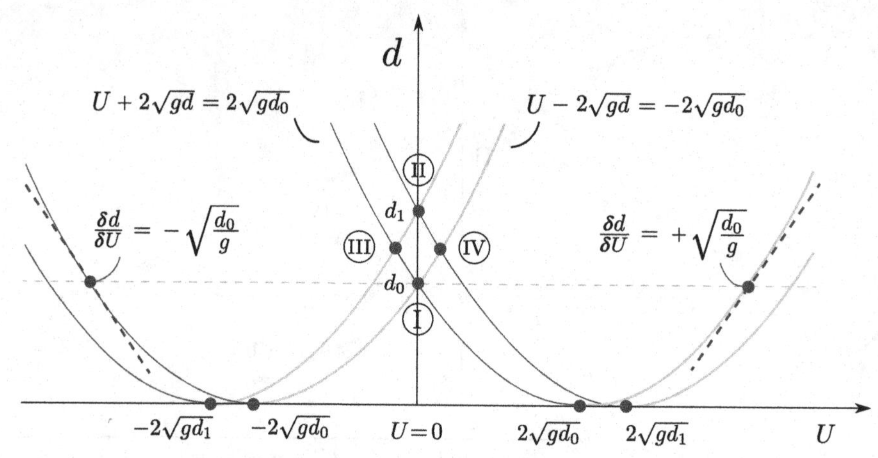

Fig. 5.13 Example: (U, d)-state diagram with parabolas $R^{\pm} = U \pm 2\sqrt{gd} = \text{constant}$

that the difference in head over the structure obtains the correct sign as the flow direction reverses (as with the resistance term in the momentum equation).

In cases of such mixed boundary conditions, it can be convenient to use (U, d) as state variables. The advantage of this is the ease of the graphical representation of the mixed boundary condition in the (U, d)-state diagram, but this is at the expense of nonlinear relations expressing the constancy of the Riemann invariants. These are now expressed as $R^{\pm} = U \pm 2\sqrt{gd} = \text{constant}$, i.e. two half-parabolas in the (U, d)-diagram (see Figure 5.13, which is the (U, d)-counterpart of the corresponding (U, c)-state diagram of Figure 5.10).

The graphical solution procedure remains effectively the same as when the value of U or d is prescribed; i.e. establish the values of these state variables from the intersection of the appropriate curve representing the boundary condition Eq. (5.21) with a line $R = \text{constant}$ for the outgoing characteristic.

Changes in the state of motion correspond to displacements along the parabolas of constant R^+ or R^-, i.e. a curved path in the (U, d)-plane. For weak disturbances of an undisturbed state (U_0, d_0), these displacements are small and follow approximately the local tangent to the parabola, given by $dU/dd = \mp\sqrt{g/d_0}$. Thus, for small steps along these local tangents, small variations in U and d are related by

$$\delta U = \mp\sqrt{\frac{g}{d_0}}\,\delta d = \mp\frac{g}{c_0}\,\delta d = \mp\frac{c_0}{d_0}\,\delta d \tag{5.22}$$

in which $c_0 = \sqrt{gd_0}$ (see also Section 3.1.2).

Absorbing Boundaries

The influence of engineering measures in water systems is often mainly local, diminishing with increasing distance from the site. This is utilized in physical or numerical model studies of the local situation by cutting off the model at a sufficiently remote boundary, such that it can reasonably be assumed to be beyond the influence of the engineering measure concerned.

Because in reality the system extends further, such a cut-off creates an artificial, open model boundary. The condition to be imposed there is that disturbances approaching that boundary from within the study area should not be reflected, so as to simulate reality in which they continue unimpeded. In these cases we speak of an *absorbing boundary* (see also Section 3.5). The method of characteristics is particularly suited to determine the appropriate condition to be imposed at an open boundary in order to make it absorbing, because disturbances travelling in opposite directions are tracked separately.

The influence from the region outside the open boundary on the motion within the model domain should be allowed to enter this domain without change. This can be achieved by prescribing the incoming Riemann invariant at the open boundary. Suppose now that a disturbance approaches this boundary from within. If it would (partially) reflect there, its reflection would add to the disturbance entering the domain from outside the model boundary. This possibility is excluded by prescribing the incoming Riemann invariant, which therefore makes the boundary absorbing.

5.5 Simple Wave

This section deals in detail with the important case of a disturbance entering a region of rest or of uniform flow, first considered in Section 3.1. Such disturbance, travelling in one direction only, is called a *simple wave*. They occur frequently in practice as a result of operation of control structures following a steady state.

5.5.1 General Solution

Initially, at $t = 0$, the flow is uniform with depth d_0 (propagation speed $c_0 = \sqrt{gd_0}$) and flow velocity U_0, indicated as state I in the (U, c)-diagram of Figure 5.14. Starting at $t = 0$, a time-varying disturbance is imposed at $s = 0$ from where it propagates into the canal, in the direction of s-positive, say ($s > 0$). We assume that the canal is prismatic and that there are no reflections nor autonomous disturbances propagating in the direction of s-negative.

The disturbance enters the domain $s > 0$ at $t = 0$ with velocity $U_0 + c_0$. Where it has not yet arrived, i.e. in the domain $s > (U_0 + c_0)t$, the initial state of uniform flow is still present.

In order to determine the motion in the region where the disturbance is present, we need (among others) negative characteristics issuing at $t = 0$ from the undisturbed domain. All of these share the same value of R^-, also in the disturbed region. It follows that *in the entire domain, including the disturbed part, all points have the same value of* R^-, given by $R_0^- = U_0 - 2c_0$. In other words, in the entire region, the two state variables U and c obey the relation

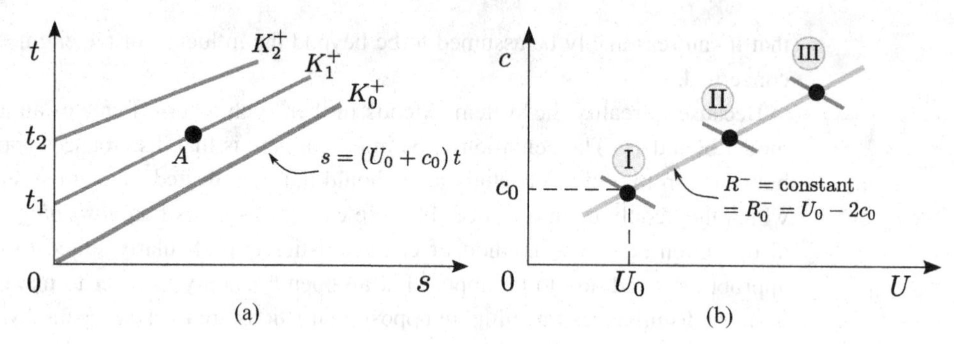

Fig. 5.14 Simple wave travelling in positive s-direction; (s, t)-diagram with characteristics (a) and state diagram (b)

$$U - U_0 = 2c - 2c_0 = 2\sqrt{gd} - 2\sqrt{gd_0} \tag{5.23}$$

This is a straight line in the (U, c)-diagram passing through point I, as shown in Figure 5.14b. It shows that the flow velocity in the simple wave increases with the local depth, i.e. with the local wave height, and how. It is shown in Box 5.2 that for low waves, Eq. (5.23) reduces to Eq. (3.8), derived in Chapter 3 for that category.

We now consider a positive characteristic K_1^+ issuing from $s = 0$ at an arbitrary time $t = t_1$, say (see Figure 5.14a). Along this characteristic, R^+ is constant = $R_1{}^+$. But, as we have seen, all points share the same value of R^-, viz. R_0^-. It follows that both R^+ and R^- are constant along K_1^+, so that the same is true for the two state variables U and c separately. Since this applies to an arbitrary positive characteristic, it applies to all of them.

This in turn implies generally that *in a simple wave, the state of motion is constant along any positive characteristic*. Since both U and c are constant along the positive characteristics, so is their sum $U + c$, i.e. the characteristic direction (ds/dt) in the (s, t)-plane. This means that *in a simple wave, the positive characteristics are straight.*

The value of $U + c$ for any positive characteristic depends on the associated value of R^+, which in turn is determined by the boundary condition in $s = 0$ at the time when the characteristic was started. Using Eq. (5.23), this can be written as

$$\frac{ds}{dt} = U + c = U_0 + 3c - 2c_0 = U_0 + 3\sqrt{gd} - 2\sqrt{gd_0} \tag{5.24}$$

It follows that points of constant U and c (or constant U and d, therefore also of constant discharge $q = Ud$) are moving with a constant speed, which is higher for points of the wave with a larger wave height (larger depth) than it is for the lower parts. Stated another way: higher parts of the wave travel faster than lower parts; the wave deforms. This can also be seen from the characteristics, because two positive characteristics with constant but different wave heights (therefore also different values of U and c) diverge or converge, which means that the distance between any two points on the wave with constant but different depths varies linearly in time.

Box 5.2 **Low simple waves**

In Chapter 3, relations were derived between variations in flow velocity and surface elevation for a low simple wave; see Eq. (3.8). Eq. (5.23) is a similar relation, valid for simple waves of arbitrary height. Let us check whether it reduces to Eq. (3.8) for low waves, i.e. $\delta d = d - d_0 \ll d_0$. To this end, we make the approximation $\sqrt{gd} = \sqrt{gd_0}\,(1 + \delta d/d_0) \simeq \sqrt{gd_0}\left(1 + \frac{1}{2}\delta d/d_0\right)$ and substitute this into Eq. (5.23), to obtain $\delta U = U - U_0 \simeq \sqrt{g/d_0}\,\delta d = c_0\,\delta d/d_0$, which is indeed the same as Eq. (3.8) (accounting for the difference in notation).

Thus, *the wave deforms as it propagates*. This was already made plausible in the context of high translatory waves in Chapter 4. The difference is that we now have proof, and a method to calculate the rate of deformation.

5.5.2 Expansion Wave

In order to investigate the wave deformation further, we choose a uniform state of rest ($U_0 = 0$, $d = \text{constant} = d_0$) in a long canal as the initial condition, and a prescribed time variation of the flow velocity in $s = 0$ as one of the boundary conditions. A second boundary condition is provided by the assumed absence of reflected waves from downstream, as in any simple wave.

We first consider a situation with *outflow* at $s = 0$ with an initially increasing outflow velocity, followed by a constant value after some time, as sketched in Figure 5.15a. The starting outflow causes a *negative wave* propagating away from the outflow boundary.

Along the characteristic $K_0{}^+$ through the point $s = 0, t = 0$, the wave speed is $U_0 + c_0$ (actually, $U_0 = 0$, but we write it for consistency in the expressions). It separates the disturbed domain from the undisturbed one, thus representing the leading edge.

Next consider a characteristic K_1^+ starting at the left boundary at some time $t = t_1$, when $U = U_1 < 0$, so $d = d_1 < d_0$, and therefore also $U_1 + c_1 < U_0 + c_0$. This implies that K_1^+ diverges from $K_0{}^+$. Thus, the point on the wave where $d = d_1$ lags more and more

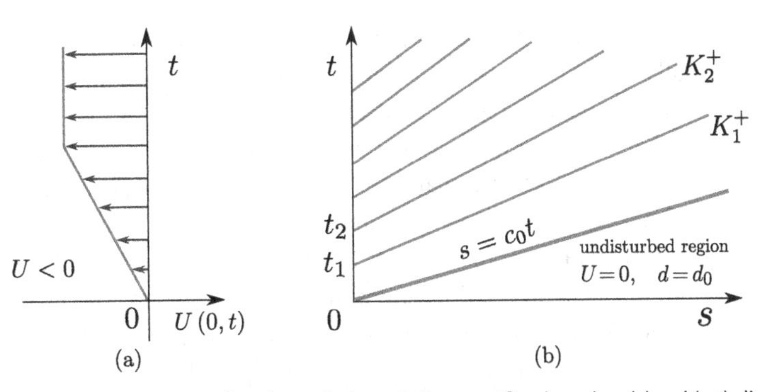

Fig. 5.15 Expansion wave travelling in positive s-direction; velocity variation at outflow boundary (a) and (s, t)-diagram with characteristics (b)

behind the leading edge. The wave, being negative, becomes less and less steep and more and more stretched as time goes on. This type of wave is called an *expansion wave*.

At some instant, the outflow velocity has become constant. From that moment on, the positive characteristics issuing at $s = 0$ have the same values of U and c; they are mutually parallel, signifying a new state of uniform flow, with a smaller depth than initially, and with a negative flow velocity equal to the velocity finally imposed at $s = 0$.

Stoker (1957) treats in detail the case of a horizontally translating vertical gate, with special attention to the case of withdrawal, which generates a negative wave. When the gate recedes at a sufficiently high velocity, the water cannot keep up with it and the bed falls dry. This is the so-called dam-break problem, which can be solved analytically in closed form using the method of characteristics.

Example 5.1 Consider a semi-infinite canal with initial depth $d_0 = 5$ m and initial velocity $U_0 = 0$. A pumping station at the (left) boundary of the canal starts withdrawing water, as a result of which an expansion wave starts travelling along the canal. Consider a point on the expansion wave in which the velocity $U_1 = -0.5$ m/s.

1. Calculate the water depth (d_1) at this point.
2. Determine the speed ($\mathrm{d}s/\mathrm{d}t|_1$) with which this point propagates along the canal.

Solution

In the undisturbed region $c_0 = \sqrt{gd_0} = 7.0$ m/s.

1. Using Eq. (5.23) (with $U_0 = 0$) we obtain $c_1 = c_0 + \frac{1}{2}U_1 = 6.75$ m/s, so $d_1 = c_1^2/g = 4.5$ m.
2. The propagation speed $\mathrm{d}s/\mathrm{d}t|_1$ equals $U_1 + c_1 = c_0 + \frac{3}{2}U_1 = 6.25$ m/s.

Notice how the smaller value of d and that of U (with respect to the initial state) both contribute to the propagation speed ($\mathrm{d}s/\mathrm{d}t$) being smaller than its undisturbed value.

5.5.3 Compression Wave

We now treat a situation of *inflow*, with a flow velocity initially increasing from zero, and finally going to a constant value, as in Figure 5.16a. The starting inflow causes a *positive wave* propagating into the adjacent canal reach.

When in a certain time interval the inflow velocity increases, so do the local values of d and c, therefore also that of $U + c$: the positive characteristics converge, and the wave becomes steeper in time. This type of wave is called a *compression wave*. At some moment, positive characteristics intersect, at which point (in space and in time) the solution becomes multivalued, with two different values of the momentary surface elevation at the same cross section. A discontinuity in surface elevation develops: a bore or *shock wave*.

Consider two positive characteristics K_1^+ and K_2^+, issuing from $s = s_0$ at times $t = t_1$ and $t = t_2 = t_1 + \Delta t$, respectively, such that $\mathrm{d}s/\mathrm{d}t|_2 > \mathrm{d}s/\mathrm{d}t|_1$. Elementary geometry shows that they intersect at a time t_i and a location s_i given by

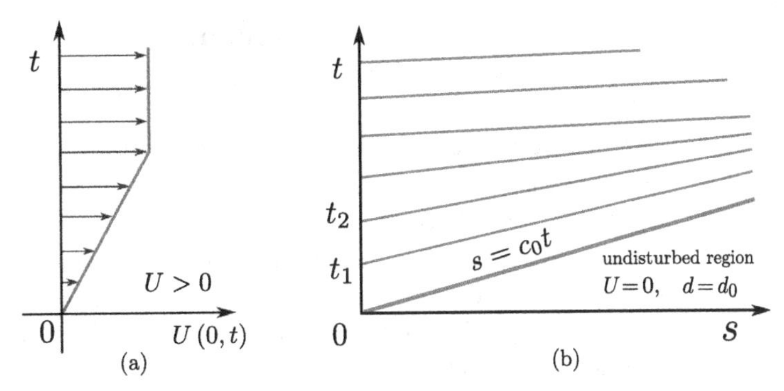

Fig. 5.16 Compression wave travelling in positive s-direction; velocity variation at inflow boundary (a) and (s, t)-diagram with characteristics (b)

$$t_i = t_1 + \frac{V_2}{V_2 - V_1}\Delta t = t_2 + \frac{V_1}{V_2 - V_1}\Delta t \tag{5.25}$$

and

$$s_i = s_0 + \frac{V_1 V_2}{V_2 - V_1}\Delta t \tag{5.26}$$

in which we have used the shorthand notation $V = \mathrm{d}s/\mathrm{d}t = U + c$.

In the above, the prescribed time variation of the flow velocity was chosen as the boundary condition, for convenience. In practice, it is more likely that the discharge is given, as at a pumping station, or a relation between surface elevation and discharge, as in a gated control structure. Such conditions make the algebra a bit more complicated without adding insight into the process of wave deformation.

Example 5.2 As in Example 5.1, but with a discharge of water from the pumping station causing a compression wave travelling along the canal. In a time interval from t_1 to $t_2 = t_1 + \Delta t = t_1 + 60$ s the velocity at the pumping station increases from $U_1 = 0.5$ m/s to $U_2 = 1$ m/s.

1. Determine the instant, with respect to t_1, at which the respective characteristics K_1^+ and K_2^+ intersect.
2. Determine the location, with respect to the pumping station, where these characteristics intersect.

Using $V = \mathrm{d}s/\mathrm{d}t = c_0 + \frac{3}{2}U$ gives $V_1 = 7.75$ m/s and $V_2 = 8.50$ m/s.

1. Instant of intersection (Eq. (5.25)): $t_i = t_1 + (V_2/(V_2 - V_1))\,\Delta t = t_1 + 680$ s
2. Location of intersection (Eq. (5.26)): $s_i = s_0 + (V_1 V_2/(V_2 - V_1))\,\Delta t = 5270$ m from the pumping station

A bore is formed when an intersection of positive characteristics *first* occurs. When and where this happens depends on the time variation of the imposed inflow velocity. The numbers used in the example were chosen arbitrarily; they do not refer to the instant and location of bore formation.

Problems

5.1 Describe the essence of the method of characteristics.

5.2 What is a characteristic?

5.3 What is the importance of characteristics?

5.4 What is the physical meaning of characteristic velocities?

5.5 Which variables determine characteristic velocities?

5.6 Considering the characteristic velocities of long waves in open water, what can be said about the relative importance of the flow velocity compared with the wave speed?

5.7 Why do the two characteristic velocities in free-surface flows have opposite signs in some cases, and the same signs in others?

5.8 What is the relevance of the answer to the previous problem for the boundary conditions? For the initial conditions?

5.9 Is the speed of propagation of a shock wave, relative to the water ahead of it (with depth d_0), larger than $\sqrt{gd_0}$, smaller, or equal to it?

5.10 A simple wave, started in $s = 0$ at $t = 0$, propagates into a prismatic canal ($s > 0$) with water at rest, where the undisturbed wave speed is c_0. Is it possible that under certain circumstances it causes disturbances in the domain $s > c_0t$?

5.11 Elaborate the examples of this chapter by making sketches of the corresponding (s, t)-diagram and the state diagram.

5.12 Choose some instants in the solutions of Problem 5.11 and sketch the corresponding longitudinal profiles of the state variables.

5.13 Choose some fixed locations in the solutions of Problem 5.11 and sketch the corresponding time variations of the state variables at those points.

5.14 The following data apply to the situation of a simple wave in a prismatic canal: $d_0 =$ 5 m, $U_0 = 0$ and

$$U(0, t) = \begin{cases} U_m \sin^2(\pi t/T) & \text{for } 0 < t < \frac{1}{2}T \\ U_m & \text{for } t \geq \frac{1}{2}T \end{cases}$$

in which $U_m = -0.7$ m/s and $T = 60$ s. The following questions should be answered using the method of characteristics.

(a) Determine the state of motion in the domain $s > c_0t$.

(b) Same for the domain $s < c_0t$, both for some instants $t < \frac{1}{2}T$ and for some instants $t > \frac{1}{2}T$.

(c) Calculate and plot the variation of U and d with s for the instants chosen in part (b).

5.15 Same as Problem 5.14, but for $U_m = +0.7$ m/s.

5.16 Same as Problem 5.14, but now the discharge per unit canal width ($q = Ud$) is given:

$$q(0, t) = \begin{cases} q_m \sin^2(\pi t/T) & \text{for } 0 < t < \frac{1}{2}T \\ q_m & \text{for } t \geq \frac{1}{2}T \end{cases}$$

where $q_m = 3.5$ m^2/s.

5.17 A canal, initially at rest with $d_0 = 5$ m, $U_0 = 0$, is connected at $s = 0$ by a movable gate to a reservoir whose surface level is 2 m above the undisturbed canal level. Initially, the gate is closed but from time $t = 0$ to $t = \frac{1}{2}T = 30$ s, the gate is gradually opened to an effective height $\mu a(t)$ and then brought to a standstill. The discharge per unit canal width (q) is related to the head difference across the gate (Δh) by

$$q = Ud = \mu a\sqrt{2g\Delta h}$$

in which the time variation of the gate opening is given as

$$\mu a = \begin{cases} \mu a_m \sin^2(\pi t/T) & \text{for } 0 < t < \frac{1}{2}T \\ \mu a_m & \text{for } t \geq \frac{1}{2}T \end{cases}$$

where $\mu a_m = 0.5$ m. The same questions apply as in Problem 5.14.

6 Tidal Basins

In the preceding chapters, the discharge and the surface elevation were treated as continuous functions of s and t. This led to a coupled system of partial differential equations in two unknowns, with wave-like solutions. Certain flow systems can be schematized in terms of separate but connected components of finite dimension, in each of which we disregard the spatial variations. In each component, either storage or transport is modelled, but not both, so that the motion within them is not wave-like. In these cases, we speak of a discrete model. One such model is presented in this chapter.

6.1 Introduction

The disregard of spatial variations in the system components considered is allowed if the dimensions (ℓ) are small compared with a typical length of the (long) waves in the domain. Stated another way, system components for which the travel time ('residence time') of long waves through them is short compared with the wave period. In such cases, phase differences within each of the system components are negligible. In other words, within each component the motion loses its wave-like character.

A good example of this category of situations is the tidal motion in a harbour basin. The water level in the basin can to a good approximation be assumed to be horizontal at all times, varying in time only. Its variation can be modelled with an ordinary differential equation instead of a partial differential equation, which simplifies the mathematics greatly.

The discrete modelling approach is utilized in the present chapter. It is relevant in itself, because numerous situations occurring in practice lend themselves to this approximation, and it is at the same time a preparation for the theory in the following chapter (on harmonic wave propagation) with respect to the linearized modelling of the flow resistance, and the use of complex algebra in the solution process. The advantage is that these building blocks, needed in the theory of Chapter 7, are introduced in the simpler context of the present chapter.

We focus on flow systems consisting of a nearly closed basin or reservoir, connected through some narrow, short opening or a channel of some length to an external body of water with a time-varying water level. The latter distinction (short vs. long) is relevant because in a short connection, inertia and wall resistance can be neglected, whereas these may be important in a long connecting channel.

An artificial, man-made example of such system is the so-called tide well, a device used for measuring tides or river stages, which may consist of a large-diameter tube, placed

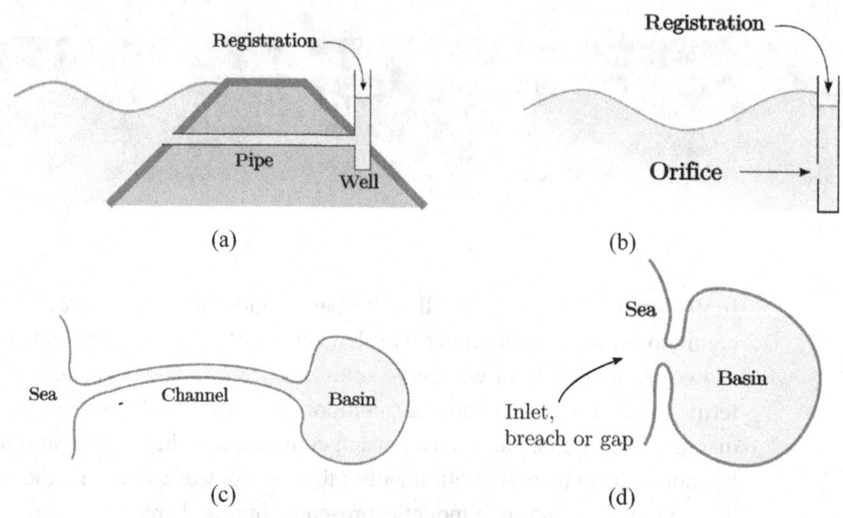

Fig. 6.1 Tide well with pipe (a) and with orifice (b), basin with channel (c) and with gap (d)

vertically and connected to an external water with varying water level. The connection may be long, such as a piece of pipe needed to bridge the distance from the external water to the location of the tide well (Figure 6.1a), or the connection may be quite short, for instance not more than an orifice in the tube wall (Figure 6.1b).

A more common, natural example is the system consisting of a basin connected to a tidal sea with a tidal channel (Figure 6.1c) or with a short connection such as an inlet, a breach in a dike or a gap in a barrier (Figure 6.1d). In fact, this is the archetype of the category of systems considered here. Below, we will for brevity use the terminology for such a tidal system, even though the theory applies more generally.

In the discrete modelling of the flow systems as described, the only function ('task') of the tidal basin is *storage*; its connection to the tidal sea has only a function of *transport*. In the basin, flow resistance and inertia are neglected, whereas in the connection these may be relevant, certainly head loss due to boundary resistance or expansion loss, but storage is not. Thus, in such a discrete model these two functions are separated, in contrast to the continuous-modelling approach in the preceding chapters, in which they were intertwined.

A similar schematization is used in the discrete model of a *mass-spring system*, well known from classical dynamics, in which kinetic energy is stored only in the mass, neglecting the mass of the spring, and all potential energy is ascribed to the spring, treating the mass as a rigid body.

6.2 Mathematical Formulation

6.2.1 Motion in the Basin

The basin is assumed to be relatively short, and it is closed except for a connection to the external body of water. There is no throughflow, so that the flow velocities in the basin are quite low. Flow resistance and inertia are negligible. Therefore, the water level in the basin can be assumed to be horizontal at all times. It can be described as a function of time only: $h_b(t)$. This is the so-called *Helmholtz mode* or *pumping mode*.

If we denote the incoming rate of flow (discharge) as Q_{in}, and the area of the free surface in the basin, available for storage, as A_b, the balance equation for the volume of water stored in the reservoir reads

$$Q_{in} = A_b \frac{dh_b}{dt} \tag{6.1}$$

Note that A_b may vary in time through its dependence on the time-varying water level: $A_b(t) = A_b(h_b(t))$.

6.2.2 Motion in the Channel

For generality, we assume initially that the connection between basin and sea consists of a channel. By letting the channel length go to zero, we cover the case of a short connection, e.g. a gap in a barrier. The main function of the channel is to convey water between the sea and the basin. Within the discrete-modelling approach, storage in the channel is neglected. This is allowable if the free-surface area in the channel is far smaller than that in the basin.

Rigid Column Approximation

Elaborating on this approximation, we start with the one-dimensional volume balance as derived in Chapter 1:

$$B\frac{\partial h}{\partial t} + \frac{\partial Q}{\partial s} = 0 \tag{6.2}$$

where B is the width of the free surface in the channel. Neglecting storage in the channel implies neglect of the first term in this equation, from which it follows that also $\partial Q/\partial s = 0$. Therefore, in the channel, Q can be considered to be a function of time only: $Q(t)$, which therefore also equals $Q_{in}(t)$.

By definition, $Q = UA_c$. For a prismatic channel, $\partial A_c/\partial s = 0$, in which case the approximation $\partial Q/\partial s = 0$ implies that also $\partial U/\partial s = 0$. This means, if in addition the channel is straight, that the distance between fluid particles is constant. The water mass in the channel is moving back and forth as a solid block. This is called the *rigid-column approximation*.

Momentum Equation

Our next task is to model the dynamics of the flow. We start with the motion *in* the channel. The transitions at both ends are considered separately. See Figure 6.2. The equation of motion for the flow in the channel (Eq. (1.21)) reads

$$\frac{\partial Q}{\partial t} + \frac{\partial}{\partial s}\left(\frac{Q^2}{A_c}\right) + gA_c\frac{\partial h}{\partial s} + c_f\frac{|Q|Q}{A_cR} = 0 \tag{6.3}$$

We integrate this over the channel length, say from $s = 0$ at the sea side to $s = \ell$ at the basin side (not counting the transitions). To do this, we must know how the various terms vary with s.

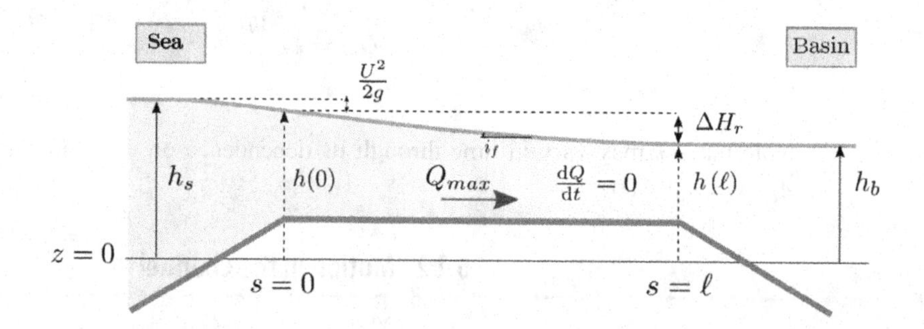

(a) Maximum flood:

$Q = Q_{max}$, $dQ/dt = 0$, $h_s - h_b = W$

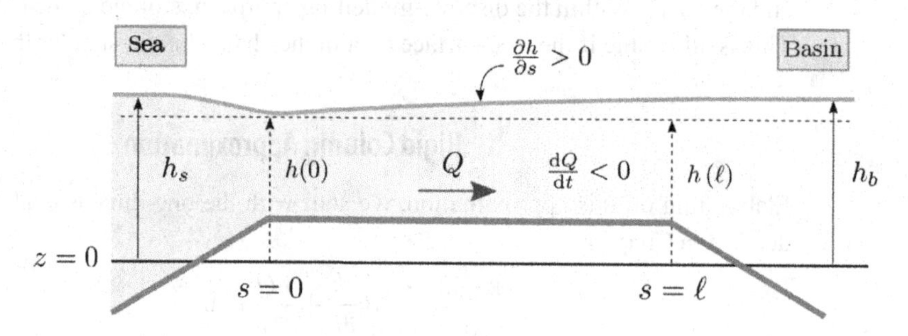

(b) Zero head difference:

$h_s - h_b = 0$, $Q > 0$, $dQ/dt = -gA_cW/\ell$

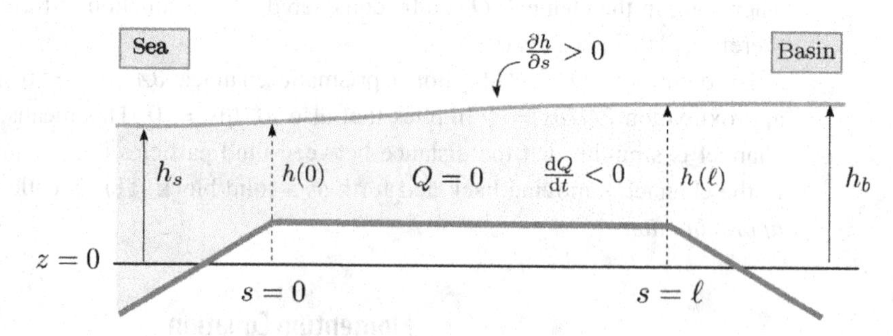

(c) Slack tide:

$Q = 0$, $W = 0$, $dQ/dt = gA_c(h_s - h_b)/\ell$

Fig. 6.2 Flow in the channel

We set $\partial Q/\partial s$ equal to zero and, assuming a prismatic channel, do the same with $\partial A_c/\partial s$, so that the second term vanishes. This also makes the resistance term independent of s, and since the same is true for the first term, it must also apply to the last remaining term, proportional to the slope of the free surface. So we obtain on integration with respect to s:

$$\ell \frac{\mathrm{d}Q}{\mathrm{d}t} + gA_c \left(h(\ell) - h(0) + \Delta H_r \right) = 0 \tag{6.4}$$

in which ΔH_r is the head loss due to boundary resistance, given by

$$\Delta H_r = c_f \frac{|Q|Q}{gA_c{}^2 R} \ell = c_f \frac{\ell}{R} \frac{|U|U}{g} \tag{6.5}$$

These equations apply to the flow in the channel.

Next, we account for the flow in both transitions. To this end, we distinguish the free-surface elevation in the channel ends, $h(0)$ and $h(\ell)$, respectively, from those at sea (h_s) and in the basin (h_b). (Of course, in practice such transitions are always gradual, and the definition of a channel end is subjective, but we must make such schematizations in the framework of a one-dimensional (s, t) model. It would require a two-dimensional (x, y, t) model to account for the transitions in a more realistic way.)

We treat the flow in the transitions as quasi-steady, i.e. at any moment adapted to the instantaneous upstream and downstream water levels, and to be modelled with the methods of classical hydraulics (see also Figure 6.2):

- When and where there is *inflow* (flood at the sea side, ebb at the basin side), the flow in the transition zone is accelerating fairly rapidly and we apply Bernoulli's law (see Chapter 9). (Boundary resistance in the short transitions is neglected.) This implies a drop in the water level equal to the velocity head in the channel ($U^2/2g$). (This drop represents not an energy loss but merely a transformation of potential energy to kinetic energy.)
- When and where there is *outflow* (ebb at the sea side, flood at the basin side), the flow is decelerating, and we can assume an expansion loss given by $\Delta H_e = U^2/2g$, implying a horizontal free surface across the transition.

For ebb as well as flood, the total head difference between sea and basin is now given by

$$h_s - h_b = h(0) - h(\ell) + \Delta H_e \tag{6.6}$$

in which ΔH_e is the expansion loss, given by $\Delta H_e = |U|U/2g$. Substituting this into Eq. (6.4) gives

$$h_s - h_b = \frac{\ell}{gA_c} \frac{\mathrm{d}Q}{\mathrm{d}t} + W \tag{6.7}$$

in which W is the total head loss due to expansion and boundary resistance:

$$W = \Delta H_e + \Delta H_r = \frac{|U|U}{2g} + c_f \frac{\ell}{R} \frac{|U|U}{g} \tag{6.8}$$

The overall head difference, i.e. the left-hand side of Eq. (6.7), accelerates the water in the channel (first term in the right-hand side) and it overcomes the losses (second term in the right-hand side). Figure 6.2 shows longitudinal profiles at a few special instants during the flood phase of the tidal cycle. (The subsequent ebb phase follows the same pattern and is not shown separately.)

Figure 6.2a applies to the moment of maximum flood current, so that $dQ/dt = 0$. The flow momentarily behaves as a steady flow, in which the available head difference ($h_s - h_b$) is spent on overcoming the losses (W).

As a result of the inflow, the water level in the basin rises; at some moment it equals the level at sea: $h_s - h_b = 0$ (Figure 6.2b). At that moment there is still flood flow, but it is being decelerated because of the adverse pressure gradient (induced by the surface slope) and the flow resistance. (Needless to say, this situation, with flow against the applied forces, is possible only on account of inertia.)

In the next phase, the water level in the basin has risen above that in the sea, strengthening the opposing pressure forces. The continuing action of the opposing forces causes the flow at some moment to change direction, i.e. slack tide occurs, at which time the flow velocity is momentarily zero, and so are the losses: $W = 0$, as shown in Figure 6.2c. (This applies to the cross-sectionally averaged velocity. The relatively low near-bottom velocities reverse earlier.) Following that moment, the ebb current starts, being accelerated by the seaward-directed pressure force.

6.2.3 Coupled System

Equations (6.1) and (6.7) form a coupled set of two first-order ordinary differential equations in h_b and Q as functions of time. These can be integrated with numerical routines, provided the initial values of h_b and Q are known, as well as the time-varying tide level at sea (h_s); see Section 11.2. In the following we will consider the corresponding second-order ordinary differential equation, which is obtained by eliminating Q (or, alternatively, h_b) from the equations.

First, we simplify the notation by introducing the dimensionless loss coefficient

$$\chi = \tfrac{1}{2} + c_f \frac{\ell}{R} \tag{6.9}$$

with which the expression for W can be written as

$$W = \chi \frac{|U|U}{g} = \chi \frac{|Q|Q}{g{A_c}^2} \tag{6.10}$$

Substituting this into Eq. (6.7) and rearranging terms we obtain

$$\frac{\ell}{gA_c}\frac{dQ}{dt} = h_s - h_b - W = h_s - h_b - \chi \frac{|Q|Q}{g{A_c}^2} \tag{6.11}$$

Substituting Eq. (6.1) into this equation, with $Q = Q_{in}$, and neglecting variations of A_c with h_b, yields

$$\frac{\ell}{g}\frac{A_b}{A_c}\frac{d^2h_b}{dt^2}+\frac{\chi}{g}\frac{A_b{}^2}{A_c{}^2}\left|\frac{dh_b}{dt}\right|\frac{dh_b}{dt}+h_b=h_s \tag{6.12}$$

This second-order ordinary differential equation for h_b has the same form as the equation for a quadratically *damped mass-spring-dashpot system*. Such system has *natural oscillations* with a *natural frequency* that in absence of damping ($\chi = 0$) has the value

$$\omega_0=\sqrt{\frac{g}{\ell}\frac{A_c}{A_b}} \tag{6.13}$$

When such a system is excited sinusoidally, it responds primarily at the forcing frequency. Initially, natural oscillations may also be excited (this is the homogeneous part of the general solution, solely dependent on the initial conditions), but these are gradually decaying due to the damping forces. In the end, the response is purely periodic at the forcing frequency (the nonhomogeneous part of the general solution), although not sinusoidal due to the nonlinearity of the damping in Eq. (6.12). In the following, we will linearize the damping, in order to simplify the mathematics. In that approximation, the response to a sinusoidal forcing is sinusoidal.

Seelig and Sorensen (1977) have applied an equation similar to Eq. (6.12) to the Pentwater Inlet-Bay system, located at the eastern shore of Lake Michigan. The system was excited by atmospherically induced seiching of Lake Michigan. The response to the measured excitation was calculated numerically. The equivalent of our loss coefficient χ was determined by trial and error, based on a comparison with the observed response. The results are shown in Figure 6.3. Given the calibrated loss coefficient, the simulated water level fluctuations agree well with the observations.

The excitation spectrum was broad but spiky, indicating the simultaneous presence of several seiching modes in Lake Michigan. It can be seen in Figure 6.3 that the inlet-bay

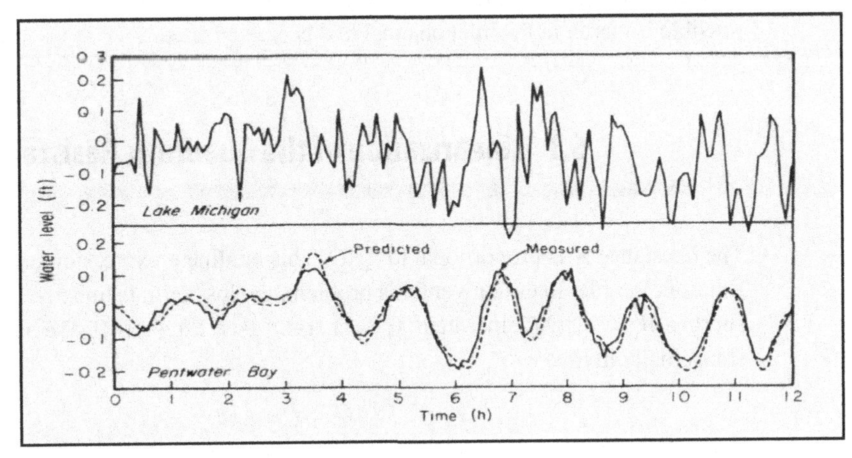

Fig. 6.3 Observed (solid) and computed (dashed) water level fluctuations in the Pentwater Inlet-Bay system, Lake Michigan; upper panel: lake level, lower panel: bay level; from Seelig and Sorensen (1977)

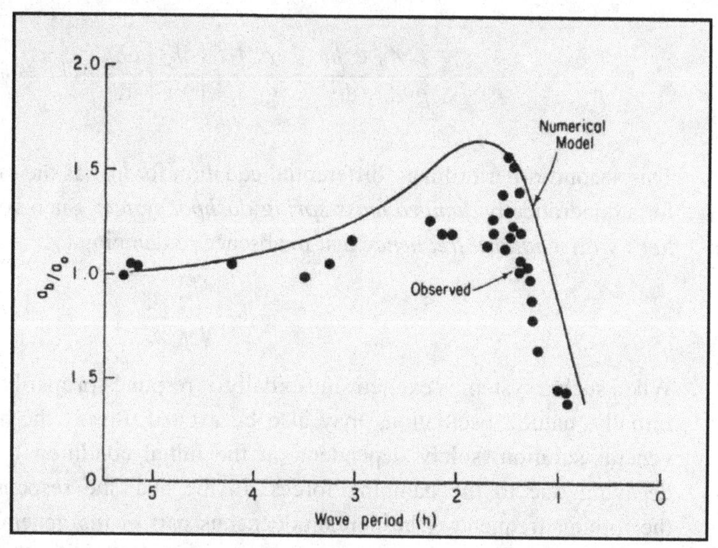

Fig. 6.4 Observed response factor (dots) of water level fluctuations in the Pentwater Bay and results of the numerical model (solid line); from Seelig and Sorensen (1977)

system acts as a filter, yielding a relatively smooth response around the apparent natural period of about 1.4 h.

Seelig and Sorensen also calculated the bay response to harmonic excitation with a 3 cm amplitude for a range of frequencies. The resulting amplitude response function (ratio of response amplitude to excitation amplitude) is shown in Figure 6.4, along with measured response values. The computed response overestimates the measured values (which Seelig and Sorensen ascribe to the pulse-like nature of the excitation), but the overall pattern is similar. Interestingly, the response shows an enhancement of the amplitude in the bay relative to the excitation amplitude. This is the result of resonance or near-resonance at frequencies at or near the natural frequency of the inlet-bay system. This would not be possible if inertia in the inlet channel had been neglected.

6.3 Linearization of the Quadratic Resistance

The resistance W is proportional to $|Q|Q$. This nonlinear expression prevents an analytical solution. In order to circumvent this problem, the loss term is linearized. Let us write it for short as $W = \lambda_1 |Q| Q$, in which $\lambda_1 = \chi / g A_c{}^2$ (see Eq. (6.10)). We assume that Q varies sinusoidally in time:

$$Q = \hat{Q} \cos \omega t \tag{6.14}$$

Figure 6.5 shows the form of the associated time variation of the resistance, plotted as $|Q|Q/\hat{Q}^2 = |\cos \omega t| \cos \omega t$.

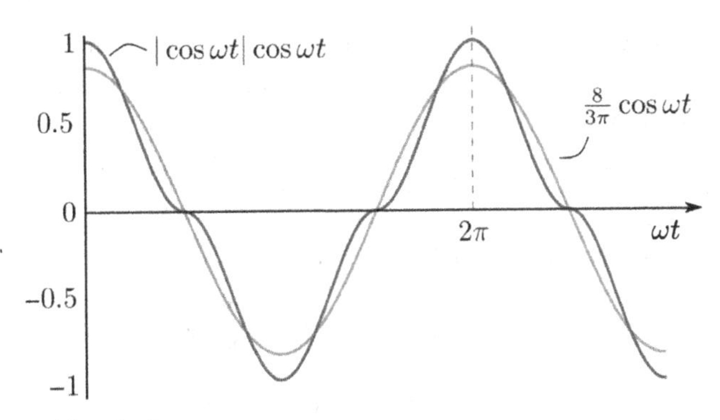

Fig. 6.5 Quadratic resistance and linearization

The quadratic form, with a modulus operator, causes a deviation from a sinusoidal variation. We want to get rid of that. To do so, we approximate the resistance term as $W = \lambda_2 Q$ instead of $W = \lambda_1 |Q| Q$, in which λ_2 is a constant to be determined. In doing so, we accept an error in the time variation, but we will choose λ_2 in such a way that the energy loss in a cycle (with duration T) has the same value in both formulations, so that the damping in the course of time is correctly represented. The rate of energy loss due to the resistance W is proportional to WQ, so that the condition to be imposed on λ_2 can be written as

$$\int_0^T WQ\,\mathrm{d}t = \int_0^T \lambda_1 |Q| Q^2 \mathrm{d}t = \int_0^T \lambda_2 Q^2 \mathrm{d}t \tag{6.15}$$

Substitution of Eq. (6.14) and carrying out the integrations yields

$$\frac{\lambda_2}{\lambda_1} = \frac{\int_0^T |\cos \omega t| \cos^2 \omega t \,\mathrm{d}t}{\int_0^T \cos^2 \omega t \,\mathrm{d}t} \hat{Q} = \frac{\int_0^{\frac{1}{4}T} \cos^3 \omega t \,\mathrm{d}t}{\int_0^{\frac{1}{4}T} \cos^2 \omega t \,\mathrm{d}t} \hat{Q} = \frac{8}{3\pi} \hat{Q} \tag{6.16}$$

With this result for λ_2, and using the definition $\lambda_1 = \chi / g A_c{}^2$, we obtain the following expression for the linearized resistance:

$$W = \frac{8}{3\pi} \chi \frac{\hat{Q}}{g A_c^2} Q \qquad \left(= \frac{8}{3\pi} \chi \frac{\hat{U}}{g} U \right) \tag{6.17}$$

This important result has first been derived by Lorentz (1926). The criterion of equivalent energy loss had been used earlier, but with an incorrect quantification (see Box 6.1). The same result is obtained by requiring that the difference between the quadratic resistance and the linear approximation be minimal in a least-square sense, or by expanding the quadratic expression in a Fourier series and retaining only the first term.

In the approximation given by Eq. (6.17), W varies linearly with Q (or U), albeit with a proportionality coefficient that contains $\hat{Q}$, which is not known beforehand. The solution can be determined iteratively by making successive, improved estimates of $\hat{Q}$ (or of $\hat{U}$) and using these as input in the next round of calculation. (It will appear that for the problem of a tidal basin the solution can in fact be determined in closed form. For more complicated situations iteration is indeed necessary; see Chapter 7.)

Using the linear approximation to W (Eq. (6.17)) in Eq. (6.11), instead of the original, quadratic expression (Eq. (6.10)), we obtain the following ordinary differential equation instead of Eq. (6.12):

$$\frac{\ell}{g}\frac{A_b}{A_c}\frac{\mathrm{d}^2 h_b}{\mathrm{d}t^2} + \tau\frac{\mathrm{d}h_b}{\mathrm{d}t} + h_b = h_s \tag{6.18}$$

in which for brevity we have introduced a resistance parameter τ defined by

$$\tau \equiv \frac{8}{3\pi}\,\chi\,\frac{A_b}{gA_c{}^2}\,\hat{Q} \tag{6.19}$$

It appears from the structure of Eq. (6.18) that τ is the *time scale* (the relaxation time) of the system response to a varying excitation. Because of the linearization of the resistance, this time scale depends partly on the unknown amplitude of the discharge ($\hat{Q}$) or of the flow velocity ($\hat{U}$).

As stated before, we neglect variations of A_b and A_c with h_b. In that approximation, Eq. (6.18) is linear with constant coefficients. This implies that the response to a sinusoidal excitation is also sinusoidal, with the same frequency.

For completeness we mention an aspect that so far has not been mentioned at all, i.e. a difference in mean water level between basin and sea. Treating A_b and A_c as constants, independent of the water level, is an approximation. Strictly speaking, the ebb flow occurs on average at a smaller depth than the flood flow, so that it experiences a higher resistance. The result is that the time-averaged water level in the basin is somewhat above that at sea (see also Section 7.8.2). This is a nonlinear effect, which we neglect in the present linear approximation. This allows us to write the instantaneous surface elevations as, respectively,

$$h_s(t) = h_0 + \zeta_s(t) \qquad \text{and} \qquad h_b(t) = h_0 + \zeta_b(t) \tag{6.20}$$

in which h_0 is the mean water level, the same in the basin as it is at sea (in the linear approximation), and ζ_s and ζ_b are the fluctuations of the water level, being zero on average. With this substitution, Eq. (6.18) is transformed into

$$\frac{\ell}{g}\frac{A_b}{A_c}\frac{\mathrm{d}^2 \zeta_b}{\mathrm{d}t^2} + \tau\frac{\mathrm{d}\zeta_b}{\mathrm{d}t} + \zeta_b = \zeta_s \tag{6.21}$$

In the following paragraphs, we present solutions (ζ_b) to this equation for a given sinusoidally varying water level at sea (ζ_s).

Box 6.1 Origin of the linearization procedure

The resistance term has been linearized by approximating it as proportional to the flow velocity U, with the requirement that the total work done in a cycle to overcome the resistance be the same in the linear approximation as in the original form. This idea seems to have been first utilized in the beginning of the twentieth century, by Dubs (1909), in order to model the oscillations in surge tanks in the context of the design of hydropower projects. Although Dubs formulated the requirement of equal energy dissipation correctly in words, his mathematical elaboration is incorrect because he in fact equates the integral of the resistance with respect to time ($\int W\,dt$, in fact a momentum and not an energy) in the two approximations, whereas the integral of the rate of work done by the resistance ($\int WU\,dt$) should have been used. The result is similar to that given by Eq. (6.17), except for the proportionality coefficient. It seems that Parsons (1918) was the first to utilize the linearization of the resistance in tidal computations. Like Dubs, he formulates a correct criterion, but his mathematical elaboration is incorrect because he assumes a parabolic variation of the flow velocity in time from zero to its maximum value, instead of a sinusoidal variation. As far as is known to the authors, Lorentz (1926) was the first to derive the correct linearization, given in Eq. (6.17). We return to this in Chapter 7.

6.4 System with Discrete Storage and Resistance

Before dealing with the full Eq. (6.21), we consider the simpler case of a short connection in order to build up the complexity gradually.

6.4.1 Governing Equation

Short connections between the basin and the sea (i.e. $\ell \to 0$), such as a gap in a barrier or in a dike, contain little mass, whose inertia we can neglect. (Moreover, if $\ell \to 0$, boundary resistance in the connection becomes negligible, so that only expansion losses remain: $\chi \to \frac{1}{2}$.) This reduces the system to one of storage in the basin and head loss in the connection. Eq. (6.21) reduces to

$$\tau \frac{\mathrm{d}\zeta_b}{\mathrm{d}t} + \zeta_b = \zeta_s \tag{6.22}$$

From a mathematical point of view, the neglect of inertia reduces the order of the equation from second order to first order.

Physically, it implies that the discharge through the connection responds instantaneously to variations in the head difference across it, being zero as soon as the head difference is zero. Because the discharge is proportional to the rate of change of the water level in the basin, this in turn implies that the maximum and minimum water levels in the basin occur at the instants when the water levels in the basin and in the sea are equal. This of course can also be seen in Eq. (6.22), since this shows that $\mathrm{d}\zeta_b/\mathrm{d}t = 0$ when $\zeta_b = \zeta_s$.

6.4.2 Nonhomogeneous Solution

We will now derive the nonhomogeneous (forced) solution of Eq. (6.22) for a sinusoidal tide at sea with water level given by

$$\zeta_s = \hat{\zeta}_s \cos \omega t \tag{6.23}$$

Since the response is also sinusoidal, with the same frequency, it can be written as

$$\zeta_b = \hat{\zeta}_b \cos(\omega t - \theta) = r\,\hat{\zeta}_s \cos(\omega t - \theta) \qquad \text{in which} \qquad r = \hat{\zeta}_b/\hat{\zeta}_s \tag{6.24}$$

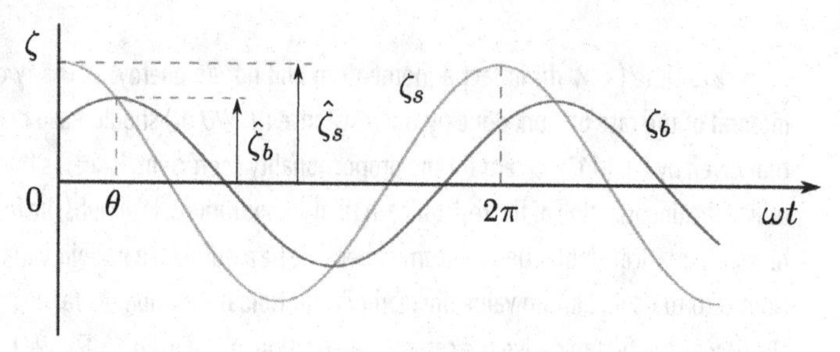

Fig. 6.6 Example solution of a linearized discrete system with storage and resistance

We have introduced the symbol r for the ratio between the two amplitudes. The amplitude $\hat{\zeta}_b$ (or the ratio r) and the phase angle θ are to be determined. Notice that θ is the phase lag of the water level in the basin behind that at sea.

Substitution of Eqs. (6.23) and (6.24) into Eq. (6.22) yields one equation in the two unknowns r and θ, but that is sufficient because it must be fulfilled for all times. Substituting two conveniently chosen phases to simplify the algebra, such as $\omega t = \frac{1}{2}\pi$ and $\omega t = \theta$, yields the results

$$\tan\theta = \omega\tau \qquad \text{and} \qquad r = \cos\theta = \frac{1}{\sqrt{1 + (\omega\tau)^2}} \tag{6.25}$$

To help visualize this result, the excitation and the response are shown in Figure 6.6, for an arbitrary value of r. The two curves satisfy the condition that the water level in the basin is at its maximum or minimum when the two curves intersect. Consider the first intersection shown in the figure, occurring at time $t = t_1$, say. At that moment, ζ_b reaches its maximum value of $\hat{\zeta}_b$, which implies that $\omega t_1 = \theta$ (see Eq. (6.24)), and ζ_s reaches the value $\hat{\zeta}_s \cos\omega t_1$ $= \hat{\zeta}_s \cos\theta$. Since this must equal $\hat{\zeta}_b$, it follows that $r = \cos\theta$.

It appears that the solution is determined by the dimensionless product $\omega\tau$ or, stated another way, by the ratio τ/T, in which τ is the time scale of the system response and $T = 2\pi/\omega$ is the excitation period. Another interpretation is that $\omega\tau = \hat{W}/\hat{\zeta}_b$, as follows from the definitions of W and τ (and using Eq. (6.26) given below). In other words, the product $\omega\tau$ represents the *relative magnitude of the resistance*.

If τ is small compared with T, or $\omega\tau \ll 1$, resistance is insignificant, so the basin level can more or less follow the relatively slowly varying water level at sea, so that $r \simeq 1$, $\cos\theta \simeq 1$, and $\theta \simeq 0$, in agreement with Eq. (6.25).

On the other hand, if the system has a time scale that is long compared with the excitation period ($\omega\tau \gg 1$), it can hardly follow the excitation, which in this case varies relatively rapidly, so the response is weak ($r \ll 1$) and the lag is large ($\cos\theta \ll 1$, or $\theta \simeq \frac{1}{2}\pi$), again in agreement with Eq. (6.25).

6.4.3 Explicit Solution

The solution has been expressed in terms of $\omega\tau$, in which τ depends on the amplitude of the discharge or the flow velocity in the connection (see Eq. (6.19)), which is not known beforehand. To deal with this problem, we substitute Eq. (6.24) into Eq. (6.1), with the result

$$\hat{Q} = A_b \, \omega \, \hat{\zeta}_b \tag{6.26}$$

Substituting this into the definition of τ (Eq. (6.19)), we obtain

$$\omega\tau = \frac{8}{3\pi} \chi \left(\frac{A_b}{A_c}\right)^2 \frac{\omega^2 \hat{\zeta}_b}{g} \tag{6.27}$$

We replace the unknown $\hat{\zeta}_b$ in the right-hand side by $r\hat{\zeta}_s$ and write the result as

$$\omega\tau = \Gamma r \tag{6.28}$$

in which Γ is defined by

$$\Gamma \equiv \frac{8}{3\pi} \chi \left(\frac{A_b}{A_c}\right)^2 \frac{\omega^2 \hat{\zeta}_s}{g} \tag{6.29}$$

Γ is a dimensionless parameter containing all independent variables playing a role in the present problem. Therefore, except for a scale factor, the solution is determined exclusively and entirely by this parameter. This also follows from Eq. (6.28), knowing that r is determined by $\omega\tau$.

If we now substitute (6.28) into (6.25), we obtain a quadratic algebraic equation in r^2, from which we finally obtain the following explicit, closed-form solution for r (therefore also for θ, since $r = \cos\theta$) as a function of the independent parameter Γ:

$$r = \cos\theta = \frac{1}{\sqrt{2}\,\Gamma}\sqrt{-1 + \sqrt{1 + 4\Gamma^2}} \tag{6.30}$$

Figure 6.7 shows this variation of r and θ as a function of Γ. Qualitatively, the influence of Γ on the solution is similar to that of $\omega\tau$, discussed above. Small values of Γ correspond to a rapid ($\theta \simeq 0$) and strong ($r \simeq 1$) system response, and large values to a slow ($\theta \simeq \pi/2$) and weak ($r \ll 1$) response. In these two limiting cases, simple approximations for r can be derived from (6.30):

$$r \simeq 1 - \frac{1}{2}\Gamma^2 \qquad \text{for} \qquad \Gamma \lesssim 10^{-1} \tag{6.31}$$

and

$$r \simeq \frac{1}{\sqrt{\Gamma}} \qquad \text{for} \qquad \Gamma \gtrsim 10^{1} \tag{6.32}$$

Keulegan (1951) presented an approximate solution to the problem considered presently, in which he used a dimensionless parameter K, the so-called coefficient of repletion, which in essence is the same as Γ (to be more explicit, $\Gamma = \frac{8}{3\pi}K^{-2}$.) However, his solution cannot be expressed in closed form.

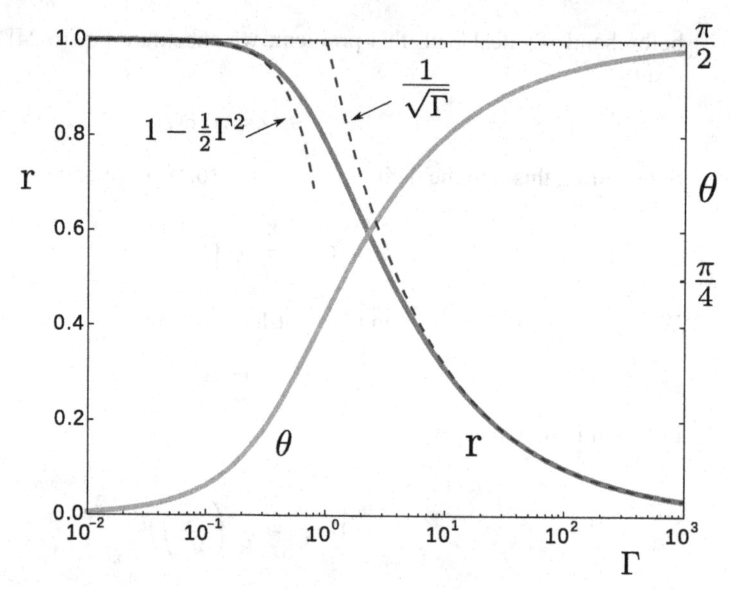

Fig. 6.7 Amplitude ratio and phase lag for a discrete system with storage and resistance as functions of the resistance parameter Γ

In the above, inertia in the connection has been neglected, as is appropriate for a gap. This may not be valid in the case of a channel. However, as we will see, it is possible for a channel of some length to contribute significantly to the overall head loss, through bed friction, while it is at the same time short enough for the inertia of the mass in it to be negligible. We will now formulate a condition to check the validity of the neglect of the inertia in the channel.

In Eq. (6.21), inertia is represented by the first term in the left-hand side, with amplitude $(\omega^2 \ell/g)(A_b/A_c)\hat{\zeta}_b$. The third term in the left-hand side of Eq. (6.21), representing the restoring force, has an amplitude $\hat{\zeta}_b$. Looking at the ratio of these terms, it follows that inertia may be neglected whenever $(\omega^2 \ell/g)(A_b/A_c) \ll 1$. See also Example 6.2.

Example 6.1 A tidal basin with length and width of the order of 10 km (A_b = 100 km^2), depths varying from 2 m to 3 m, is connected to the sea (M$_2$-tide, $\hat{\zeta}_s$ = 0.75 m) by a gap in a barrier under construction with A_c = 3000 m^2. Demonstrate that the small basin approximation is valid. Next, calculate:

1. the surface amplitude in the basin (ζ_b)
2. the phase lag (θ)
3. the maximum discharge in the gap ($\hat{Q}$).

Solution

Tidal wavelength (without friction): $L = T\sqrt{gd}$, using T = 44,700 s (M$_2$-tide) and d = 2 m (lower bound), $L \simeq 200$ km $\gg$ basin dimensions ($\approx$ 10 km), as required for the small basin approximation. Loss coefficient: $\chi = 1/2$ (gap), yielding (using given data) $\Gamma = 0.71$. We now obtain:

1. $r = 0.85 \rightarrow \hat{\zeta}_b = r\hat{\zeta}_s = 0.64\,\mathrm{m}$
2. $\theta = \arccos r = \arccos 0.85 = 31°$
3. $\hat{Q} = A_b\omega\hat{\zeta}_b = 9.0 \times 10^3\,\mathrm{m^3/s}$.

The parameter $\omega\tau = \Gamma r \simeq 0.6 < 1$, indicating that the influence of resistance is considerable.

6.5 System with Discrete Storage, Resistance and Inertia

This section deals with the complete Eq. (6.21), representing the effects of inertia, resistance and storage. We repeat it here for convenience:

$$\frac{\ell}{g}\frac{A_b}{A_c}\frac{\mathrm{d}^2\zeta_b}{\mathrm{d}t^2} + \tau\frac{\mathrm{d}\zeta_b}{\mathrm{d}t} + \zeta_b = \zeta_s \tag{6.33}$$

Equations of this form describe a forced, damped linear mass–spring system. As stated above, the natural frequency of this system in absence of damping is

$$\omega_0 = \sqrt{\frac{g}{\ell}\frac{A_c}{A_b}} \tag{6.34}$$

For harmonic forcing with frequency ω, the solution depends on $\omega\tau$ (the relative damping, as before) but also on the ratio of the forcing frequency to the natural frequency. It can be found in texts on dynamics, and reads

$$\frac{\hat{\zeta}_b}{\hat{\zeta}_s} = r = \frac{1}{\sqrt{(1 - \omega^2/\omega_0{}^2)^2 + \omega\tau}} \tag{6.35}$$

and

$$\tan\theta = \frac{\omega\tau}{1 - \omega^2/\omega_0{}^2} \tag{6.36}$$

As before, the solution is implicit because τ depends in part on the unknown amplitude $\hat{Q}$ or $\hat{U}$. Using the same approach as in the preceding section, we can make the solution explicit in terms of the same Γ as before ($\Gamma = \omega\tau/r$), with the result

$$r = \frac{1}{\sqrt{2}}\frac{1}{\Gamma}\sqrt{-\left(1 - \omega^2/\omega_0{}^2\right)^2 + \sqrt{(1 - \omega^2/\omega_0{}^2)^4 + 4\Gamma^2}} \tag{6.37}$$

Once r is determined from this equation, the phase lag θ follows from

$$\tan\theta = \frac{\omega\tau}{1 - \omega^2/\omega_0{}^2} = \frac{\Gamma r}{1 - \omega^2/\omega_0{}^2} \tag{6.38}$$

For $\omega/\omega_0 \ll 1$, i.e. a slow excitation, with consequently a small role of inertia, this solution reduces to that in the preceding section, where inertia was neglected *a priori*. The result given by Eqs. (6.37) and (6.38) seems to have been presented first by Mehta and Özsoy (1978).

The preceding expressions for r and θ show an explicit dependence on the excitation frequency ω, but there is an additional frequency dependence through the parameter Γ, which contains a factor ω^2. We remove this hidden dependence by defining

$$\Gamma_0 \equiv \Gamma\frac{\omega_0^2}{\omega^2} = \frac{8}{3\pi}\chi\left(\frac{A_b}{A_c}\right)^2\frac{\omega_0^2\,\hat{\zeta}_s}{g} \tag{6.39}$$

Insertion of $\Gamma = \Gamma_0\,(\omega/\omega_0)^2$ in Eqs. (6.37) and (6.38) turns these equations into response functions of the relative frequency ω/ω_0 with Γ_0 as a frequency-independent parameter. Figure 6.8 shows the result for the amplitude response for a chosen set of values of Γ_0. The most striking feature is the occurrence of *resonance* for small or moderate values of Γ_0, manifesting itself in high r-values for $\omega/\omega_0 \simeq 1$. Reference is made to Figure 6.4, which

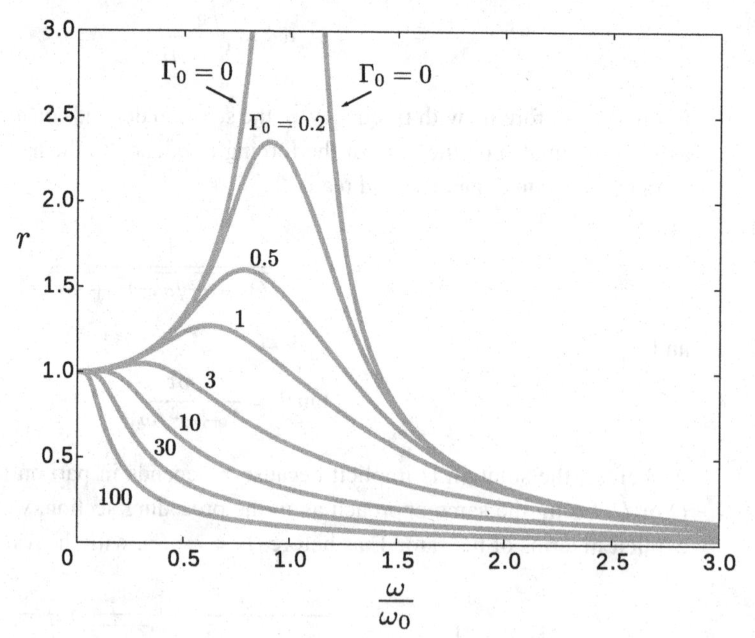

Fig. 6.8 Amplitude ratio of a discrete system with inertia, storage and resistance as functions of the resistance parameter Γ_0 and the frequency ratio ω/ω_0

shows a similar pattern in the computed and measured values for the Pentwater Inlet-Bay system.

With increasing values of Γ_0, the maximum response decreases and shifts to lower frequencies. For large values of Γ_0, i.e. a relatively high damping as a result of a narrow connection and/or a large head loss, the effects of resonance are suppressed.

The parameter Γ_0 has a simple geometric interpretation. To show this, we insert the expression for the natural frequency, Eq. (6.34), into Eq. (6.39), with the result

$$\Gamma_0 = \frac{8}{3\pi}\chi \frac{A_b \hat{\zeta}_s}{A_c \ell} \tag{6.40}$$

The latter fraction in this expression is a volume ratio, viz. the ratio of the amplitude of the volume variation in the basin, in the case of 100% response, to the volume of water in the canal.

Example 6.2 Same as in Example 6.1, except that now the connection consists of a short channel with length $\ell = 600$ m, conveyance cross section $A_c = 3000\,\text{m}^2$, hydraulic radius $R = 6$ m and friction coefficient $c_f = 0.004$. Demonstrate that the inertia term can be neglected. Next, calculate:

1. the surface amplitude in the basin (ζ_b)
2. the phase lag (θ)
3. the discharge in the gap ($\hat{Q}$).

Solution

Importance of inertia (relative to the restoring force): $\left(\omega^2\ell/g\right)(A_b/A_c) \approx 0.04$, acceptably small to neglect it in a first approximation. Including the boundary resistance in the channel, the resistance coefficient $\chi = 1/2 + c_f\ell/R = 0.9$. We now obtain:

1. $\Gamma = 1.28 \rightarrow r = 0.73 \rightarrow \hat{\zeta}_b = 0.55$ m
2. $\theta = \arccos 0.73 = 43°$
3. $\ddot{Q} = A_b\,\omega\,\zeta_b = 7.7 \times 10^3\ \text{m}^3/\text{s}$.

The channel is sufficiently short to allow the neglect of inertia, yet sufficiently long for the bed resistance to have a significant effect.

Empirical verification of the preceding results has been presented by Terra (2005), based on laboratory experiments. Details of the set-up and the results are presented in Box 6.2. Terra used three groups of excitation amplitudes in his experiments, of about 0.6 mm, 2 mm and 5.5 mm, respectively. Within each of these groups, the actual amplitude varied considerably due to a frequency dependence. (The amplitude range of each of the three groups is indicated in Figure 6.9.) For a comparison with the theory, the value of Γ_0 is required, which is proportional to the excitation amplitude. Instead of using a separate value of Γ_0 for each individual excitation amplitude, the smallest and the largest amplitude of each of the three groups was used, thus defining an upper bound and a lower bound to the theoretical

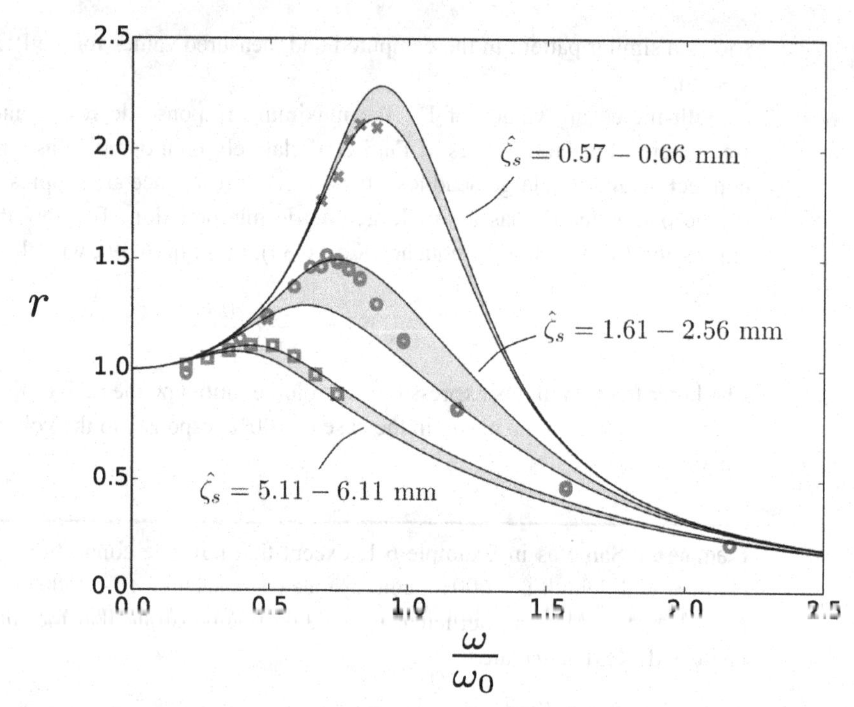

Fig. 6.9 Experimental verification of linearized channel–basin system, measured (markers) and theoretical range of the response factor for three groups of excitation amplitude; data from Terra (2005)

response values for each of the three groups. The results are shown in Figure 6.9. The agreement is quite good.

Box 6.2

Experimental verification of basin response

The water in a laboratory basin ($A_b = 0.916$ m^2) was connected by a pipe (length $\ell = 441$ mm, diameter $D = 76.4$ mm) to a 'sea' in which the free surface was forced to oscillate harmonically for a range of frequencies and amplitudes. Terra (2005) fitted the theoretical model to the data and so estimated the values of the natural frequency ω_0 and the parameter $\Gamma_0/\hat{\zeta}_s$ (in our notation). Instead, we use no fitting but make *a priori* estimates of the necessary parameters, as follows. The pipe extended some distance into the water of the bay and that of the 'sea', so that during inflow contraction occurs to half the pipe cross-sectional area (Borda), followed by expansion with an accompanying head loss of $U^2/2g$ (Carnot). This comes in addition to the well-known head loss at outflow of equal magnitude. Therefore, we use $\chi = 1 + c_f\ell/R$ instead of Eq. (6.9). Using $c_f = 0.002$, we obtain $\chi = 1.05$. To calculate the natural frequency, added mass must be taken into account, which is expressed by using an effective pipe length ℓ_{eff} in Eq. (6.34), which is larger than the geometric pipe length by an amount of the order of the pipe radius (as is known in acoustics for open organ pipes). Using $\ell_{eff} = \ell + \frac{1}{2}D$, we obtain $\omega_0 = 0.320$ rad/s and finally $\Gamma_0 = (0.371\ \text{mm}^{-1})\,\hat{\zeta}_s$. The minimum and the maximum values of $\hat{\zeta}_s$ for each of the three amplitude groups have been applied in the computation of the theoretical upper and lower bounds of the response factor.

6.6 Solution through Complex Algebra

In preparation for the following chapter, we are once more going to derive the solution given above for the system with storage and resistance, now using the representation of sinusoidal functions as complex quantities. The advantage of this formulation is that the amplitude and the phase angle are represented in terms of a single variable, the time variation can be factored out, and the partial differential equation reduces to an algebraic equation from which the solution can be more easily found and represented graphically.

6.6.1 Complex Representation

A real, harmonically varying quantity such as $A = \hat{A}\cos(\omega t + \alpha)$ can be represented as the real part of a complex quantity $\tilde{A}\exp(i\omega t)$, in which $\tilde{A}$ is the so-called complex amplitude of A, given by $\tilde{A} = \hat{A}\exp(i\alpha)$. The complex amplitude can be presented as a point in the complex plane; see Figure 6.10. The modulus of $\tilde{A}$ equals $\hat{A}$, i.e. the (real) amplitude of A, and its argument is the time-independent part of the phase of $\tilde{A}$, or $\arg\tilde{A} = \alpha$.

The time variation of A is obtained by multiplying $\tilde{A}$ with $\exp(i\omega t)$, i.e. by rotating the corresponding vector in the complex plane over an angle ωt, followed by taking the real part, i.e. the projection on the real axis. This time variation can be carried out at any moment if desired, but it is not necessary to show this in a graph. It is sufficient to know that it can be done, provided $\tilde{A}$ is known.

We will apply the complex representation to find the harmonic (i.e., sinusoidal) solution of the model with discrete storage and resistance, described by Eq. (6.22), which is repeated here for convenience:

$$\tau\frac{\mathrm{d}\zeta_b}{\mathrm{d}t} + \zeta_b = \zeta_s \tag{6.41}$$

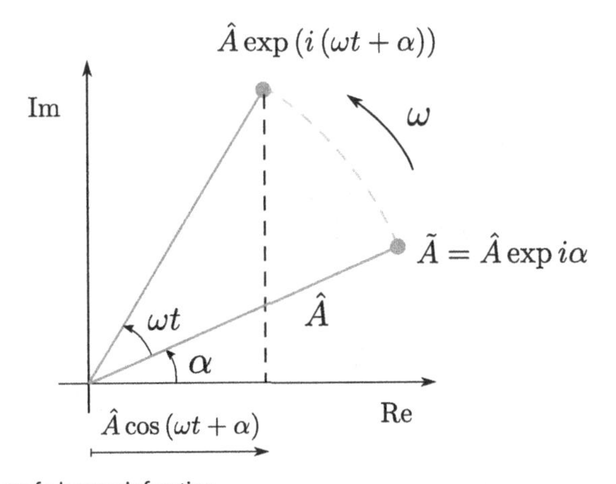

Fig. 6.10 Complex representation of a harmonic function

Each time-varying sine function is represented as the real part of a complex quantity:

$$\zeta_s(t) = \text{Re}\left(\tilde{\zeta}_s \exp(i\omega t)\right) \quad \text{and} \quad \zeta_b(t) = \text{Re}\left(\tilde{\zeta}_b \exp(i\omega t)\right) \tag{6.42}$$

In view of Eq. (6.42), the first term in Eq. (6.41) is given by

$$\tau \frac{\mathrm{d}\zeta_b}{\mathrm{d}t} = \text{Re}\left(i\omega\tau\tilde{\zeta}_b \exp(i\omega t)\right) \tag{6.43}$$

We substitute this and Eq. (6.42) into Eq. (6.41). Each term in that equation is the real part of a complex, time-varying quantity, but the presence of the time factor $e^{i\omega t}$ requires that Eq. (6.41) be fulfilled by the corresponding complex, time-independent amplitudes. Moreover, because the time factor is common to all terms, it can be factored out. The result is

$$i\omega\tau\tilde{\zeta}_b + \tilde{\zeta}_b = \tilde{\zeta}_s \tag{6.44}$$

Here, we see two great advantages of using the complex representation: the time dependence is represented as a common multiplier that can be factored out, and the differential equation (6.41) is replaced by the algebraic Eq. (6.44).

6.6.2 Solution

For given $\tilde{\zeta}_s$, we need to find $\tilde{\zeta}_b$. First, we do this graphically because that gives a good insight. After that the solution is determined purely algebraically.

For the graphical solution procedure, we plot $\tilde{\zeta}_b$ in the complex plane as a vector with length $|\tilde{\zeta}_b|$, i.e. the amplitude $\hat{\zeta}_b$, at an angle $\arg\tilde{\zeta}_b$ with the real axis (see Figure 6.11). The modulus and the argument of $\tilde{\zeta}_b$ have been chosen arbitrarily. The quantity $i\omega\tau\tilde{\zeta}_b$ has been plotted in Figure 6.11 as well. It stands at a right angle to $\tilde{\zeta}_b$, because of the factor i, and its modulus is a factor $\omega\tau$ larger than $\hat{\zeta}_b$ (or smaller, as the case may be); in the figure, an arbitrarily chosen value of $\omega\tau$ of about 0.5 was used. (Note that $i = \exp(i\pi/2)$, so multiplying with i means turning over 90°.) According to Eq. (6.44), the sum of $i\omega\tau\tilde{\zeta}_b$ and $\tilde{\zeta}_b$, i.e. the resultant of the corresponding two vectors in the figure (the hypotenuse),

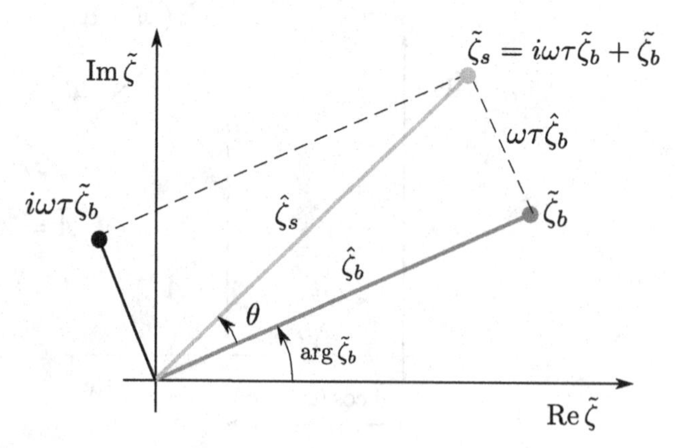

Fig. 6.11 Complex amplitudes

equals $\tilde{\zeta}_s$. The angle enclosed between this resultant and $\tilde{\zeta}_b$ represents the phase difference between the two, which we had denoted as θ, as indicated in the figure. We see at once in the resulting right triangle that $\tan\theta = \omega\tau$ and that $\cos\theta = \hat{\zeta}_b/\hat{\zeta}_s = r$, the same as was found above.

For the analytical solution method, we define a complex-amplitude ratio $\tilde{r}$ as

$$\tilde{r} \equiv \tilde{\zeta}_b/\tilde{\zeta}_s \tag{6.45}$$

For the real amplitudes, we have $|\tilde{r}| = |\tilde{\zeta}_b/\tilde{\zeta}_s| = |\tilde{\zeta}_b|/|\tilde{\zeta}_s| = \hat{\zeta}_b/\hat{\zeta}_s = r$, and for the phases we have $\arg\tilde{r} = \arg\left(\tilde{\zeta}_b/\tilde{\zeta}_s\right) = \arg\tilde{\zeta}_b - \arg\tilde{\zeta}_s = -\theta$. It follows from Eq. (6.44) and the definition of $\tilde{r}$ that

$$\tilde{r} = \frac{1}{1 + i\omega\tau} \tag{6.46}$$

so that

$$r = |\tilde{r}| = \left|\frac{1}{1 + i\omega\tau}\right| = \frac{1}{|1 + i\omega\tau|} = \frac{1}{\sqrt{1 + (\omega\tau)^2}} \tag{6.47}$$

and

$$\arg\tilde{r} = -\theta = -\arctan(\omega\tau) \qquad \text{or} \qquad \tan\theta = \omega\tau \tag{6.48}$$

The preceding simple example illustrates the advantages and the potential of the complex representation for solving a linear problem with sinusoidal forcing and response. This is a good preparation for the application in the following chapter, which deals with the more complicated problem of one-dimensional *propagation* of long waves, including damping.

Problems

6.1 What is the so-called small-basin approximation?

6.2 Under which conditions is the small-basin approximation valid?

6.3 Can a variation in the (undisturbed) depth in the basin affect the behaviour of the system? If so, how (through which mechanism)?

6.4 Same question as in Problem 6.3, now with respect to the bed friction coefficient c_f.

6.5 What is the so-called rigid-column approximation?

6.6 Under which conditions is the rigid-column approximation valid?

6.7 Explain why the water in a small basin, connected to a tidal sea by a gap in a barrier, reaches its highest level at the instant when this level equals that at sea.

6.8 Which physical criterion is applied in the linearization of the quadratic resistance? Why?

6.9 Which physical process is lost in the linearization of the quadratic resistance?

6.10 What is the meaning of the dimensionless parameters $\omega\tau$ and Γ as used in this chapter?

6.11 Derive Eq. (6.30), given Eq. (6.25) and the definition $\Gamma = \omega\tau/r$.

6.12 Argue why the amplitude ratio r is nearly 1 for small Γ and goes to zero for large Γ (without using the analytical solution).

6.13 A cylindrical tide well with an inner diameter of 1 m, placed in a harbour, should be able to measure harbour oscillations (seiches) with a period of about 10 min and an amplitude of 0.5 m with a damping of at most 1% (r at least 0.99). Calculate the required diameter of the orifice that connects the tide well with the surrounding water.

6.14 For the orifice diameter calculated in the answer to Problem 6.13, calculate the damping of wind waves given that these have a period of 8 s and an amplitude of 0.3 m at the location of the tide well.

6.15 Argue for each of the following parameters how a 20% increase in their value could affect the discharge through the channel between basin and sea: A_c, B_c, R_c, c_f, ℓ, A_b, $\hat{\zeta}_s$, ω. Think of possible feedback! Verify your estimates through numerical calculation of the discharge for a self-chosen example. (Note that the number of calculations required is less than the number of parameters.)

6.16 A tidal basin of approximately 10 km × 20 km in plan is connected by a channel to a sea with M_2-tide with a tidal range of 2 m. The channel is 10 km long and has a cross-sectionally averaged mean depth of about 9 m, a free-surface width of 600 m and a conveyance cross section of 5×10^3 m^2 with $c_f = 0.004$. Ignore inertia at first.

(a) Calculate the amplitude of the tidal elevation in the basin.
(b) Calculate the phase lag of the basin tide behind the tide at sea.
(c) Calculate the maximum discharge through the channel.
(d) Sketch in one plot the time variation of the tides at sea and in the basin as well as the discharge through the channel and interpret the result.

Now take inertia into account.

(e) Calculate the natural frequency.
(f) Verify whether the neglect of inertia in the parts (a)–(c) was allowed.
(g) Regardless of the answer to part (f), take inertia into account and calculate the amplitude of the tidal elevation in the basin and the phase lag of the basin tide behind the tide at sea. Why is this lag greater with inertia taken into account than when it is neglected?

6.17 Using complex algebra, and starting from Eqs. (6.33) and (6.34), derive the results of Eqs. (6.35) and (6.36).

6.18 Given $\zeta_s = \hat{\zeta}_s \cos(\omega t - \alpha)$ with $\hat{\zeta}_s = 1$ m and $\alpha = 212°$, calculate $\tilde{\zeta}_s$. Plot this in the complex plane and calculate its real part and its imaginary part.

6.19 Given $\tilde{\zeta}_s = (0.9 - 0.8i)$ m, plot this and give the corresponding expression for $\zeta_s(t)$. Calculate ζ_s for $\omega t = 0$ and $\omega t = \frac{1}{2}\pi$.

6.20 Suppose that $\omega\tau = 0.5$. Plot $\tilde{r}$, equal to $1/(1 + i\omega\tau)$, and calculate r and $\arg \tilde{r}$.

7 Harmonic Wave Propagation

This chapter deals with low periodic long waves and oscillations such as tides and seiches, taking into account inertia and resistance. The aim of this chapter is to provide insight into the dynamics of these waves, in particular with respect to the effects of resistance, which were mainly ignored so far. Using linear approximations, valid for low waves, explicit yet simple analytical solutions are obtained in the form of complex exponential functions, representing damped harmonic (sinusoidal) waves or oscillations. Solutions will be derived for uniform as well as non-uniform channels, which are used subsequently to analyze periodic wave motion in channels that are closed at one end or at both ends. The chapter concludes with a discussion of nonlinear influences, such as the variation of the cross-sectional parameters with the free-surface elevation, or the quadratic nature of the resistance.

7.1 Introduction

A typical situation occurs when an initial state of rest is affected by continual periodic disturbances at a boundary, which propagate into the domain considered. Given enough time, a periodic state of motion is established in the entire domain, without memory of the initial situation. In such cases the motion is unsteady within each wave cycle, but the cycles themselves do not vary in time, as is approximately the case for the variation of amplitudes and phases in seiches and tidal waves. The principal features of such waves can reasonably well be represented with a linear model, assuming a harmonic variation of the state variables involved.

To the authors' knowledge, analytical modelling of the one-dimensional propagation of long harmonic waves was pioneered by Parsons (1918), for the prediction of the tidal flows in the then future Cape Cod Canal. Parsons' linearization of the quadratic friction was incorrect (see Box 6.1), and the system considered by him was simple, but his work was nevertheless a commendable achievement. Yet it does not seem to have found a recognizable follow-up. In fact, it is not even mentioned in the review of tidal computations with the harmonic method by Ippen and Harleman (1966).

Apparently without being aware of Parsons' work, a Dutch committee chaired by the Nobel prize–winning physicist Hendrik Antoon Lorentz developed and used a more extensive analytical model for tidal computations for the purpose of predicting the changes that were to be expected in the tides as a result of the enclosure in 1932 of the previous Zuiderzee, a large tidal basin in The Netherlands. For this purpose, Lorentz (1926) developed the correct linearization of the quadratic friction now named after him. The resulting model was successfully applied to a network of 26 sections representing the tidal inlets and the inner network of tidal channels. This so-called *harmonic method* soon found its way into civil engineering practice and has been applied to numerous projects through the world.

The approach in the present chapter is a continuation of the preceding chapter in the use of the linearization of the quadratic resistance and the complex algebra formulation. The difference is that we are now dealing with progressive and standing waves. We start with a description of damped, harmonic progressive waves in terms of the complex formalism, prior to a derivation of the characteristic features of these waves, in particular the propagation speed and the rate of damping.

7.2 Complex Representation of Damped Progressive Harmonic Waves

We assume a periodic wave progressing in the positive s-direction in a prismatic channel. The wave period is T, the angular frequency (i.e. the phase change per unit time) $\omega = 2\pi/T$, the wavelength L, the wave number (i.e. the change of phase per unit distance) $k = 2\pi/L$, and the propagation speed (or the phase speed, i.e. the speed of points of constant phase) $c = L/T = \omega/k$. The surface elevation ζ has a location-dependent amplitude $\hat{\zeta}$:

$$\zeta(s,t) = \hat{\zeta}(s)\cos(\omega t - ks + \alpha) \tag{7.1}$$

In complex form, this can be written as

$$\zeta(s,t) = \mathrm{Re}\left(\tilde{\zeta}(s)\exp(i\omega t)\right) \tag{7.2}$$

in which

$$\tilde{\zeta}(s) = \hat{\zeta}(s)\exp(i(-ks+\alpha)) \tag{7.3}$$

Notice that in this complex formulation a *separation of variables* has taken place, i.e. the dependences on s and on t occur in separate factors instead of combined as in $\cos(\omega t - ks)$. This simplifies the analysis.

As we saw in the preceding chapter, the complex amplitude $\tilde{\zeta}$ at a fixed point (constant s) can be represented as a point in the complex plane (see Figure 7.1a). Its argument is the phase of $\zeta(s,t)$ at that point at time $t = 0$. The time variation is given by $\exp(i\omega t)$.

In contrast with the discrete approximation used in Chapter 6, in which the amplitudes and phases were treated as constants, they are now continuously varying in space, i.e. as a function of s. Starting at $s = 0$, $\arg\tilde{\zeta} = \alpha$, i.e. the phase at time $t = 0$, as shown in Figure 7.1b. The phase at $t = 0$ at an arbitrary other point can be found by multiplication of $\tilde{\zeta}(0)$ with $\exp(-iks)$, i.e. by rotation over an angle $-ks$ (Figure 7.1b).

Choosing a continuous succession of points, i.e. a continuous variation of s, we obtain a continuously varying sequence of values of $\tilde{\zeta}$, generating a smooth curve in the complex

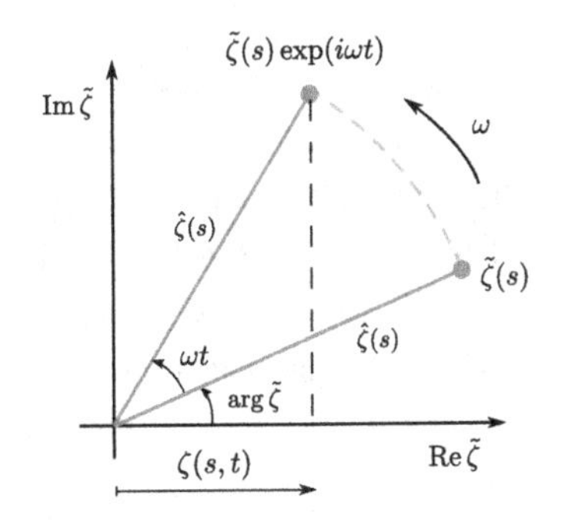

$$\begin{aligned}\zeta(s,t) &= \mathrm{Re}\left(\tilde{\zeta}(s)\exp(i\omega t)\right)\\ &= \mathrm{Re}\left(\hat{\zeta}(s)\exp\left(i\omega t + \arg\tilde{\zeta}(s)\right)\right)\\ &= \hat{\zeta}(s)\cos\left(\omega t + \arg\tilde{\zeta}(s)\right)\end{aligned}$$

(a)

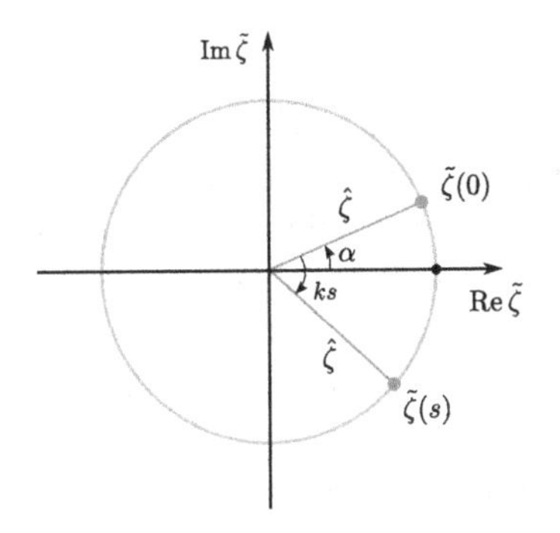

hodograph of $\tilde{\zeta}(s) = \hat{\zeta}\exp\left(i\left(\alpha - ks\right)\right)$
with $\hat{\zeta}$ and k constant

(b)

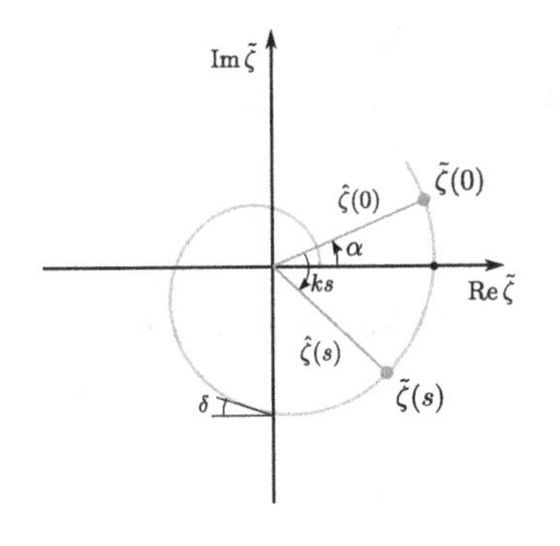

hodograph of $\tilde{\zeta}(s) = \hat{\zeta}(0)\exp(-\mu s)\exp\left(i\left(\alpha - ks\right)\right)$
in which $\tan\delta = \mu/k$ (plotted for $\tan\delta = 0.2$)

(c)

Fig. 7.1 Complex plane with hodographs; (a) variation in time at a fixed point; (b) phase shift as a function of distance in a progressive wave of constant amplitude; (c) variation of amplitude and phase as functions of distance in a damped progressive wave

plane, a so-called *hodograph*. For constant amplitude, the hodograph is a circle, shown in Figure 7.1b.

A special case of amplitude variation is the exponential decay with s, as occurs in systems with linear(ized) resistance. Denoting the damping modulus with μ, this can be written as $\hat{\zeta}\,(s) = \hat{\zeta}\,(0)\exp\left(-\mu s\right)$. In this case, Eq. (7.3) yields

$$\tilde{\zeta}\,(s) = \hat{\zeta}\,(0)\,\exp\left(-\mu s\right)\,\exp\left(i\left(-ks + \alpha\right)\right) = \tilde{\zeta}\,(0)\,\exp\left(-ps\right) \tag{7.4}$$

in which $\tilde{\zeta}\,(0) = \hat{\zeta}\,(0)\exp\left(i\alpha\right)$ and we have introduced the complex propagation constant

$$p = \mu + ik \tag{7.5}$$

Box 7.1 The convergence angle of the logarithmic spiral

When propagating over a small distance Δs from s_1 to s_2, a phase change occurs given by $k\Delta s$, corresponding to a rotation of the associated vector in the complex plane over a small angle $k\Delta s$; see Figure 7.1c. The amplitude, represented by the length of the radius to the origin, thereby reduces by a factor $\exp(-\mu s_2) / \exp(-\mu s_1) = \exp(-\mu(s_2 - s_1)) = \exp(-\mu\Delta s)$. The relative amplitude reduction is therefore given by $1 - \exp(-\mu\Delta s)$, which for small Δs can be approximated as $\mu\Delta s$. The angle between the tangent and the normal to the radius, i.e. the angle of convergence, is therefore given by $\arctan(\mu/k)$, which is constant along the spiral.

In the right-hand side of Eq. (7.4), $\tilde{\zeta}(0)$ is the initial complex amplitude, and $\exp(-ps)$ describes its spatial variation. The rate of change of the real amplitude is given by μ, and that of the phase is given by k. These two are combined in the single complex parameter p.

In the case of exponential decay, the hodograph of $\tilde{\zeta}(s)$ takes the form of a logarithmic spiral, characterized by a constant angle of convergence (i.e. the angle between the tangent at a point of the spiral and the normal to the radius connecting that point with the origin, designated as δ in Figure 7.1c). See Box 7.1.

The values of ω, μ and k are not independent because the wave motion must fulfill the balances of mass and momentum. Only one of them can be chosen freely, normally the frequency. Using the balance equations, an expression for $p(\omega)$, the so-called *complex dispersion equation*, will be derived in the following sections. From this we also know $\mu(\omega)$ and $k(\omega)$ and from the latter the phase speed $c = \omega/k$.

7.3 Formulation and General Solution

This section presents the basic equations and their solutions for harmonic motion.

7.3.1 Formulation

We start from the linearized one-dimensional balance equations for mass and momentum. The quadratic resistance is linearized as in Chapter 6, and the advection of momentum is neglected, as is allowed for low waves. The latter restriction also allows us to neglect variations in the cross section that occur as a result of variations in flow depth. With this, the linearized balance equations for mass and momentum are written as follows:

$$B\frac{\partial \zeta}{\partial t} + \frac{\partial Q}{\partial s} = 0 \tag{7.6}$$

and

$$\frac{\partial Q}{\partial t} + gA_c\frac{\partial h}{\partial s} + \kappa Q = 0 \tag{7.7}$$

The parameter κ is a shorthand factor (dimension: 1/time) in the expression for the linearized resistance, defined by (see also Eq. (6.17)):

$$\kappa = \frac{8}{3\pi} c_f \frac{\hat{Q}}{A_c R} = \frac{8}{3\pi} c_f \frac{\hat{U}}{R} \tag{7.8}$$

In the following derivations and applications, κ is assumed to be a known constant. Since it contains the amplitude $\hat{Q}$ or $\hat{U}$, which are not known beforehand, iteration is necessary.

Eqs. (7.6) and (7.7) form a set of two coupled partial differential equations in ζ and Q, with constant coefficients (in the linear approximation). Such equations allow (complex) exponential solutions.

For harmonic motion, the first term in the momentum balance is of the order of $\omega\hat{Q}$, and the resistance term is of the order $\kappa\hat{Q}$. It follows that the ratio of resistance to inertia is of the order κ/ω, written as σ for short:

$$\sigma \equiv \frac{\kappa}{\omega} = \frac{8}{3\pi} c_f \frac{\hat{Q}}{\omega A_c R} = \frac{8}{3\pi} c_f \frac{\hat{U}}{\omega R} \tag{7.9}$$

Apart from the factor $8/3\pi$ (about 0.85), this is the same σ as in Chapter 2, of which some typical values have been given in Table 2.1.

Eliminating Q between Eqs. (7.6) and (7.7), and using the property of a prismatic channel that $\partial A_c/\partial s = 0$, we obtain

$$\frac{\partial^2 \zeta}{\partial t^2} - c_0^2 \frac{\partial^2 \zeta}{\partial s^2} + \kappa \frac{\partial \zeta}{\partial t} = 0 \tag{7.10}$$

in which c_0^2 is the long-wave speed in absence of resistance:

$$c_0 = \sqrt{\frac{g A_c}{B}} \tag{7.11}$$

Eq. (7.10) is the linearized wave equation (see Chapter 3) with (linearized) resistance. Because of this, and with reference to Chapter 3, we expect solutions that in general consist of two waves, travelling in opposite directions, decaying as they propagate.

7.3.2 General Solution

We seek solutions for $\zeta(s, t)$ in the form

$$\zeta(s,t) = \mathrm{Re}\left(\tilde{\zeta}(s)\exp(i\omega t)\right) \tag{7.12}$$

We substitute this into Eq. (7.10), drop the time factor $\exp(i\omega t)$, and obtain

$$\frac{\mathrm{d}^2\tilde{\zeta}}{\mathrm{d}s^2} + \frac{\omega^2 - i\omega\kappa}{c_0^2}\tilde{\zeta} = 0 \tag{7.13}$$

Substitution of $\kappa = \sigma\omega$ and of $k_0 = \omega/c_0$, the wave number in absence of resistance, yields

$$\frac{d^2\tilde{\zeta}}{ds^2} + k_0{}^2(1 - i\sigma)\tilde{\zeta} = 0 \tag{7.14}$$

This is an ordinary differential equation with constant coefficients. Such equations have exponential solutions. Therefore, we pose

$$\tilde{\zeta}(s) = \exp(Ps) \tag{7.15}$$

and substitute this into (7.14), which results in

$$P^2 + k_0^2(1 - i\sigma) = 0 \tag{7.16}$$

This is the so-called *complex dispersion equation.* For given frequency (contained in k_0) and relative resistance (σ), it determines the complex propagation constant P, which governs the spatial variation, as expressed in Eq. (7.15).

Notice the reduction in mathematical complexity in the various steps: we started with a partial differential equation, Eq. (7.10). Due to the restriction to harmonic solutions and the complex representation, this reduced to an ordinary differential equation, Eq. (7.14). Finally, assuming a (complex) exponential solution, the integration constant P is determined by an algebraic equation (7.16).

Eq. (7.16) is of second degree. It has two opposite, complex roots P_1 and $P_2 = -P_1$, which will be designated as p and $-p$, respectively, in which

$$p = ik_0\sqrt{1 - i\sigma} \tag{7.17}$$

The general solution for $\tilde{\zeta}(s)$ can then be written as

$$\tilde{\zeta}(s) = C^+\exp(-ps) + C^-\exp(ps) = \tilde{\zeta}^+(s) + \tilde{\zeta}^-(s) \tag{7.18}$$

Eq. (7.18) is the general solution of Eq. (7.14). If we include the time factor $\exp(i\omega t)$, it represents two waves travelling in opposite directions. The two (complex) integration constants C^- and C^+, each containing amplitude and phase information, are to be determined from the (two) boundary conditions, as we will see further on. If either C^- or C^+ is zero, the propagation is in one direction only. Usually, there is some reflection somewhere, in which case there is propagation of two wave systems in opposite directions.

7.3.3 Solution of the Dispersion Equation

Our next task is to derive a more explicit solution of Eq. (7.17). The propagation constant p is in general complex. (If $\sigma = 0$ it is purely imaginary.) We write $p = \mu + ik$, in which μ and k are real, representing damping and propagation, respectively (see Section 7.2).

In order to determine μ and k, we have to separate the right-hand side of Eq. (7.17) into its real part and its imaginary part, for which we need (among others) the modulus and the argument of $\sqrt{1-i\sigma}$. The argument of this square root is $-\frac{1}{2}\arctan\sigma$. To shorten the notation, we introduce an auxiliary variable, an angle δ, such that

$$\tan 2\delta \equiv \sigma = \frac{8}{3\pi} c_f \frac{\hat{U}}{\omega R} \tag{7.19}$$

where δ is a dimensionless measure of the resistance/inertia ratio, just as σ. It can be seen as a kind of friction angle. In terms of δ, we have $\arg\sqrt{1-i\sigma} = -\frac{1}{2}\arctan\sigma = -\delta$. Furthermore, $|1-i\sigma| = \sqrt{1+\sigma^2} = \sqrt{1+\tan^2 2\delta} = 1/\cos 2\delta$, so that altogether we have

$$p = ik_0 \frac{\exp(-i\delta)}{\sqrt{\cos 2\delta}} \tag{7.20}$$

or

$$p = i\frac{\cos\delta - i\sin\delta}{\sqrt{\cos 2\delta}} k_0 = \frac{\sin\delta + i\cos\delta}{\sqrt{\cos 2\delta}} k_0 \tag{7.21}$$

so that finally

$$\mathrm{Im}\,p = k = \frac{\cos\delta}{\sqrt{\cos 2\delta}} k_0 = \frac{k_0}{\sqrt{1-\tan^2\delta}} \quad \text{and} \quad \mathrm{Re}\,p = \mu = \frac{\sin\delta}{\sqrt{\cos 2\delta}} k_0 = k\tan\delta \tag{7.22}$$

The preceding algebraic steps have been visualized in Figure 7.2. (Note: it follows directly from Eq. (7.22) that $\delta = \arctan(\mu/k)$, which is equal to the angle of convergence of the logarithmic spiral representing the hodograph of $\tilde{\zeta}$; see Box 7.1.)

Finally, we write the phase speed $c = \omega/k$ in a more explicit form, using the preceding results:

$$c = \frac{\omega}{k} = \frac{\omega}{k_0}\sqrt{1-\tan^2\delta} = c_0\sqrt{1-\tan^2\delta} \tag{7.23}$$

or

$$c = \sqrt{\frac{gA_c}{B}}\sqrt{1-\tan^2\delta} \tag{7.24}$$

Apparently, resistance not only dampens the waves but also slows them down.

As far as the authors are aware, the results given in Eqs. (7.22) and (7.24) have first been derived by Parsons (1918) in his model of tidal propagation through the Cape Cod Canal (though with a numerically different linearized resistance factor, and in an algebraically different appearance in terms of σ instead of our auxiliary angle δ).

7.3.4 Solution for the Discharge

We express the discharge in complex form, analogous to Eq. (7.12). Substiuting this and the general solution for $\tilde{\zeta}(s)$ (Eq. (7.18)) in the continuity equation (7.6) yields the following expression for the associated discharge, assuming a zero net value:

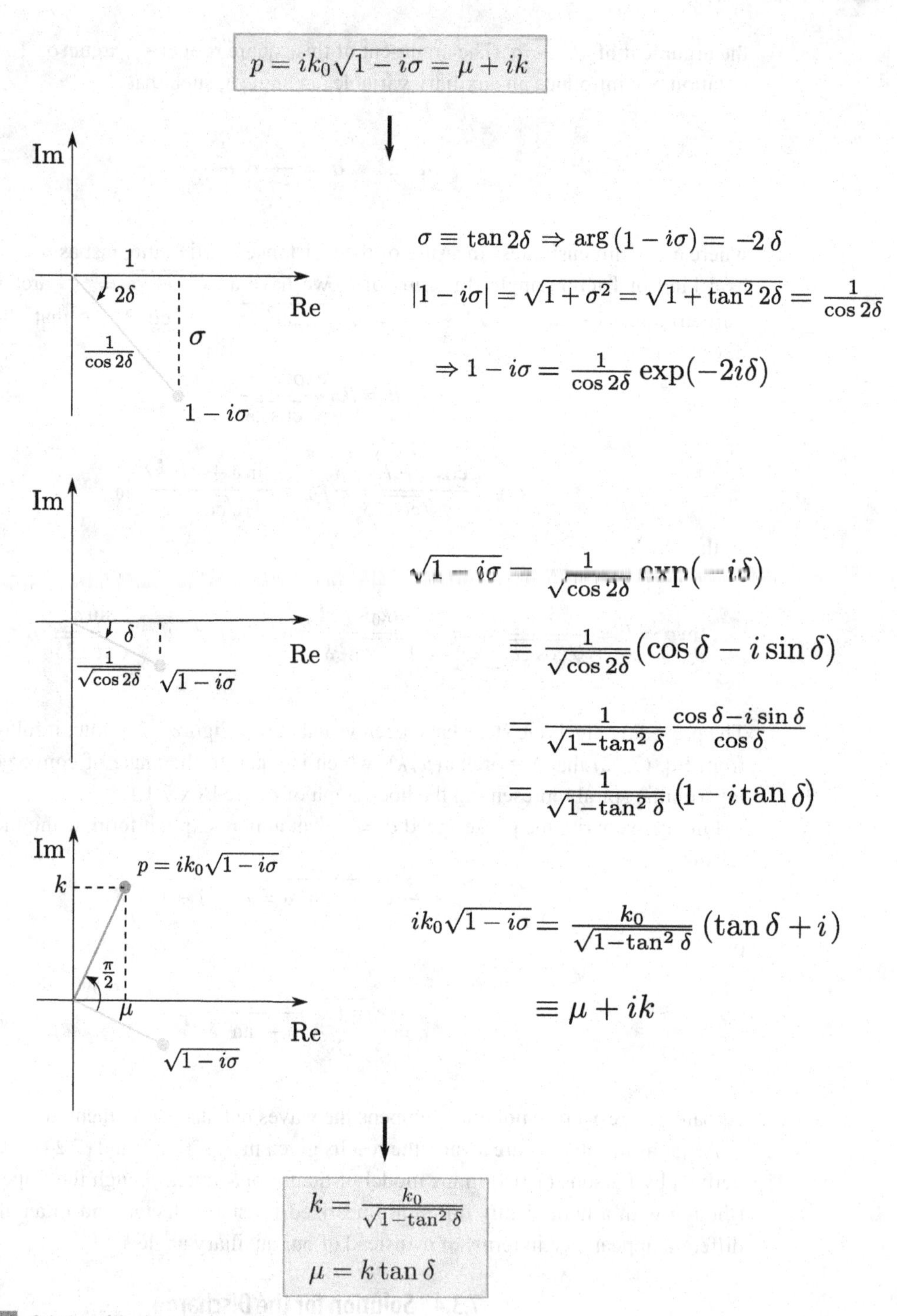

$p = ik_0\sqrt{1-i\sigma} = \mu + ik$
Im
Re
1
2δ
σ
$\frac{1}{\cos 2\delta}$
$1 - i\sigma$
$\sigma \equiv \tan 2\delta \Rightarrow \arg(1 - i\sigma) = -2\,\delta$
$|1 - i\sigma| = \sqrt{1+\sigma^2} = \sqrt{1+\tan^2 2\delta} = \frac{1}{\cos 2\delta}$
$\Rightarrow 1 - i\sigma = \frac{1}{\cos 2\delta}\exp(-2i\delta)$
Im
Re
δ
$\frac{1}{\sqrt{\cos 2\delta}}$
$\sqrt{1-i\sigma}$
$\sqrt{1-i\sigma} = \frac{1}{\sqrt{\cos 2\delta}}\exp(-i\delta)$
$= \frac{1}{\sqrt{\cos 2\delta}}(\cos\delta - i\sin\delta)$
$= \frac{1}{\sqrt{1-\tan^2\delta}}\frac{\cos\delta - i\sin\delta}{\cos\delta}$
$= \frac{1}{\sqrt{1-\tan^2\delta}}(1 - i\tan\delta)$
Im
Re
$p = ik_0\sqrt{1-i\sigma}$
k
$\frac{\pi}{2}$
μ
$\sqrt{1-i\sigma}$
$ik_0\sqrt{1-i\sigma} = \frac{k_0}{\sqrt{1-\tan^2\delta}}(\tan\delta + i)$
$\equiv \mu + ik$
$k = \frac{k_0}{\sqrt{1-\tan^2\delta}}$
$\mu = k\tan\delta$

Fig. 7.2 Solution of the dispersion equation

$$\tilde{Q}(s) = \frac{i\omega B}{p} \left(C^+ \exp(-ps) - C^- \exp(ps) \right) \tag{7.25}$$

or

$$\tilde{Q}(s) = \frac{i\omega B}{p} \left(\tilde{\zeta}^+(s) - \tilde{\zeta}^-(s) \right) \tag{7.26}$$

The complex factor $i\omega B/p$ can be reworked as follows:

$$\frac{i\omega B}{p} = \frac{i\omega B}{\mu + ik} = \frac{\omega}{k} \frac{B}{1 - i\mu/k} = \frac{Bc}{1 - i \tan \delta} \tag{7.27}$$

Since $\arg(1 - i \tan \delta) = -\delta$, and $|1 - i \tan \delta| = \sqrt{1 + \tan^2 \delta} = 1/\cos \delta$, we finally obtain

$$\frac{i\omega B}{p} = Bc \cos \delta \exp(i\delta) \tag{7.28}$$

Summarizing, the general solution for $\tilde{\zeta}$, Eq. (7.18), is

$$\tilde{\zeta}(s) = C^+ \exp(-ps) + C^- \exp(ps) = \tilde{\zeta}^+(s) + \tilde{\zeta}^-(s) \tag{7.29}$$

and the corresponding discharge is

$$\tilde{Q}(s) = \frac{i\omega B}{p} \left(C^+ \exp(-ps) - C^- \exp(ps) \right) = Bc \cos \delta \exp(i\delta) \left(\tilde{\zeta}^+ - \tilde{\zeta}^- \right) \tag{7.30}$$

This general solution will be interpreted and elaborated in the following sections, first for the simple case of a unidirectional wave system, so as to facilitate the understanding of the meaning of the results, to be followed by various cases of bi-directional waves.

7.4 Unidirectional Propagation

The preceding results apply to the spatial variation of the complex amplitudes of the surface elevation and the discharge, the time variation having been set aside temporarily. We will now reintroduce it:

$$\zeta(s, t) = \mathrm{Re} \left(\tilde{\zeta}(s) \exp(i\omega t) \right) \tag{7.31}$$

This is still the general solution. In the following, we will examine the case of a purely progressive wave.

7.4.1 Physical Interpretation

We restrict the following discussion to ζ^+:

$$\zeta^+(s,t) = \text{Re}\left(\tilde{\zeta}^+(s)\exp(i\omega t)\right) = \text{Re}\left(C^+\exp(-ps)\exp(i\omega t)\right) \tag{7.32}$$

Substituting $p = \mu + ik$ and expanding C^+ into its modulus and argument, we obtain

$$\zeta^+(s,t) = \text{Re}\left(|C^+|\exp(-\mu s)\exp(i(\omega t - ks + \arg C^+))\right) \tag{7.33}$$

or

$$\zeta^+(s,t) = \hat{\zeta}^+(s)\cos\left(\omega t - ks + \arg C^+\right) \tag{7.34}$$

in which $\hat{\zeta}^+(s) = |C^+|\exp(-\mu s)$. The associated discharge is

$$Q^+(s,t) = Bc\,\hat{\zeta}^+(s)\,\cos\delta\,\cos(\omega t - ks + \arg C^+ + \delta) \tag{7.35}$$

Inspecting these equations, we note the following items:

- $\arg C^+$ is the initial phase (when $\omega t = 0$) of ζ^+ in $s = 0$.
- The phase varies in s and t as $(\omega t - ks)$, implying that we deal with a wave progressing in the positive s-direction (that is the reason for the superscript + in C^+, ζ^+ and Q^+) with speed $c = \omega/k$, the so-called phase speed, because, observed at this speed, the phase is constant.
- $|C^+|$ is the amplitude of ζ^+ in $s = 0$.
- $|C^+|\exp(-\mu s)$ is the amplitude of ζ^+ as a function of s: exponential damping in the direction of propagation.
- In absence of damping ($\delta = 0$), $Q(s,t) = Bc\,\zeta(s,t)$, the same as was found in Chapter 3 for purely progressive waves without resistance.
- The discharge is a factor $\cos\delta$ smaller than $Bc\,\zeta$, and it is an angle δ ahead in phase relative to the surface elevation. The fact that resistance reduces the discharge for a given surface elevation, i.e. for a given driving force due to the slope of the free surface, is fairly obvious. The advance in phase of the discharge, by an angle δ, is due to the fact that for a given periodic forcing the flow acquires less momentum in the presence of resistance than it does without resistance, and therefore responds faster to the oscillatory driving forces.

The preceding items refer to a purely progressive wave propagating in the positive s-direction, i.e. proportional to C^+, with $C^- = 0$. Needless to say, they are equally applicable to a wave propagating in the negative s-direction, apart from a change of sign in Q^-. As anticipated, *the general solution of the wave equation* (7.10) *for harmonic motion consists of two wave systems, propagating in opposite direction, each exponentially decaying in its propagation direction.*

Example 7.1 Consider, respectively, an M_2-tide (T = 12 h 25 min) and a seiche (T = 1 h) in a channel in an estuary with $c_f = 0.004$, hydraulic radius R = 11 m, depth d = 12 m, conveyance width B_c = 350 m and storage width $B = 2B_c$ = 700 m. We consider the

Table 7.1 Example: unidirectional propagation of tidal waves and seiches

Quantity	Equation	Tide	Seiche
T	(given)	745 min	60 min
$\hat{U}$	(given)	1.2 m/s	0.6 m/s
ω	$\omega = 2\pi/T$	1.405×10^{-4} rad/s	1.745×10^{-3} rad/s
c_0	$c_0 = \sqrt{gA_c/B}$	7.7 m/s	7.7 m/s
k_0	$k_0 = \omega/c_0$	1.8×10^{-5} rad/m	2.3×10^{-4} rad/m
L_0	$L_0 = c_0 T = 2\pi/k_0$	343 km	28 km
σ	$\sigma = \frac{8}{3\pi} c_f \frac{\hat{U}}{\omega R}$	2.64	0.106
δ	$\delta = \frac{1}{2} \arctan \sigma$	34.6°	3.02°
$\tan \delta$	$\tan \delta$	0.69	0.053
$\sqrt{1 - \tan^2 \delta}$	$\sqrt{1 - \tan^2 \delta}$	0.72	1.00 (0.998...)
k	$k = k_0/\sqrt{1 - \tan^2 \delta}$	2.5×10^{-5} rad/m	2.3×10^{-4} rad/m
μ	$\mu = k \tan \delta$	1.8×10^{-5} m^{-1}	1.2×10^{-5} m^{-1}
c	$c = c_0 \sqrt{1 - \tan^2 \delta}$	5.5 m/s	7.7 m/s
L	$L = 2\pi/k = cT$	247 km	28 km

incident component of each, with $\hat{U} = 1.2$ m/s for the tide and $\hat{U} = 0.6$ m/s for the seiche, and calculate the corresponding phase speed and damping. See Table 7.1 for the results.

The phase speed of the tide (5.5 m/s) is only about one-half of $\sqrt{gd}$ (or $\sqrt{gA_c/B_c}$), which is almost 11 m/s. The difference is due to two factors: the width of the free surface (available for storage) is twice the conveyance width B_c, which slows down the wave speed by a factor $1/\sqrt{2}$, or about 70%, and resistance reduces it further with a factor $\sqrt{1 - \tan^2 \delta}$, which is roughly another 70%. The phase speed of the higher-frequency seiches is hardly affected by resistance, in agreement with the general conclusion in Chapter 2 about the relative impact of resistance (Table 2.1).

Although σ (and therefore δ) is far smaller for the seiche than it is for the tide, their μ-values, i.e. the relative damping per unit propagation distance, are comparable. This is because μ is proportional not only to $\tan \delta$, but also to the wave number k, which is almost a factor of ten larger (shorter wavelength) for the seiches than it is for the tides. Therefore, the smaller damping of the seiches per wavelength, determined by the value of δ, is more or less compensated by the larger number of wavelengths in a given space interval. Over a propagation distance Δs of 10 km, say, the tidal amplitude is reduced with a factor $\exp(-\mu \Delta s)$, or about 0.84, whereas the seiche amplitude over that same distance is reduced with a factor of about 0.89.

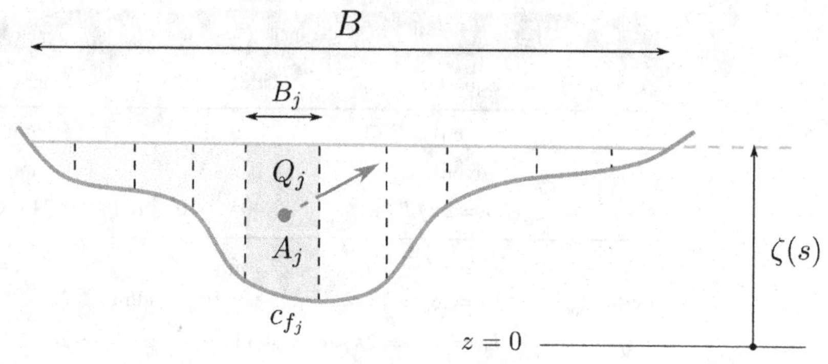

Fig. 7.3 Cross section of a compound channel

7.4.2 Propagation in Compound Channels

In the derivations in this and other chapters, allowance was made for the possibility that only a part of the total channel cross section contributes substantially to the transport; the remainder was assumed to be available for storage only. We are now generalizing this notion to channels with a compound cross section, consisting of (or rather, schematized to) a number ($j = 1, 2, 3, ...$) of distinct, adjacent parts, each with its own characteristic width (B_j), depth (d_j), area ($A_j = B_j d_j$) and resistance coefficient (c_{f_j}). See Figure 7.3. The total width of the free surface is $B = \Sigma_j B_j$, the total cross-sectional area is $A = \Sigma_j A_j$, and the mean depth is $d = A/B$. The values of these parameters define an equivalent rectangular cross section. As before, the free-surface elevation is assumed to be laterally uniform.

The flow in a compound channel is commonly calculated on the basis of the assumption that lateral momentum exchange between adjacent parts can be neglected, so that they are dynamically decoupled. For steady, uniform flow, the discharge in each part (Q_j) can then be determined by applying the conventional uniform-flow results to each part separately (for the given, common bed slope). The result can be used to calculate an equivalent bed friction coefficient ($c_{f,eq}$) for the equivalent rectangular cross section:

$$Q = \Sigma_j A_j \sqrt{\frac{g d_j i_b}{c_{fj}}} = A \sqrt{\frac{g d i_b}{c_{f,eq}}} \tag{7.36}$$

It is customary to use the equivalent rectangular cross section together with the equivalent resistance coefficient also in calculation of unsteady flow, but this is in fact incorrect because it ignores phase differences between the flow velocities in the respective parts. In the following, we present a model for harmonic wave propagation in a compound channel that takes such phase differences within the cross section into account.

The continuity equation retains its original form as derived for a non-compound channel, given by Eq. (7.6):

$$B \frac{\partial \zeta}{\partial t} + \frac{\partial Q}{\partial s} = 0 \tag{7.37}$$

The momentum balance (Eq. (7.7)) is applied to each subchannel separately:

$$\frac{\partial Q_j}{\partial t} + gA_j \frac{\partial \zeta}{\partial s} + \kappa_j Q_j = 0 \qquad \text{for} \quad j = 1, 2, 3, \ldots \tag{7.38}$$

We expand ζ as follows:

$$\zeta(s,t) = \mathrm{Re}\left(C_\zeta \ \exp\left(i\omega t + Ps\right)\right) \tag{7.39}$$

in which C_ζ is a complex proportionality constant and P is the complex propagation constant. Likewise, we write

$$Q_j(s,t) = \mathrm{Re}\left(C_{Q_j} \ \exp\left(i\omega t + Ps\right)\right) \tag{7.40}$$

Substitution of these expressions into Eqs. (7.37) and (7.38) yields

$$i\omega B C_\zeta + P \sum_j C_{Q_j} = 0 \tag{7.41}$$

and

$$C_{Q_j}\left(i\omega + \kappa_j\right) + gA_j P C_\zeta = 0 \qquad \text{for} \quad j = 1, 2, 3, \ldots \tag{7.42}$$

Substituting C_{Q_j} from the latter equation into Eq. (7.41) yields

$$i\omega B - P^2 \sum_j \frac{gA_j}{i\omega + \kappa_j} = 0 \tag{7.43}$$

With the substitution $\tan 2\delta_j = \sigma_j = \kappa_j/\omega$, and after some algebra, as in the transformation of Eq. (7.17) into Eq. (7.20), we finally obtain the following expression for the complex propagation constant P of the compound channel:

$$P^2 = -\frac{\omega^2 B/g}{\sum_j A_j \cos 2\delta_j \exp 2i\delta_j} \tag{7.44}$$

This equation has two opposite roots, say p and $-p$, in which $p = \mu + ik$, determining the damping modulus (μ) and the wave number (k), and therefore also the propagation speed $c = \omega/k$. The difference with the single-channel case is that it is not possible to obtain explicit expressions for these constants because of the summation of complex quantities. Nevertheless, they are fully determined through Eq. (7.44).

A part of the profile that is very shallow and rough was up to now supposed to contribute to storage only but not to transport. With the formulation for a compound channel being available, this rough schematization is no longer necessary. However, it can be seen as a special case of the more general formulation by letting the resistance factor of the shallow portion (κ_1, say), go to infinity. This implies that the corresponding Q_1 vanishes (see Eq. (7.42)). Another way to see this is through Eq. (7.44), since $\kappa_1 \rightarrow \infty$ implies that $\sigma_1 = \tan 2\delta_1 \rightarrow \infty$, so $\cos 2\delta_1 \rightarrow 0$. Therefore, the corresponding portion of the cross section does not contribute to the sum in the denominator in the right-hand side of Eq. (7.44), whereas the contribution B_1 to the storage width remains unaffected.

It is shown in Box 7.2 that for a non-compound channel, Eq. (7.44) reduces to Eq. (7.16), which was derived ab initio for that simple geometry.

Box 7.2 Non-compound channels

In the case of a non-compound channel, Eq. (7.44) should reduce to the expression for the single channel given by Eq. (7.16). This is not obvious at first sight, but by substituting $k_0 = \omega/c_0 = \omega/\sqrt{gA_c/B}$ and $1 - i\sigma = \exp(-2i\delta)/\cos 2\delta$ into Eq. (7.16), we obtain $P^2 = -\left(\omega^2 B/g\right) / \left(A_c \cos 2\delta \exp 2i\delta\right)$. It appears that the product in the denominator for the single-channel case is simply applied to each distinct portion of the compound channel and then summed to obtain the denominator for the compound-channel case (Eq. (7.44)).

7.5 Bi-directional Wave Propagation

In the linearization of the resistance, the amplitude of the discharge and that of the flow velocity have been assumed constant. Even in a prismatic channel, this is not the case because of the wave decay due to resistance. In practice, a long channel is subdivided in sections of a limited length, such that in each of them the amplitude does not vary too much. The above equations are applied to each of these sections, while the various solutions are coupled by demanding continuity of discharge and surface elevation at the junctions. In the preceding example, a length of some 10 km appears acceptable in view of the decay numbers given above.

In this section, we will derive relationships between the surface elevation and the discharge at one end of a prismatic channel section, expressed in terms of those at the other end (Figure 7.4). These relationships should be valid for arbitrary combinations of two wave systems travelling in opposite directions, not necessarily unidirectional waves.

7.5.1 Relation between the Complex Amplitudes at the Ends of a Prismatic Section

Our starting point is the general solution for the surface elevation and the discharge, given in Eq. (7.29) and (7.30), which are repeated here for convenience:

$$\tilde{\zeta}(s) = C^+ \exp(-ps) + C^- \exp(ps) \tag{7.45}$$

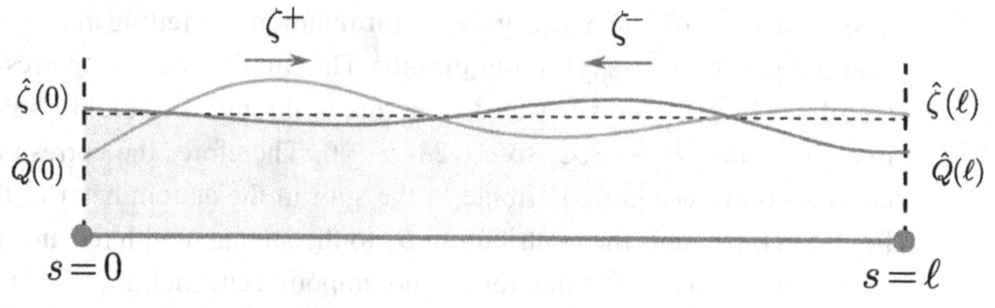

Fig. 7.4 Bi-directional wave propagation in a section

and

$$\tilde{Q}(s) = \frac{i\omega B}{p}\left(C^+ \exp(-ps) - C^- \exp(ps)\right) \tag{7.46}$$

For easier algebraic manipulation, we eliminate temporarily the factor $i\omega B/p$ by introducing a new complex variable $\tilde{Z}$, with the dimension of a length

$$\tilde{Z}(s) \equiv \frac{p}{i\omega B}\tilde{Q}(s) = C^+ \exp(-ps) - C^- \exp(ps) \tag{7.47}$$

It then follows that

$$\tilde{\zeta}_0 = C^+ + C^- \qquad \text{and} \qquad \tilde{Z}_0 = C^+ - C^- \tag{7.48}$$

where we have written $\tilde{\zeta}_0$ for $\tilde{\zeta}(0)$, and likewise for $\tilde{Z}$. These two equations determine the integration constants C^+ and C^- in terms of $\tilde{\zeta}_0$ and $\tilde{Z}_0$:

$$C^+ = \frac{\tilde{\zeta}_0 + \tilde{Z}_0}{2} \qquad \text{and} \qquad C^- = \frac{\tilde{\zeta}_0 - \tilde{Z}_0}{2} \tag{7.49}$$

Substituting these integration constants into Eq. (7.45) and (7.46), we can express the values of $\tilde{\zeta}$ and $\tilde{Z}$ at $s = \ell$ in terms of those at $s = 0$:

$$\tilde{\zeta}_\ell = \frac{\tilde{\zeta}_0 + \tilde{Z}_0}{2}\exp(-p\ell) + \frac{\tilde{\zeta}_0 - \tilde{Z}_0}{2}\exp(p\ell) \tag{7.50}$$

$$\tilde{Z}_\ell = \frac{\tilde{\zeta}_0 + \tilde{Z}_0}{2}\exp(-p\ell) - \frac{\tilde{\zeta}_0 - \tilde{Z}_0}{2}\exp(p\ell) \tag{7.51}$$

or

$$\tilde{\zeta}_\ell = \tilde{\zeta}_0 \cosh p\ell - \tilde{Z}_0 \sinh p\ell \tag{7.52}$$

$$\tilde{Z}_\ell = -\tilde{\zeta}_0 \sinh p\ell + \tilde{Z}_0 \cosh p\ell \tag{7.53}$$

(See Box 7.3 for the hyperbolic functions of complex argument.) This is the key result of this section, i.e. two equations relating the four complex amplitudes of surface elevation and discharge at both ends of the channel section. Given any two of these, the other two are determined. These relations play an important role in the calculations for a set of channel sections in series or in a network, where it is necessary to transfer information on amplitudes and phases from junction to junction.

A set of equations essentially similar to Eqs. (7.52) and (7.53) has been derived by Parsons (1918), although with a different, incorrect linearization of the resistance, as noted above. He used them to predict the tidal currents in the Cape Cod Canal, given the Atlantic Ocean tidal surface elevations at both ends of the canal.

Note: although we have referred to channel section 'ends' in the preceding text, no special conditions were imposed there. Therefore, the relations in fact apply to any pair of cross sections of a prismatic channel. In other words, '$p\ell$' in the equations above might just as well be replaced by 'ps', indicating an arbitrary location.

Example 7.2 We present a simple application of this result to a system as considered in Chapter 6, viz. a small basin, closed except for a connection by a channel to a tidal sea. The water in the channel is excited sinusoidally at the channel mouth ($s = 0$, say), while at the other end ($s = \ell$) it in turn excites a Helmholtz mode in the basin with complex surface elevation amplitude $\tilde{\zeta}_b$. In contrast to what was done in Chapter 6, we take not only inertia but also storage in the channel into account, allowing for wave propagation in the channel with the associated phase lags.

Solution

The volume balance for the basin reads $Q(\ell, t) = A_b \, d\zeta_b/dt$, implying that $\tilde{Q}_\ell = i\omega A_b \tilde{\zeta}_b$. Ignoring velocity-head effects in the transition between the channel and the basin, we have $\tilde{\zeta}_b = \tilde{\zeta}_\ell$, so that

$$\tilde{Q}_\ell = i\omega A_b \tilde{\zeta}_b = i\omega A_b \tilde{\zeta}_\ell = \frac{i\omega B}{p} \tilde{Z}_\ell \tag{7.54}$$

Substitution of this result into Eqs. (7.52) and (7.53) yields the following result for the ratio ($\tilde{r}$) of the complex amplitude in the basin to that at the channel mouth:

$$\tilde{r} = \frac{\tilde{\zeta}_b}{\tilde{\zeta}_0} = \left(\frac{pA_b}{B} \sinh p\ell + \cosh p\ell \right)^{-1} \tag{7.55}$$

In Chapter 6, it was assumed that the channel was short so that the storage in it could be neglected. In that approximation, the motion in the channel was not wave-like. Instead, the water was seen to oscillate back and forth as a rigid column. Here, we have gone beyond that approximation by modelling wave propagation in the channel.

As the channel length goes to zero, $\tilde{r} \to 1$. This seems to contradict the results presented in Chapter 6 for a short connection such as a gap or a breach. The apparent discrepancy is resolved if one realizes that here we have modelled boundary resistance only, which goes to zero in proportion to the channel length, whereas in Chapter 6 expansion losses were included as well, which are even dominant for short connections such as a gap.

Box 7.3 Hyperbolic functions of complex argument

It was shown above that the superposition of two exponentially decaying waves, travelling in opposite directions, gives rise to hyperbolic functions (cosh and sinh) of complex argument. To facilitate computation, these are expressed in terms of their real and imaginary parts in the following. By definition, $\cosh ps = (\exp ps + \exp(-ps))/2$. Substituting $p = \mu + ik$ and collecting the real and imaginary parts, we obtain $\cosh ps = \cosh \mu s \cos ks + i (\sinh ps \sin ks)$. Likewise, since $\sinh ps = (\exp ps - \exp(-ps))/2$, we find $\sinh ps = \sinh \mu s \cos ks + i (\cosh \mu s \sin ks)$. Taking the squares of these expressions, we obtain for the moduli, after some reworking: $|\cosh ps|^2 = \sinh^2 \mu s + \cos^2 ks = \cosh^2 \mu s - \sin^2 ks$ and $|\sinh ps|^2 = \sinh^2 \mu s + \sin^2 ks = \cosh^2 \mu s - \cos^2 ks$, in which we have used $\cos^2 ks + \sin^2 ks = 1$ and $\cosh^2 \mu s - \sinh^2 \mu s = 1$.

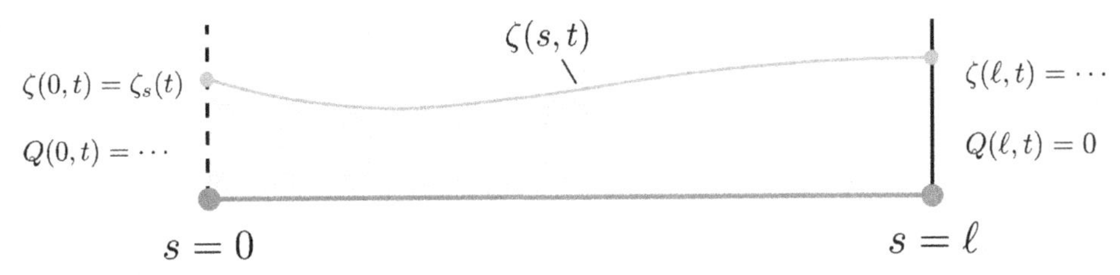

Fig. 7.5 Standing wave in a semi-closed basin

7.5.2 Response Function of a Semi-Closed Prismatic Basin

Consider a semi-closed prismatic basin, with its mouth at $s = 0$, where it is subjected to harmonic excitation from the sea, and with a closed end at $s = \ell$; see Figure 7.5. We will investigate the variation of the response as a function of the frequency of the harmonic excitation. This yields the so-called *response function*.

At the open end, $\zeta(0, t)$ is prescribed as a sine function with known frequency, amplitude and phase. At the closed end, the boundary condition is $Q(\ell, t) = 0$. Since this is just a special case of the situation considered in the preceding section, we can immediately obtain the required response from those results. Setting $\tilde{Z}_\ell = 0$ in Eqs. (7.52) and (7.53), and returning to the complete expression for the complex discharge ($\tilde{Q} = (i\omega B/p)\,\tilde{Z}$), we obtain

$$\tilde{\zeta}_\ell = \frac{\tilde{\zeta}_0}{\cosh p\ell} \tag{7.56}$$

$$\tilde{Q}_0 = \frac{i\omega B}{p}\,(\tanh p\ell)\,\tilde{\zeta}_0 \tag{7.57}$$

Note that Eq. (7.56) also follows from Eq. (7.55) by setting $A_b = 0$.

According to Eq. (7.56), the ratio of the amplitude at the closed end to that at the mouth, the so-called *response factor*, is given by

$$r = \frac{\hat{\zeta}(\ell)}{\hat{\zeta}(0)} = \frac{1}{|\cosh p\ell|} = \frac{1}{\sqrt{\sinh^2 \mu\ell + \cos^2 k\ell}} \tag{7.58}$$

This response factor is determined by two constants for the given channel: $\mu\ell$ and $k\ell$, determining the damping and the phase change of a wave progressing over the length of the basin. These two parameters can for given ℓ be calculated from $k_0\ell$ and σ, which can therefore be used as independent parameters in the calculations. The graph in Figure 7.6 is based on this. It consists of a set of curves, each showing the variation of r as a function of $k_0\ell$ (proportional to the frequency) for constant σ according to Eq. (7.58). The response function shown in Figure 7.6 gives at a glance insight into the dynamics of wave propagation and reflection in a semi-closed basin.

For basins that are very short compared with the wavelength ($k_0\ell \ll 1$), the amplitude at the closed end is virtually the same as it is at the mouth, signifying that the water level in the basin rises and falls in unison with the external tide, the so-called Helmholtz mode

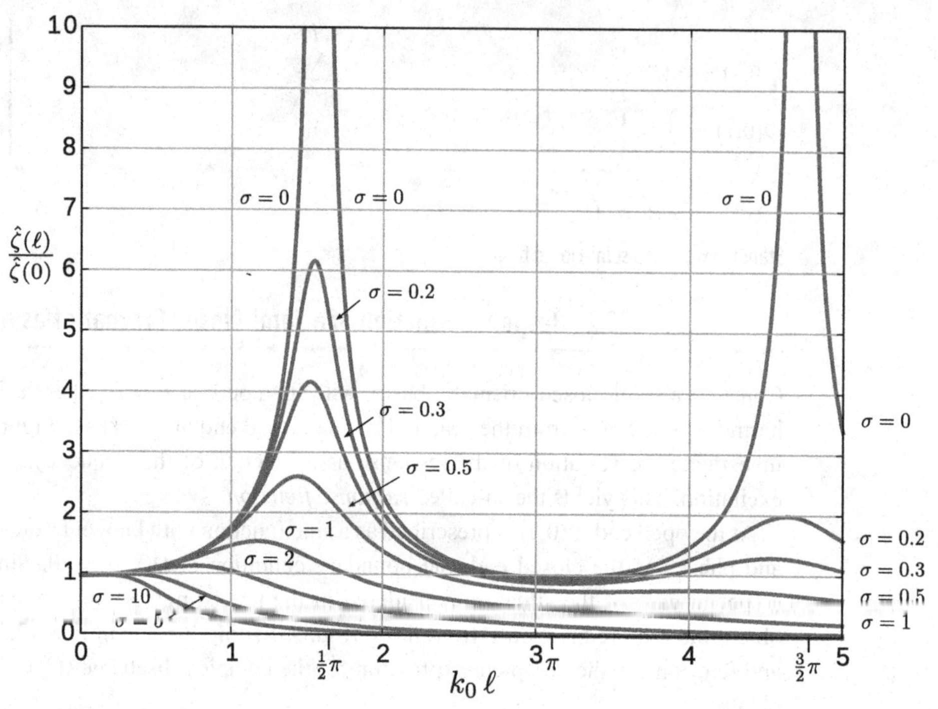

Fig. 7.6 Response factor for a basin closed at one side

that we have encountered in Chapter 6. Notice that in this range the value of σ has no influence, as expected, since resistance is negligible in a short, semi-closed basin.

For small to moderate values of σ, the response shows a high peak near $k_0\ell = \frac{1}{2}\pi$, signifying quarter-wave resonance. The influence of σ is significant, causing a strong reduction in the response peak with increasing σ and a slight lowering of the resonance frequency, which is the result of the influence of σ (or δ) on the phase speed. This is because quarter-wave resonance occurs when $k\ell = \frac{1}{2}\pi$, or $k_0\ell = \frac{1}{2}\pi\sqrt{1 - \tan^2\delta}$, therefore for lower values of $k_0\ell$ when σ increases.

For values of $k_0\ell$ near $\frac{3}{2}\pi$, there is another set of response peaks for small to moderate values of σ, signifying three-quarter-wave resonance. These are less pronounced than those near $k_0\ell = \frac{1}{2}\pi$ because damping is more important for these longer basins (and/or shorter waves).

The complex amplitude of the discharge at the mouth follows from Eq. (7.57). Substitution of Eq. (7.28) into this equation yields the following expression for the real amplitude of the discharge:

$$\frac{\hat{Q}_0}{Bc\,\hat{\zeta}_0\,\cos\delta} = \left|\frac{\sinh k\ell}{\cosh k\ell}\right| = \sqrt{\frac{\sinh^2 \mu\ell + \sin^2 k\ell}{\sinh^2 \mu\ell + \cos^2 k\ell}} \tag{7.59}$$

Because of the sine and cosine functions of $k\ell$ in this expression, the discharge amplitude at the mouth is not a monotonic function of the relative basin length. Shortening of a basin may therefore increase the discharge at the mouth (see Box 7.4).

Box 7.4 The Zuiderzee closure (1932)

A significant example where the shortening of a tidal basin increased the discharge through its mouth occurred when the previous Zuiderzee in the central part of The Netherlands was closed by the construction of an enclosure dam in 1932. This shortened considerably the tidal basin, extending from the North Sea to the closed end of the basin, but counter to intuitive expectations the discharge through the tidal inlets feeding the tides in the basin increased significantly, more than 25% in the most nearby inlet. The reason was that the tide in the shorter, remaining basin was nearer to quarter-wave resonance. This unexpected behaviour was correctly predicted by the Lorentz Committee, using the harmonic method described above (though in a more complicated version, to reflect the network topology of the tidal channels), which was specifically developed for that purpose (see Figure 7.11 below).

7.6 Propagation in Non-Uniform Channels

7.6.1 Abrupt Channel Transition

In Chapter 4, we considered the partial reflection of low translatory waves at a location where a discrete change in the channel geometry occurs. Here we do the same for harmonic waves, using the same approach. Only the algebra is a bit different.

Two long, prismatic channel sections with mutually different geometry are connected at a point $s = 0$ (Figure 7.7). From section 1 ($s < 0$), a harmonic wave is approaching the transition and is partially reflected there, as well as partially transmitted into section 2 ($s > 0$), from where it does not reflect back to the transition. For given channel geometry and incident wave parameters at the site of the transition, we have to make a first estimate of the values of $\hat{U}$ or $\hat{Q}$ in both sections near the transition, so that we can start the calculation with an estimated σ or δ. If necessary the calculations can be repeated with improved estimates to obtain better results.

We write the following expressions for the incident waves (subscript i) and the reflected waves (subscript r) in channel 1:

$$\tilde{\zeta}_i = C_i \exp(-p_1 s) \qquad \text{and} \qquad \tilde{\zeta}_r = C_r \exp(p_1 s) \tag{7.60}$$

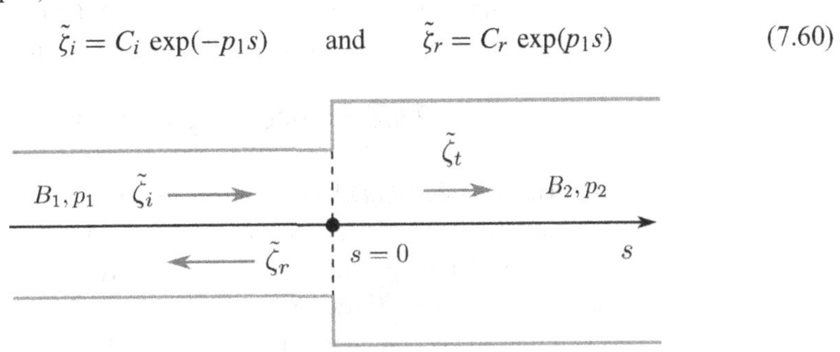

Fig. 7.7 Transition in channel geometry

and the following for the transmitted waves (subscript t) in section 2:

$$\tilde{\zeta}_t = C_t \exp(-p_2 s) \tag{7.61}$$

The associated discharges are given by

$$\tilde{Q}_i = (i\omega B_1/p_1)\, C_i \exp(-p_1 s) \qquad \text{and} \qquad \tilde{Q}_r = -(i\omega B_1/p_1)\, C_r \exp(p_1 s) \tag{7.62}$$

and

$$\tilde{Q}_t = (i\omega B_2/p_2)\, C_t \exp(-p_2 s) \tag{7.63}$$

At the transition, where $s = 0$, we must have continuity in ζ and in Q, which implies

$$C_i + C_r = C_t \tag{7.64}$$

and

$$\frac{i\omega B_1}{p_1}(C_i - C_r) = \frac{i\omega B_2}{p_2} C_t \tag{7.65}$$

For brevity, we introduce a complex transition ratio

$$\tilde{\gamma} = \frac{i\omega B_2/p_2}{i\omega B_1/p_1} = \frac{B_2 c_2}{B_1 c_1} \frac{\cos\delta_2}{\cos\delta_1} \exp(i(\delta_2 - \delta_1)) \tag{7.66}$$

(see Eq. (7.28)) as well as a complex coefficient for reflection and one for transmission:

$$\tilde{r}_r = \frac{\tilde{\zeta}_r(0)}{\tilde{\zeta}_i(0)} = \frac{C_r}{C_i} \qquad \text{and} \qquad \tilde{r}_t = \frac{\tilde{\zeta}_t(0)}{\tilde{\zeta}_i(0)} = \frac{C_t}{C_i} \tag{7.67}$$

It then follows from Eqs. (7.64) and (7.65) that

$$\tilde{r}_r = \frac{1 - \tilde{\gamma}}{1 + \tilde{\gamma}} \qquad \text{and} \qquad \tilde{r}_t = \frac{2}{1 + \tilde{\gamma}} = 1 + \tilde{r}_r \tag{7.68}$$

These results have a structure identical to those obtained in Chapter 4 for low translatory waves, the difference being that here the transition ratio and the coefficients are complex, bearing both amplitude and phase information. They also show that at the transition the reflected wave is not in phase with the incident wave, because $\tilde{r}_r$ is complex. The only exception to this is when the resistance angle δ does not change at the transition, in which case $\tilde{\gamma}$ is real (equal to $B_2 c_2 / B_1 c_1$, the same as in Eq. (4.6)), and therefore also $\tilde{r}_r$ and $\tilde{r}_t$. In that case, the results are fully identical to those of Chapter 4.

7.6.2 Exponentially Varying Cross Section

Alluvial tidal rivers generally have a funnel shape, in which the cross section increases in seaward direction. Examples are the Western Scheldt estuary in The Netherlands, shown in Figure 7.8, the Gironde estuary in France, the river Thames in the United Kingdom, and the Delaware River in the United States.

This widespread phenomenon is related to tidal damping, which decreases the discharge amplitude in the propagation direction of the tidal wave. This influences the *net* sediment

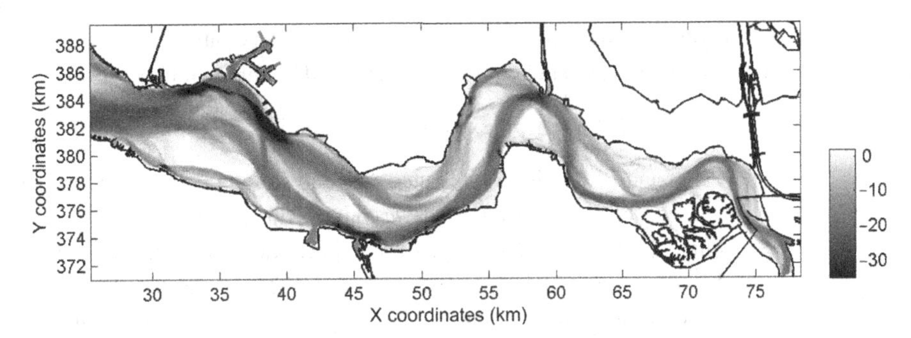

Fig. 7.8 Bathymetry of the Western Scheldt estuary, 2014; bed level in metres with repect to mean sea level; data from the Dutch Ministry of Public Works

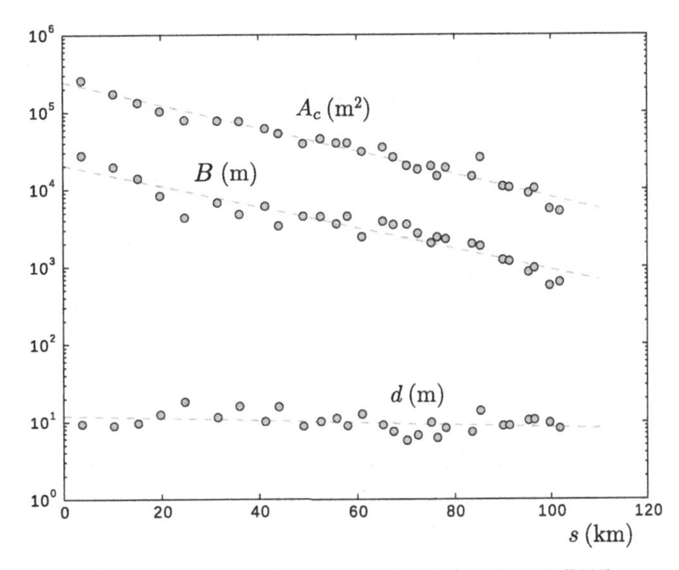

Fig. 7.9 Conveyance area, storage width and depth in the Western Scheldt; data from Savenije (2005)

transport capacity, to which the river will respond by narrowing its cross section in the inland direction. The diminishing discharge is then compensated by a proportional decay of the cross-sectional area, giving the typical funnel shape. In this way the system tends to a morphological equilibrium characterized by a uniform distribution of the maximum tidal current velocities and surface elevation amplitudes along the river. A detailed account of this self-regulating mechanism is given by Savenije (2005).

Most alluvial tidal rivers are in this state of (near-)equilibrium. Deviations from this state, if any, are mostly related to changing environmental conditions (e.g. sea level rise, fresh water depletion) or human interventions (e.g. dredging works, barrier construction). For a broad range of conditions, it has been found that the storage width and conveyance area of such rivers converge exponentially, at the same rate, while the depth is nearly constant. Figure 7.9 illustrates this for the Western Scheldt estuary; similar geometrical properties have been observed in tidal rivers around the world.

For an analytical treatment of tidal wave propagation in converging rivers, we can therefore schematize the conveyance area and storage width as, respectively,

$$A_c(s) = A_{c,0}\exp(-2\lambda s) \qquad \text{and} \qquad B(s) = B_0\exp(-2\lambda s) \tag{7.69}$$

where $A_{c,0}$ and B_0 are the conveyance area and storage width, respectively, at the mouth of the river, λ is a *convergence parameter*, having the dimension of L^{-1}, and s is a streamwise coordinate, positive in inland direction and zero at the mouth. The factor 2 in Eq. (7.69) is introduced with a view to the algebra that will follow shortly. We further assume that the resistance of the tidal river can be characterized by a constant linear relative resistance σ, suggested by the small spatial variations of depth and velocity amplitude commonly observed in nature; see also Eq. (7.9). Using this assumption, Ippen and Harleman (1966) analyze the tidal motion in the Delaware estuary by using Green's law combined with exponential frictional damping. We take a different approach, starting from the dispersion equation, which, in contrast to Green's law, gives closed-form solutions for the complex amplitudes of the surface elevation and discharge; see also Toffolon and Savenije (2011).

Dispersion Equation

Substitution of the expressions in Eq. (7.69) into the linearized continuity and momentum equations, eliminating Q, and assuming periodic solutions with frequency ω results in the following linear ordinary differential equation for the complex amplitude of the surface elevation:

$$\frac{\mathrm{d}^2\tilde{\zeta}}{\mathrm{d}s^2} - 2\lambda\frac{\mathrm{d}\tilde{\zeta}}{\mathrm{d}s} + k_0^2(1 - i\sigma)\tilde{\zeta} = 0 \tag{7.70}$$

where $k_0 = \omega/\sqrt{gA_c/B}$ is the wave number in the absence of resistance and convergence. In view of Eq. (7.69), this wave number is constant in the entire estuary. Elementary solutions are therefore of the form $C\exp(Ps)$, where C is a constant and P is a solution of the corresponding *complex dispersion equation*

$$P^2 - 2\lambda P + k_0^2(1 - i\sigma) = 0 \tag{7.71}$$

After some rearrangement, the two solutions for P can be written as

$$P_{1,2} = \pm ik_0\sqrt{1 - \lambda^2/k_0^2 - i\sigma} + \lambda \tag{7.72}$$

This result is similar to Eq. (7.17) for prismatic channels, and reduces to the latter if λ vanishes. The two roots are associated with wave components travelling in the positive (P_2) and negative direction (P_1), respectively. Their imaginary parts, differing in sign only, constitute the wave number k. Their real parts, differing in magnitude if $\lambda \neq 0$, constitute the damping modulus μ.

Analytical expressions for the wave number k and the damping modulus μ may be derived by following a procedure similar to that for the prismatic channel. We define the *convergence number* $\alpha \equiv \lambda/k_0$, which measures the change of the cross section over the wavelength in absence of resistance, and an auxiliary angle δ as the *non-negative* argument for which

$$\tan 2\delta = \frac{\sigma}{1-\alpha^2} \qquad (0 \leq \delta \leq \pi) \tag{7.73}$$

Using these definitions, the expression under the square root operator in Eq. (7.72) is rewritten in exponential form as

$$1 - \lambda^2/k_0^2 - i\sigma = \frac{1-\alpha^2}{\cos 2\delta} \exp(-2i\delta) \tag{7.74}$$

With this result, Eq. (7.72) can be rewritten as

$$P_{1,2} = \pm ik_0 \sqrt{\frac{1-\alpha^2}{\cos 2\delta}} (\cos\delta - i \sin\delta) + \lambda \tag{7.75}$$

Separating the real and imaginary parts gives for the wave numbers

$$k_{1,2} = k = k_0 \sqrt{\frac{1-\alpha^2}{\cos 2\delta}} \cos\delta = k_0 \sqrt{\frac{1-\alpha^2}{1-\tan^2\delta}} \tag{7.76}$$

and for the damping moduli

$$\mu_{1,2} = k \tan\delta \pm \lambda \tag{7.77}$$

where the subscripts 1 and 2 correspond to the waves travelling in the negative (seaward) and positive direction (landward), respectively. It can be verified that the above solution is continuous at $\alpha = 1$, despite the denominator in Eq. (7.73) going to zero for $\alpha \to 1$.

Unidirectional Wave Propagation

An infinitely long estuary involves only the incident wave component, propagating in the positive s-direction (no reflected wave), with complex surface elevation amplitude given by

$$\tilde{\zeta}(s) = \tilde{\zeta}_0 \exp(-(\mu_2 + ik)s) \tag{7.78}$$

The corresponding solution for the complex discharge amplitude can be derived by following the same steps as for the prismatic channel. The result, written in terms of the flow velocity, is

$$\tilde{U}(s) = \frac{\tilde{Q}(s)}{Bd} = \frac{c}{1 + i(\lambda/k - \tan\delta)} \frac{\tilde{\zeta}(s)}{d} \tag{7.79}$$

Figure 7.10 shows the wave number and damping parameter in dimensionless form (by scaling with the wave number k_0), as functions of the convergence number α for a range of values of the relative resistance σ.

Figure 7.10a shows that the wave number k decreases with increasing convergence, implying that the funnel shape of an estuary increases the phase speed $c = \omega/k$. The latter may even exceed the wave speed in the absence of resistance and convergence ($c_0 = \sqrt{gA_c/B}$). For every value of σ there is one value of λ for which $c = c_0$. In the absence of resistance, the wave number is zero for $\lambda > k_0$; the corresponding solutions are standing waves. However, these are unlikely to occur in reality.

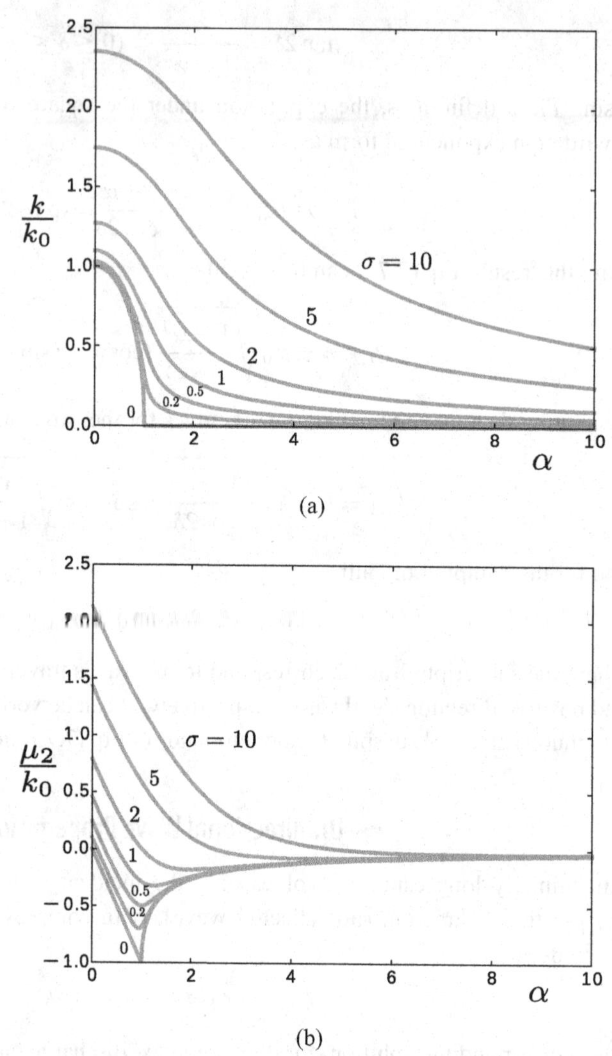

Fig. 7.10 Infinitely long tidal river: wave number (a) and damping modulus (b) as functions of the relative rate of convergence α for a set of values of the relative resistance σ

Inspection of Figure 7.10b shows that the damping modulus in general decreases with increasing convergence. It may even become negative, in which case the amplitude of the tidal wave is amplified in inland direction. For every σ there is one value of λ for which the damping modulus vanishes, in which case the surface elevation amplitude is constant along the estuary. It is shown in the next section that this value of σ is the same as the one for which $k = k_0$ and, furthermore, that this combination of parameter values corresponds to morphological equilibrium.

In the absence of resistance, the damping modulus is $\mu_2 = -\lambda$, resulting in an amplitude variation proportional to $\exp(\lambda s)$. In view of the width variation given in Eq. (7.69), this implies that $\hat{\zeta}$ varies in proportion to $B^{-1/2}$, in agreement with Green's law.

Morphological Equilibrium

It can be seen in Eq. (7.76) that $k = k_0$ if $\alpha = \tan\delta$. Substituting this into Eq. (7.77), we find $\mu_2 = k\tan\delta - \lambda = k_0\tan\delta - \lambda = k_0(\tan\delta - \alpha) = 0$. In other words, if $\lambda/k_0 = \tan\delta$, we have $k = k_0$ (also $c = c_0$) and $\mu_2 = 0$, signifying that under this critical condition the accelerating and enhancing effects of the convergence just compensate for the retarding and damping effects of resistance.

In the critical condition referred to above, $c = c_0$ and $\lambda/k = \tan\delta$, in which case Eq. (7.79) reduces to

$$\tilde{U}(s) = \frac{c_0}{d}\tilde{\zeta}(s) \tag{7.80}$$

It follows that under these conditions the flow velocity varies in phase with the surface elevation. Moreover, since in the critical condition the amplitude of the surface elevation is uniform along the estuary, so is the amplitude of the flow velocity. This solution therefore represents a state of morphological equilibrium. The shape of such an estuary, characterized by the equilibrium convergence parameter λ_e, is obviously related to the resistance. Indeed, substitution of $\alpha = \tan\delta = \lambda_e/k_0$ into Eq. (7.73), and using the formula for the tangent of a double angle, gives

$$\lambda_e = \tfrac{1}{2}\sigma k_0 \tag{7.81}$$

Substituting Eq. (7.9) for σ, we find that the following proportionality holds:

$$\lambda_e \propto c_f \frac{B_0}{RA_{c,0}}\hat{\zeta}_0 \tag{7.82}$$

which relates the geometrical parameters of an estuary in equilibrium to the resistance coefficient and the tidal amplitude. Roughly speaking, a shallow tidal river subject to a large vertical tidal range narrows rapidly; deep estuaries are likely to be more elongated, the more so if the tidal range is small.

7.7 Propagation in Networks

It has been stated at numerous places in the above that a long channel may have to be divided into a number of adjacent sections, forming a one-dimensional chain, with sections in series. In other cases, the tidal channels in the study region form a network structure. In all such cases, there are internal nodes at the channel junctions, and nodes providing a link to the external world, usually a tidal sea or a closed end.

The primary unknowns are the complex amplitudes of the surface elevation and the discharge at the nodes of the system, i.e. four (complex) unknowns for each section, or on average two per node for each section, or $2N$ unknowns per node if there are N sections joined at a node. As we have seen in Section 7.5, there are two algebraic equations for each channel section describing the propagation therein, or (on average) one per node

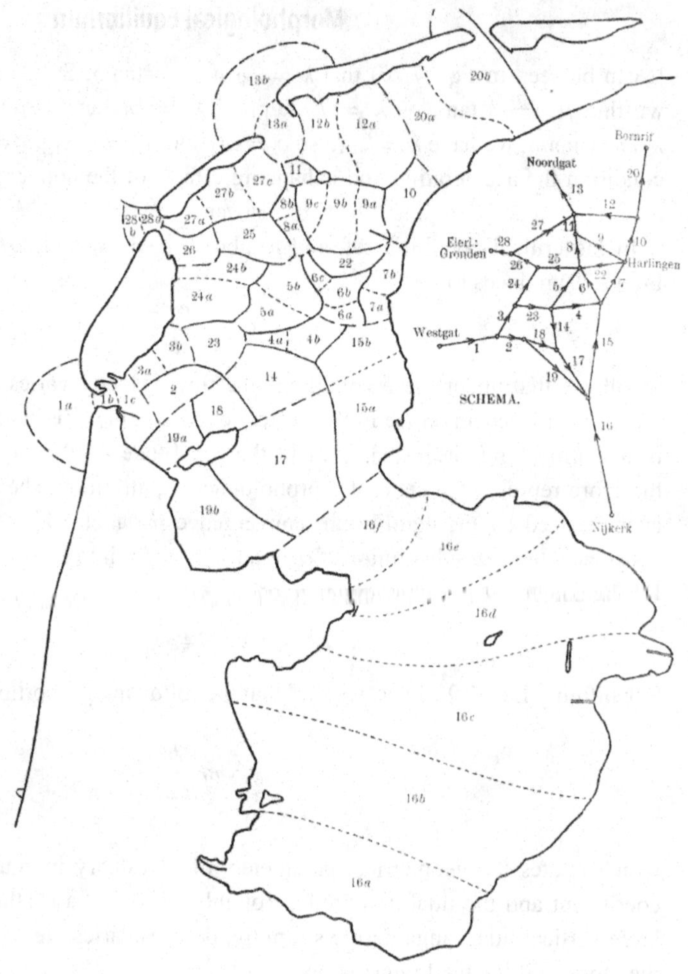

Fig. 7.11 Harmonic network model of the Zuiderzee, The Netherlands, as used by the Lorentz Committee; from Lorentz (1926)

for each section, or N equations per node. For each node there is one more equation describing that the sum of the discharges towards the node is zero (neglecting storage at the node), and $(N-1)$ equations describing that the water levels of the N sections at the node are equal (neglecting velocity-head effects). At external nodes, boundary conditions are imposed. Altogether, the number of available independent equations is equal to the number of unknowns, as required for a well-posed problem, so that the solution is uniquely determined.

The tidal computations of the Lorentz Committee, performed for the prediction of the tides after the construction of the enclosure dam in the previous Zuiderzee, utilized a network of 26 channels. Figure 7.11 shows the previous Zuiderzee and the subdivision into compartments representing tidal channels and adjacent flats. The inset presents the corresponding network of compound channels. Boundary conditions were provided by the

M_2-tides at the four inlets connecting the basin with the North Sea, and the condition of zero discharge was imposed at the southern boundary of the Zuiderzee. As mentioned previously, the committee successfully predicted the changes in the tides that were the result of the construction of the enclosure dam. The harmonic method has been applied since then in many projects. Needless to say, numerical codes have long since replaced the analytical harmonic method in practical projects, but the method is still useful for insight into the wave dynamics and for preliminary computations.

7.8 Nonlinear Effects

In this section, we briefly touch on a few nonlinear effects, which were systematically ignored in the above. We distinguish two kinds of nonlinearities in the equations describing the flow: terms that are nonlinear in the dependent variables, e.g. quadratic terms, and geometric nonlinearities, in particular the variation of the wet cross section with the water level. In contrast to a linear system, the response of a nonlinear system to harmonic forcing is not sinusoidal, although still periodic. Higher harmonics can be excited, as well as a nonzero mean response.

7.8.1 Tidal Wave Deformation

Let us take the quadratic advective term in the momentum balance, $\partial(Q^2/A_c)/\partial s$, as an example. If Q varies in time as $\cos \omega t$, then Q^2 varies in time as $\cos^2 \omega t = (1 + \cos 2\omega t)/2$. We see that the square introduces both a higher harmonic, at a frequency 2ω, and a nonzero mean value. Likewise, a factor such as $|Q|Q$, occurring in the resistance term, introduces third, fifth and higher odd harmonics if Q itself varies sinusoidally. The presence of higher harmonics causes deformation of the wave profile.

Nonlinear effects increase with increasing ratio of wave height to water depth. They are strongest in shallow water. That is why the nonlinearly generated higher harmonics of the M_2-tide, i.e. the components M_4, M_6 etc., are called 'shallow-water tides'.

Geometric nonlinearities often play an important role in tidal propagation, because the cross section available for flow can vary greatly between high water and low water, even to the extent of falling dry.

Figure 7.12 gives a series of tidal curves, showing the variation in time of the surface elevation at a sequence of stations along the Western Scheldt estuary and river. Station 'V' (Vlissingen) is the most seaward station, 'A' is Antwerp, and 'G' is Gentbrugge, far inland. The increased deformation of the tidal curves for the more inland stations is obvious.

7.8.2 Mean Slope of the Free Surface

The tide in estuaries is usually mainly progressive, so that the surface elevation and the discharge are in phase for most of the time. This means that for flood flow the cross section available for conveyance is larger than it is for ebb flow. As a consequence, ebb flow

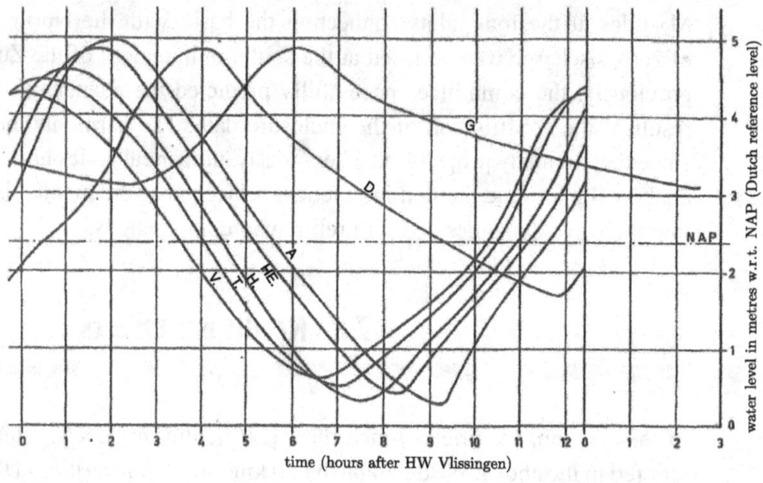

Fig. 7.12 Tidal wave propagation in the Western Scheldt, The Netherlands; records are shown for a sequence of stations, ranging from the seaward location at Vlissingen (V), via Antwerp (A) to the far inland station Gentbrugge (G); from the Dutch Department of Public Works

velocities are on average higher than flood flow velocities, and so is the corresponding resistance (which is proportional to $|U|U/d$). As a result, the mean water level slopes down towards the sea, even in purely oscillatory flow in the absence of a net discharge. This slope of the mean water level is clearly visible in Figure 7.12 (where it is partly due to a net discharge).

An approximate analytical expression for the mean surface slope can be derived by time-averaging the momentum balance, given by Eq. (1.21). For periodic motions, as considered here, the time average of $\partial Q/\partial t$ vanishes. For low waves, the advective term is neglected from the outset. The mean resistance is the only remaining contribution:

$$\overline{\frac{\partial h}{\partial s}} = \frac{d\overline{h}}{ds} = -\overline{i_f} \tag{7.83}$$

in which the overbar denotes the time average over an integer number of cycles. We use the linearized version of the friction slope i_f, given by

$$i_f = \frac{8}{3\pi} c_f \frac{\hat{U} U}{gR} = \frac{8}{3\pi} c_f \frac{\hat{Q} Q}{g A_c^2 R} \tag{7.84}$$

In order to express the influence of the varying water level on the conveyance cross section, we write $A_c = B_c d$, neglect variations of B_c with the water level and replace the hydraulic radius R by the flow depth d. Taking the time average, we obtain

$$\overline{i_f} = \frac{8}{3\pi} c_f \frac{\hat{Q}}{g {B_c}^2} \overline{(Q/d^3)} \tag{7.85}$$

Writing $d = d_0 + \zeta$, assuming $|\zeta| \ll d_0$, and restricting ourselves to purely oscillatory flow ($\overline{Q} = 0$), we approximate $\overline{Q/d^3}$ as follows:

$$\overline{Q/d^3} = \overline{Q/d_0^3\,(1+\zeta/d_0)^3} \approx d_0^{-3}\,\overline{Q\,(1-3\zeta/d_0)} = -3\,d_0^{-4}\,\overline{Q\zeta} \tag{7.86}$$

We assume sinusoidal variations of ζ and Q, with phase angles α and β at a fixed location, as in

$$\zeta = \hat{\zeta}\,\cos(\omega t + \alpha) \qquad \text{and} \qquad Q = \hat{Q}\,\cos(\omega t + \beta) \tag{7.87}$$

This yields

$$\overline{Q\zeta} = \tfrac{1}{2}\,\hat{Q}\hat{\zeta}\,\cos(\alpha - \beta) \tag{7.88}$$

This shows explicitly that a mean slope of the water surface is the result of ζ and Q being in phase ($\alpha - \beta = 0$, as in a purely progressive wave without resistance), or partly so ($\alpha - \beta \neq 0$ and $\alpha - \beta \neq \frac{1}{2}\pi$). Only when ζ and Q are exactly 90° out of phase, as in a pure standing wave, does the resistance not contribute to a mean slope of the water surface.

Substituting Eqs. (7.86) and (7.88) into Eq. (7.85), we obtain

$$\overline{i_f} = -\frac{4}{\pi}\,c_f\,\frac{\hat{Q}^2}{gB_c{}^2 d_0{}^3}\,\frac{\hat{\zeta}}{d_0}\,\cos(\alpha - \beta) = -\frac{4}{\pi}\,c_f\,\frac{\hat{U}^2}{gd_0}\,\frac{\hat{\zeta}}{d_0}\,\cos(\alpha - \beta) \tag{7.89}$$

(Note that the factor $c_f\hat{U}^2/gd_0$ is the friction slope in uniform flow with velocity $\hat{U}$ in a depth d_0.) Finally, considering a purely progressive wave, we substitute Eqs. (7.34) and (7.35) into Eq. (7.89) and perform some straightforward mathematical manipulations, in which we also use Eq. (7.24) for the phase velocity c. The result for the mean surface slope can be written as

$$\frac{d\bar{h}}{ds} = -\overline{i_f} = \frac{4}{\pi}\,c_f\,\cos\delta\,\cos 2\delta\,\frac{B}{B_c}\left(\frac{\hat{\zeta}}{d_0}\right)^3 \tag{7.90}$$

Notice that this implies a rise of the mean water surface elevation in the direction of propagation of the wave. Because of the third power, the mean slope soon becomes negligible for decreasing, small values of $\hat{\zeta}/d_0$, but for finite-amplitude waves in shallow water, the resistance can cause a non-negligible mean surface slope.

Numerical results obtained with Eq. (7.90) are only approximate, considering its assumptions, and should therefore be interpreted with care. Importantly, it provides insight into the causative factors, as is always the case for analytical relations.

Problems

7.1 Mention two effects of resistance on the relation between variations in the discharge and in the surface elevation.

7.2 An M_2-tide in a channel with a width of the free surface B of 500 m has a damping modulus $\mu = 2 \times 10^{-5}$ /m and a wave number $k = 1.5\mu$. The integration constants are given as $C^+ = (1.2\,\text{m})\exp(i\frac{1}{6}\pi)$ and $C^- = (0.3\,\text{m})\exp(-i\frac{1}{4}\pi)$. Plot the values of these constants in the complex plane and construct the corresponding two hodographs of the + wave and the − wave for a self-chosen sequence of values of s. Construct the hodograph for the total surface elevation. Compare the angle of convergence with the theoretical value.

7.3 For the data in Problem 7.2, construct the hodographs for $\tilde{Q}^+$, for $\tilde{Q}^-$, and for the total discharge $\tilde{Q} = \tilde{Q}^+ + \tilde{Q}^-$.

7.4 Using the data listed in Example 7.1, calculate the values of the dependent variables in the left column of Table 7.1.

7.5 Calculate the wave number and the damping modulus for propagation in the channel in Problem 6.16.

7.6 Using the result of Problem 7.5, verify whether the rigid-column approximation is justified for this case.

7.7 Suppose that the channel of Problem 6.16 is closed at one end and that the M_2-tide at the mouth has a surface elevation amplitude of 1 m. Calculate the surface elevation amplitude at the closed end, estimate it from Figure 7.6 and compare the results. Calculate the value of the amplitude of the discharge at the mouth.

7.8 Do the same as in Problem 7.7 for a seiche with a period of 45 min and a surface elevation amplitude at the mouth of 0.3 m.

7.9 The Bay of Fundy, where the world's highest tides occur, has a length of about 270 km. The width does not vary too much, in contrast with the depth, but we will ignore the latter variations for this problem and put the depth at a constant value of 75 m. We take c_f= 0.0025. If the M_2-tide at the mouth has a tidal range of 5 m, estimate the tidal range at the closed end, using Figure 7.6.

7.10 Using Figure 7.6, estimate the amplification factor for a seiche with a period of 20 min in a semi-closed harbour basin with a length of 3 km and a depth of 15 m; the surface elevation amplitude at the mouth is 0.25 m. Check the influence of the offshore amplitude on the amplification factor for this case. Do the same for the preceding problem and compare both sensitivities.

7.11 Mention two sources of nonlinearity of the long-wave equations.

7.12 Derive an analytical expression for the complex amplitude of the discharge in an infinitely long tidal river with an exponentially converging cross section.

7.13 Derive an analytical expression for the complex surface elevation amplitude in an exponentially converging tidal river with the entrance in $s = 0$ and with a closed end in $s = \ell$.

7.14 Explain why a mean slope of the free surface occurs in a purely oscillatory tide (no net discharge).

8 Flood Waves in Rivers

Flood waves, which are resistance dominated, constitute the final category in the progression of wave types considered so far, ranging from the rapidly varying translatory waves, for which resistance could be neglected, to the intermediate category of tidal waves with their mixed character to the present case of flood waves, in which resistance is dominant.

8.1 Introduction

Flood waves are in essence humps of water traveling downriver. Stated in more detail, they are transient increases and decreases of discharge and water level ('stage') in a river caused by temporarily enlarged run-off in the catchment area due to heavy rainfall or snow melt, which travel downriver as a wave.

The temporal evolution and dynamics of flood waves differ greatly between the upper and lower reaches of a river. In the upper reaches, characterized by relatively steep slopes and a small catchment area, the response to an increased run-off can be quite fast, with rapid variations in flow rate and water level, even to the extent that inhabitants along the river border are taken by surprise, sometimes with fatal consequences.

In contrast, flood waves in the lower river reaches are slow processes, in many cases taking place over several days, due to the larger catchment area, the existence of tributaries and the greater propagation distance from upstream. Run-off peaks in different parts of the larger catchment area do not in general occur simultaneously, so that the maximum of their sum is less than the sum of the individual maxima. Moreover, as we will see, the internal dynamics of the flood wave cause it to flatten and to elongate as it propagates. As a result, the variations in flow rate and water level in the lower reaches are gradual, even such that inertia is insignificant (see Table 2.1 and Box 8.1).

In engineering practice, numerical models are used to simulate flood waves. These are based on the complete equations of De Saint-Venant (the long-wave equations), including inertia (see Chapter 11). This is needed in order to cover a wide range of occurrences of flood waves. Moreover, they should be able to cover the estuarine reaches of rivers, where tides penetrate, for which inertia and resistance are of comparable magnitude. And they should be able to simulate properly the effects of the operation of weirs, which can induce rapidly varying translatory waves superimposed on the slowly varying flood waves.

Box 8.1 Estimation of the relative inertia in a flood wave

We take the flood wave in the river Rhine shown in Figure 2.2 as an example for a comparison of the order of magnitude of the inertia in the momentum balance ($\partial U/\partial t$) with the down-slope gravity force per unit mass (or the acceleration), $g\,\partial h/\partial s$. The rate of rise of the water level in the steepest part of the rating curve is about 2 m/24 h. The increase in flow velocity would be about 0.5 m/s per 24 h, corresponding to an acceleration of about 5×10^{-6} m/s^2. This is roughly a factor of 200 less than $g\,|\partial h/\partial s|$, if we equate the slope of the free surface to the bed slope, which for the lower Rhine is about 10^{-4}.

For an analytical study of flood waves, which is pursued here to obtain a fundamental insight into their behaviour, we consider their most elementary form, in which inertia is neglected, leaving a balance between the driving force and the resistance. The remainder of this chapter is based on that assumption.

8.2 Quasi-Steady Approximation

The neglect of inertia leads to the so-called *quasi-steady approximation*: although the flow is unsteady on a long time scale, the relations between dependent variables are at any time assumed to be the same as for steady flow. In other words, the flow is instantaneously adjusted to (slow) variations in the driving force. It has no memory, so to speak.

We use the continuity equation in the form given by Eq. (1.20):

$$B\frac{\partial h}{\partial t}+\frac{\partial Q}{\partial s}=0 \tag{8.1}$$

In the momentum balance (Eq. (1.21)), we neglect the inertia terms, in which case it reduces to

$$gA_c\frac{\partial h}{\partial s}+c_f\frac{Q^2}{A_cR}=0 \tag{8.2}$$

We have replaced $|Q|Q$ by Q^2 because the flow is unidirectional. From a mathematical point of view, the neglect of inertia in Eq. (1.21) has important consequences because it reduces the set of coupled balances of mass and momentum from being of second order in time to first order (though still second order in space). This implies that dynamic wave propagation in two directions is not possible within this simplified model.

In previous chapters, dealing with translatory waves, tides etc., there was no preferred flow direction. The undisturbed reference state was one of rest. A bed slope, if present, played no particular role. This is quite different for river flow, where the reference state is one of uniform, downward flow. For that category, it is meaningful to bring the bed slope explicitly into account: $i_b\equiv-\partial z_b/\partial s$, in which z_b is the cross-sectionally averaged bed elevation above the reference plane $z=0$. Writing d for the average depth in the conveyance cross section, we have $h=z_b+d$ (see Figure 8.1). Substituting this into Eq. (8.2) gives

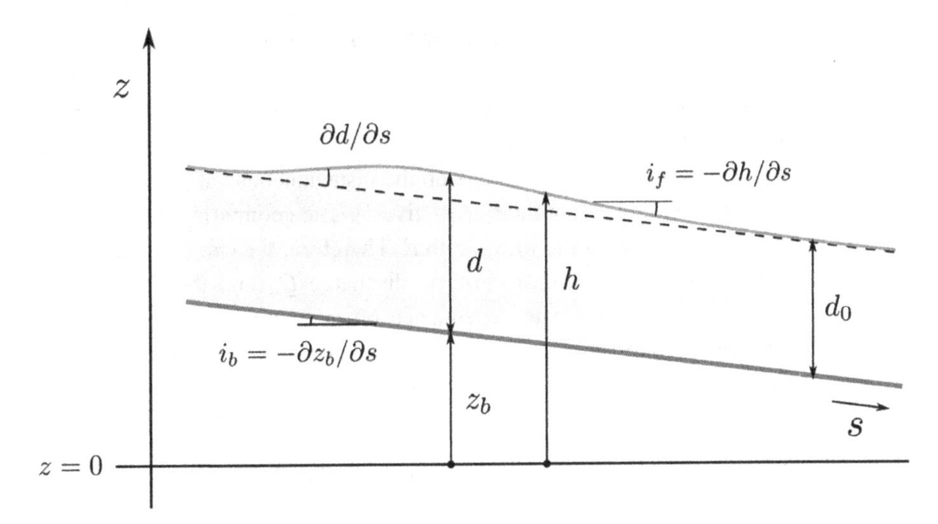

Fig. 8.1 Longitudinal profile of flood wave

$$\frac{\partial d}{\partial s} - i_b + c_f \frac{Q^2}{g{A_c}^2 R} = 0 \tag{8.3}$$

Eq. (8.3) forms the basis of the analyses in the remainder of this chapter, using approximations of increasing complexity.

8.3 Quasi-Uniform Approximation

We assume an initial state of uniform flow that is gradually being disturbed by a flood wave.

In *uniform flow*, the free-surface slope and the friction slope both equal the bed slope, and the depth gradient $\partial d/\partial s$ is zero, in which case the discharge and the flow velocity are given by

$$Q = Q_u \equiv A_c \sqrt{\frac{gRi_b}{c_f}} \quad \text{and} \quad U = U_u \equiv \frac{Q_u}{A_c} = \sqrt{\frac{gRi_b}{c_f}} \tag{8.4}$$

These equations express two messages: they *define* Q_u and U_u, and they state that in uniform flow the actual discharge Q and flow velocity U equal Q_u and U_u, respectively. The latter does not hold in nonuniform flow, but in that case Q_u and U_u are still given by the expressions in the respective right-hand sides of Eq. (8.4).

If the flood wave is sufficiently low and/or elongated, the difference between the free-surface slope and the bed slope is small. We will initially neglect it. This is a so-called *quasi-uniform approximation*, so named because the flow, though varying in space on a large scale, is treated as if it were uniform as far as the relations between local variables are concerned.

8.3.1 Formulation and General Solution

In the quasi-uniform approximation, the value of the discharge is taken to be the value in uniform flow, or $Q = Q_u$ as defined in Eq. (8.4). This is an algebraic equation, describing the dependence of the discharge on the instantaneous values of the local geometric profile variables A_c and R (not their derivatives!). The geometric profile variables are monotonic, unique functions of the flow depth d. Therefore, we can consider Eq. (8.4) as an implicit relation between the uniform-flow discharge Q_u and the depth d, which is written as $Q_u = Q_u(d(s,t))$. (The dependence on g, i_b and c_f given in Eq. (8.4) is not written explicitly.) Substitution of $Q = Q_u$ and of $Q_u = Q_u(d(s,t))$ into Eq. (8.1), in which we replace $\partial h/\partial t$ by $\partial d/\partial t$ (which is allowed since $h = z_b + d$ and $\partial z_b/\partial t = 0$), yields

$$\frac{\partial d}{\partial t} + \frac{1}{B}\frac{\mathrm{d}Q_u}{\mathrm{d}d}\frac{\partial d}{\partial s} = 0 \tag{8.5}$$

The left-hand side of this equation has the structure of a total derivative of d for an observer moving with a velocity $\mathrm{d}s/\mathrm{d}t = (1/B)\mathrm{d}Q_u/\mathrm{d}d$, and the equation states that an observer moving at this speed sees no change in the local value of d (see Section 3.2.3 for an introduction of the total derivative). Neither would this observer see changes in the other local geometric variables such as R, A_c or A (assuming a prismatic channel), nor in the discharge Q (within the quasi-uniform flow approximation $Q = Q_u$). This can be expressed mathematically as follows:

$$\frac{\mathrm{d}d}{\mathrm{d}t} = 0, \quad \frac{\mathrm{d}Q}{\mathrm{d}t} = 0 \quad \text{provided} \quad \frac{\mathrm{d}s}{\mathrm{d}t} = \frac{1}{B}\frac{\mathrm{d}Q_u}{\mathrm{d}d} \tag{8.6}$$

It follows from these relations that the specified value of $\mathrm{d}s/\mathrm{d}t$ is the speed of propagation of the flood wave. Denoting this as c_{HW} ('HW' for high water), we have:

$$c_{\mathrm{HW}} \equiv \frac{1}{B}\frac{\mathrm{d}Q_u}{\mathrm{d}d} \tag{8.7}$$

It is noted that in finite-difference form, Eq. (8.7) can be written as $\delta Q = Bc_{\mathrm{HW}}\,\delta h$, the same as was found for translatory waves (Eq. (3.23)), except for the difference in propagation speed.

We see that flood waves can only propagate downstream (at a low speed, as we will see), as a consequence of the neglect of inertia in the momentum balance. For the latter reason, flood waves are said to belong to the category of *kinematic waves*, as distinct from the dynamic waves considered in preceding chapters, which – thanks to inertia – can travel both downstream and upstream.

8.3.2 The High-Water Wave Speed

In order to obtain a more transparent, explicit expression for c_{HW}, we substitute $A_c = B_c d$ into Eq. (8.4), as well as the approximation $R = d$, with the result

$$Q_u = B_c d \sqrt{\frac{g d i_b}{c_f}} \tag{8.8}$$

so that, assuming a constant value of c_f,

$$\frac{\mathrm{d}Q_u}{\mathrm{d}d} = \frac{3}{2} B_c \sqrt{\frac{g d i_b}{c_f}} = \frac{3}{2} B_c U \tag{8.9}$$

With this, we find the following important expression for the propagation speed of the flood wave:

$$c_{\mathrm{HW}} \equiv \frac{1}{B} \frac{\mathrm{d}Q_u}{\mathrm{d}d} = \frac{3}{2} \frac{B_c}{B} U \tag{8.10}$$

We see that this speed is of the order of magnitude of the flow velocity. In the lower river reaches, where the Froude number usually is much less than 1, the flow velocity U is much less than the classical long-wave speed of $\sqrt{gd}$, and so is c_{HW}. For $B_c/B < 2/3$, c_{HW} is even less than the flow velocity! When at a high river stage the flood plains are submerged, B can be much larger than B_c, which causes the propagation speed of the flood wave to become considerably less than the flow velocity in the main (conveyance) channel (see Section 8.4.3).

For inertia-dominated waves in a channel, the wave speed is also affected by the ratio B_c/B, but to a smaller extent. This can be seen from the expression for that speed, viz. $c = \sqrt{gA_c/B}$. This can be written as $c = \sqrt{B_c/B}\,\sqrt{gd}$. We see that for a given depth the wave speed is here proportional to the square root of B_c/B, implying a lower sensitivity than for resistance-dominated flood waves.

In the above, we have treated c_f as a constant, independent of the depth, in which case U varies in proportion to the square root of the depth (see Eq. (8.4)). Had we used a Strickler- or Manning-type of resistance law (see Chapter 9), U would vary with the $\frac{2}{3}$-power of the depth, and the discharge with the $\frac{5}{3}$-power, in which case the coefficient $\frac{3}{2}$ in Eq. (8.10) would have to be replaced by $\frac{5}{3}$.

8.3.3 Kinematic Wave Behaviour

According to the preceding results, it is possible for an observer to follow a point of constant depth or discharge, provided the observer travels at the appropriate speed. This also applies to the point with the highest surface elevation, or river stage, which implies that this maximum remains constant: the flood wave does not diminish in height according to this simple model. Nor does the flood wave lengthen in this model, since two points with equal depth, one upstream of the maximum stage, the other downstream, would travel at the same speed.

The fact that the propagation speed increases with depth (through its proportionality to U) causes deformation of the wave, since points of larger depth travel faster than points of smaller depth. It follows that the leading part of the flood wave steepens and the trailing part flattens. If this continues for a sufficient time and distance, the assumption that the change in surface slope is negligible compared with the slope in uniform flow, i.e. the bed slope, becomes untenable. In the following section we will consider the effect of this variation in slope, first qualitatively and thereafter quantified in an extended mathematical model.

8.4 Influence of Variable Free-Surface Slope

In the preceding section, the variations in the slope of the free surface, which accompany the passage of a flood wave, were ignored. Actually, the slope varies, being larger than the slope in uniform flow (i_b) in the leading part of a flood wave, i.e. at rising river stage at a fixed point, whereas it is less than i_b in the trailing part, i.e. at falling river stage. This can be seen in Figure 2.2. Prior to the arrival of the flood wave, the difference in water surface elevation between the upper and lower gauging stations, some 100 river-km apart, is constant in time at about 10 m, corresponding to a slope of about 1 in 10^4, equal to the bed slope. When the flood wave is at its maximum, the difference has increased to about 13 m, corresponding to a slope of about 1.3×10^{-4}.

At a given river stage at a fixed point, the free-surface slope is larger when the water level rises than when it falls. The same then applies to the discharge. This is a manifestation of the phenomenon of *hysteresis*. The hysteresis is illustrated with a rating curve in Figure 8.2, i.e. a plot of the discharge versus the simultaneously occurring river stage at the same location as a function of time during the passage of a flood wave. It can be seen that the maximum discharge occurs ahead of the maximum stage. This is because the enhanced slope of the free surface ahead of the cross section of maximum stage enhances the flow

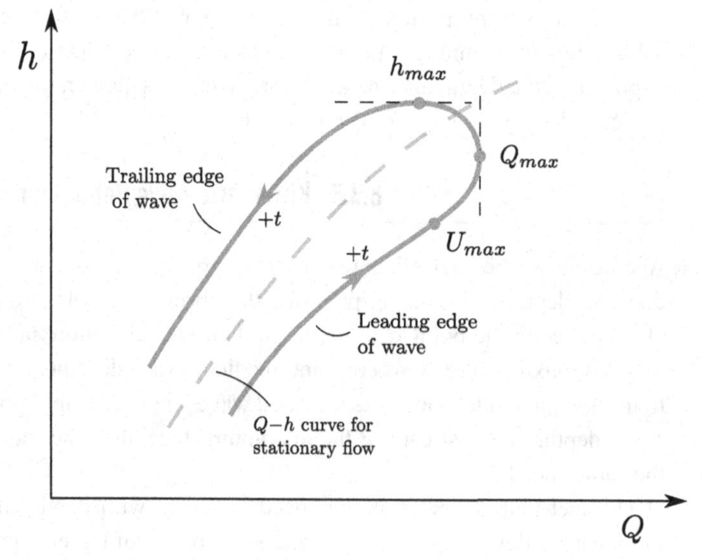

Fig. 8.2 Example Q–h curve with hysteresis

rate, which more than compensates for the larger depth at maximum stage, which must do with a smaller slope, viz. i_b.

The occurrence of hysteresis in a flood wave has important consequences for the evolution of the wave as it travels downriver. As noted above, the discharge is greater ahead of the peak than it is following the peak, for two cross sections with the same stage. Considering a (moving) control volume between two such sections, we see that mass is flowing out of it at the downstream section at a higher rate than it is flowing into it at the upstream section. In other words, the mass contained between those two cross sections on either side of the crest of the flood wave decreases in time. As a result, the surface elevations between these moving cross sections decrease as the wave propagates downriver. This means that *the flood wave decreases in height and* (consequently) *increases in length as it propagates downriver.*

8.4.1 Diffusion Model for Flood Waves

In order to model this phenomenon mathematically, we start with Eq. (8.3), recasting it into an expression for the discharge:

$$Q = A_c \sqrt{\frac{gR}{c_f}} \sqrt{i_b - \frac{\partial d}{\partial s}} = Q_u \sqrt{1 - \frac{1}{i_b}\frac{\partial d}{\partial s}} \tag{8.11}$$

We have substituted the expression for Q_u, i.e. the discharge for the given depth if the flow were uniform, given by Eq. (8.4).

We take account of a variable surface slope on the assumption that it deviates by a relatively small amount from the uniform-flow value, i.e. the bed slope i_b. This implies that the depth gradient, though nonzero, is small compared with the bed slope ($|\partial d/\partial s| \ll i_b$). This allows us to approximate the square root as in $\sqrt{1-\epsilon} \simeq 1 - \frac{1}{2}\epsilon$ if $\epsilon \ll 1$. Using this approximation, we obtain

$$Q = Q_u \left(1 - \frac{1}{2i_b}\frac{\partial d}{\partial s}\right) \tag{8.12}$$

With this approximation, we have obtained a linearized expression for the effect of the variable free-surface slope on the discharge, with a correction to $Q = Q_u$ that is proportional to the depth gradient $\partial d/\partial s$. Considering the depth of flow as a measure of volume 'concentration', i.e. volume per unit horizontal area, we can say that the correction to the volume transport is proportional to the gradient of the volume concentration. This is typical for *diffusive transport* (see Box 8.2).

Box 8.2 Diffusive transport

The classical example of diffusion is heat transport in a continuous medium, which in good approximation is proportional to the gradient of the temperature, the latter being a measure of the heat concentration. Such transport takes place in the direction of decreasing temperature. Differences in temperature are gradually erased as time goes on. That is typical for diffusion (see Chapter 10). A similar process applies in the case of flood waves, although the physical mechanism behind it is drastically different.

Our next task is to express the effect of diffusion on the flow depth mathematically. In view of the volume balance, Eq. (8.1), we need the derivative of Q with respect to s, to be determined from Eq. (8.12):

$$\frac{\partial Q}{\partial s} = \frac{\partial Q_u}{\partial s}\left(1 - \frac{1}{2i_b}\frac{\partial d}{\partial s}\right) - \frac{Q_u}{2i_b}\frac{\partial^2 d}{\partial s^2} \tag{8.13}$$

We neglect the small, second term between parentheses, which is consistent with the approximation already made in the transformation of Eq. (8.11) into Eq. (8.12). Furthermore, since Q_u can be considered a function of d only (for given geometry), we can express $\partial Q/\partial s$ as follows:

$$\frac{\partial Q}{\partial s} = \frac{\mathrm{d}Q_u}{\mathrm{d}d}\frac{\partial d}{\partial s} - \frac{Q_u}{2i_b}\frac{\partial^2 d}{\partial s^2} \tag{8.14}$$

We substitute this into Eq. (8.1), replace $\partial h/\partial t$ by $\partial d/\partial t$, and divide by B, with the result

$$\frac{\partial d}{\partial t} + \frac{1}{B}\frac{\mathrm{d}Q_u}{\mathrm{d}d}\frac{\partial d}{\partial s} - \frac{Q_u}{2i_b B}\frac{\partial^2 d}{\partial s^2} = 0 \tag{8.15}$$

The factor multiplying the first-order derivative of d can be recognized as the wave speed c_{HW}. The factor multiplying the second-order derivative of d is called the diffusivity, to be denoted as K:

$$K \equiv \frac{Q_u}{2i_b B} \tag{8.16}$$

With these substitutions, Eq. (8.15) is transformed into

$$\frac{\partial d}{\partial t} + c_{\mathrm{HW}}\frac{\partial d}{\partial s} - K\frac{\partial^2 d}{\partial s^2} = 0 \tag{8.17}$$

This is a so-called (one-dimensional) *advection–diffusion equation* (see Chapter 10), a standard equation in mathematical physics. The first two terms represent the displacement (advection) of the longitudinal surface profile, with velocity c_{HW}, as in the quasi-uniform approximation. The third term adds the effect of diffusion. It is proportional to $\partial^2 d/\partial s^2$, i.e. the curvature of the free surface, which results in a spatial smoothing of the profile, elongating and flattening the flood wave as time progresses. This can be seen as follows.

Consider a control volume between two cross sections in a reach where the free surface is concave (hollow) upwards. In that case, the discharge at the most upstream cross section (inflow) is larger than it is at the downstream cross section (outflow), causing a net inflow into this control volume and therefore a rise of the free surface.

Where the free surface is convex upward, i.e. concave downward, the opposite occurs, with net outflow and a lowering of the free surface as a result. The overall effect is that bumps are lowered and troughs are filled. In other words, the effect is to smoothen the longitudinal profile of the free surface.

8.4.2 Elementary Solution

Solutions of the one-dimensional advection–diffusion equation have been derived in the classical literature in physics for a range of initial and boundary conditions. Without derivation, we present just one of them here, for the case of the spreading of an initially concentrated mass in an infinitely long river reach. (The analogous case for heat diffusion would be the spreading of heat in a long conducting rod following a localized initial heating.)

The advection–diffusion equation is of first order in time and of second order in space. Therefore, one initial condition and two boundary conditions are required for a well-posed problem. We choose the following conditions:

Initial condition: uniform flow with depth $d =$ constant $= d_0$ and flow velocity $U = U_0$, to which at time $t = 0$ a volume V is added abruptly, concentrated at the point $s = 0$. Note: addition of a volume V means addition of a volume per unit width equal to V/B. This ratio has the meaning of an area in the longitudinal profile of the surface elevation.

Boundary conditions: for $s \to \pm\infty$, $d = d_0$ at all times.

For simplicity, we treat c_{HW} and K as given constants, equal to their initial values. That is acceptable to obtain an impression of the effects of the diffusion term.

Without proof, we state that the solution of Eq. (8.17) with the above stated initial and boundary conditions is given by:

$$d(s,t) = d_0 + \frac{V/B}{\sqrt{2\pi}\,\sigma_s(t)} \exp\left(-\frac{(s - c_{\text{HW}}t)^2}{2{\sigma_s}^2(t)}\right) \tag{8.18}$$

in which

$$\sigma_s(t) = \sqrt{2K_0 t} \quad \text{and} \quad K_0 = \frac{Q_0}{2i_b B} = \frac{U_0 d_0}{2i_b}\frac{B_c}{B} \tag{8.19}$$

It can be shown through back substitution that these expressions satisfy Eq. (8.17) as well as the initial and boundary conditions. For a derivation and discussion, see for instance, Strauss (1992).

The solution has been plotted for two instances $t = t_1$ and $t = t_2 = 4t_1$ in Figure 8.3. We note the following features of Eq. (8.18), also shown in the figure:

1. The first term in the right-hand side of Eq. (8.18) is the undisturbed depth; the second term represents the flood wave. At any instant $t > 0$, the longitudinal profile of the flood wave is the classical bell-shaped Gauss curve, with total area V/B and 'standard deviation' (here: the distance from the location of the maximum to the locations of the inflection points) σ_s as specified in Eq. (8.19).
2. σ_s increases in time proportional to $\sqrt{t}$: the wave elongates. At $t = t_2 = 4t_1$, we have $\sigma_s(t_2) = 2\sigma_s(t_1)$, so the 'length' of the wave is doubled compared with the situation at $t = t_1$.

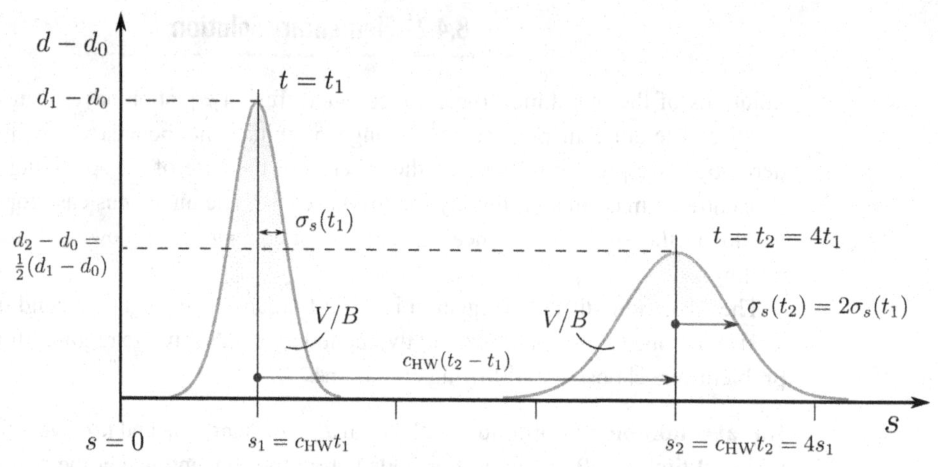

Fig. 8.3 Elementary solution of the advection–diffusion equation

3. At an instant $t = t_0$, say, the spatially maximal depth occurs where $\partial d(s, t_0)/\partial s = 0$, i.e. at $s = c_{HW}t_0$, where the exponential function has its maximum value of unity. This shows that c_{HW} is the speed with which this maximum travels downriver.
4. The value of this spatial maximum decreases in time proportional to $1/\sigma_s$, or proportional to $1/\sqrt{t}$: the wave becomes lower. At $t = t_2 = 4t_1$, its maximum height is halved, compared with the situation at $t = t_1$. Note that the product of the height and the 'length' of the flood wave is constant, as required in view of volume conservation.
5. At a fixed point $s = s_0$, say, the depth (or river stage) does not vary in time according to a Gauss curve. This is because the wave becomes lower and longer as it passes the point considered, so that the rising branch corresponds to a smaller value of σ_s than the falling branch. Therefore, the rise occurs faster than the fall.
6. The spatially maximal depth passes the fixed location $s = s_0$ at time $t = s_0/c_{HW}$, but since it is continually falling, it is at the same location preceded by higher values. This implies that the temporal maximum depth occurring at s_0 occurs earlier than this and has a larger value than the spatial maximum at time $t = s_0/c_{HW}$ (this can be verified by solving for t from $\partial d(s_0, t)/\partial t = 0$).

8.4.3 Observations

Figure 8.4 shows three plots of river stage versus time, measured in the river Rhine and one of its branches (the Waal) in The Netherlands, following the bombardment in World War II (May 1943) by the RAF of the Möhne dam in the river Ruhr in Germany (with the aim to harm German capacity of electricity production, needed in the war industry; see Brickhill (1951)). As a result of this bombardment, the dam failed and an amount of water of about 110×10^6 m^3 flowed from the Möhne reservoir via the Ruhr river into the Rhine near the city of Ruhrort, some 140 km downstream of the dam, where the height of the resulting flood wave had been reduced to roughly 1.5 m.

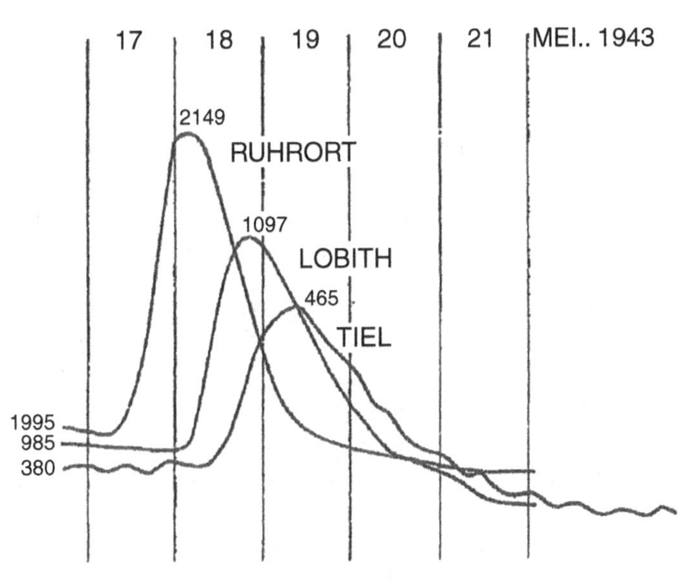

Fig. 8.4 Water level recordings in the Rhine/Waal river following the collapse of the Möhne dam. The numbers represent the height of the water level above mean sea level, expressed in centimetres; from Wemelsfelder (1947)

Because of the abrupt release, it is meaningful to compare the solution given by Eq. (8.18) with the observations.

The downstream progression of the flood wave, the lowering of the peak height and the asymmetry between the rising branch and the falling branch are clearly present in the measured data. The propagation speed determined from the observations varied somewhat per river section, but in all cases it was of the order of 1 m/s, in fact somewhat less than the flow velocity, due to the presence of groynes. As a result, the debris, which resulted from the damage caused in the Ruhr valley by the raging flood waters, floating downstream with the main current, arrived in The Netherlands ahead of the high water! In fact, these observations led to the development of the diffusion model by Schönfeld (1948).

8.5 Discussion

The analytical model described above is meant mainly to provide insight into the special features of the resistance-dominated flood waves in rivers, not so much as a tool for practical applications. We mention in particular the following aspects:

- the relatively low propagation speed (of the order of the flow velocity)
- the important influence of the ratio of storage width to conveyance width on the propagation speed (more than it is for inertia-dominated waves)
- the decrease of the flood wave height and increase of the flood wavelength during the propagation

- the occurrence of hysteresis in the relation between the local discharge and surface elevation; i.e. for the same river stage the discharge is greater when the water rises than when it falls
- at a fixed location, the maximum discharge occurs ahead of the maximum surface elevation
- the asymmetry in the local variation of the river stage with time, the rising branch being steeper and (therefore) of shorter duration than the falling branch.

The classical flood wave model described above allows wave propagation in the downstream direction only, at a low speed, of the order of magnitude of the flow velocity. These are so-called kinematic waves.

The theory for kinematic flood waves rests on the assumption that the inertia is negligible compared with the resistance, but its predictions are partly in conflict with this assumption. Take the case of the pulse-type addition of a volume of water at one particular cross section. The theory predicts a response in the form of a family of Gauss curves spreading in time. This implies an infinitely fast spreading of the wave (the tails of the Gauss curves), which in turn implies large accelerations, in conflict with the basic assumption that these would be negligible.

A more complete theory, in which inertia is not neglected, has positive and negative characteristics, corresponding to so-called dynamic waves, which can travel both downstream and upstream (assuming subcritical flow) at the speed $U \pm \sqrt{gA_c/B}$. However, the mass associated with these is very small, in practice insignificant. The bulk of the water mass flows downstream with the much lower speed derived for the kinematic flood wave. Yet the dynamic waves cannot be ignored in numerical models because their speed is the speed at which information travels; numerical models should be able to deal with that.

Problems

8.1 What is the difference between the theories for so-called dynamic waves and kinematic waves?

8.2 What is the principal difference in the behaviour of the categories of dynamic and kinematic waves?

8.3 What is the so-called quasi-steady approximation?

8.4 What is the so-called quasi-uniform approximation?

8.5 Choose a few characteristic instants in the river stage records in Figure 2.2 and estimate the associated values of h and $\partial h/\partial t$.

8.6 Estimate for the instances in Problem 8.5 the value of the acceleration and verify the validity of the quasi-steady approximation, assuming a bed slope of 10^{-4}.

8.7 Verify for the instances in Problem 8.5 the validity of the quasi-uniform flow approximation.

8.8 Prove that in the uniform-flow approximation the height and the length of the flood wave are constant as it propagates downstream.

8.9 Prove that in the uniform-flow approximation the flood wave deforms as it propagates downstream.

8.10 Explain why theoretically the crest of a flood wave decreases in height during propagation when the quasi-uniform flow approximation is not made.

8.11 What is the so-called hysteresis in flood waves? Explain why it occurs.

8.12 The propagation speed of a flood wave can be less than the flow velocity. What is the condition for this to happen?

8.13 Because of the failure of a dam, a volume of 10^6 m^3 of water is suddenly released in a river, at $t = 0$ in $s = 0$, say. Initially, the flow is uniform; the bed slope is $i_b = 1.5 \times 10^{-4}$, $c_f = 0.005, d = 3$ m, $B_c = B = 50$ m.

(a) Calculate the value of the diffusivity K (using the initial values of the flow parameters).

(b) Calculate the maximum height of the resulting flood wave at $t = 5$ h, and the location s_{max} where this occurs.

(c) Explain why the maximum height from the previous question is less than the maximum height occurring in $s = s_{max}$ during the passage of the flood wave, and verify this with calculations for a few chosen instants near the time $t = 5$ h.

9 Steady Flow

The preceding chapters have dealt with various types of long waves, in a sequence from rapidly varying translatory waves to slowly varying flood waves in lowland rivers, slow enough to be modelled as quasi-stationary. In the present chapter, we continue this line of development by considering steady flow. Within this class, we again deal with rapidly (spatially) varying flow first, in particular flow in control structures. This is followed by a section on gradually varying steady flow, the subject of the so-called backwater curves. Last, we deal with steady, uniform flow, mainly for a summary of expressions for flow resistance.

9.1 Rapidly Varying Flow

So far, we considered the propagation of various classes of long waves in canals, tidal channels and similar water courses, without obstructions of any kind. This no longer holds in cases where control structures are installed and operated, e.g. water intakes or weirs of various kinds for different functions such as

- discharge measurement
- control of discharge
- control of water level or flow depth.

From a hydrodynamics point of view, flows through or over a control structure belong to the category of *rapidly varying flows*, with length scales of the order of the flow depth. Because of these short scales, vertical accelerations are important, precluding the use of an assumption of hydrostatic pressure. The long-wave approach is locally not applicable.

9.1.1 Scaling Analysis

Up to a distance of a few depths upstream of a control structure, as well as beyond distances of some twenty flow depths downstream from it, the streamlines are more or less parallel, so that the long-wave approximation is applicable there. The structure determines the boundary conditions required for the calculation of these long waves. This is illustrated in Figure 9.1 for the case of a vertical, movable gate above a sill.

This chapter presents the key characteristic features of the flow in, over or through control structures, with respect to both the internal flow and the relations between the discharge and the water levels on either side.

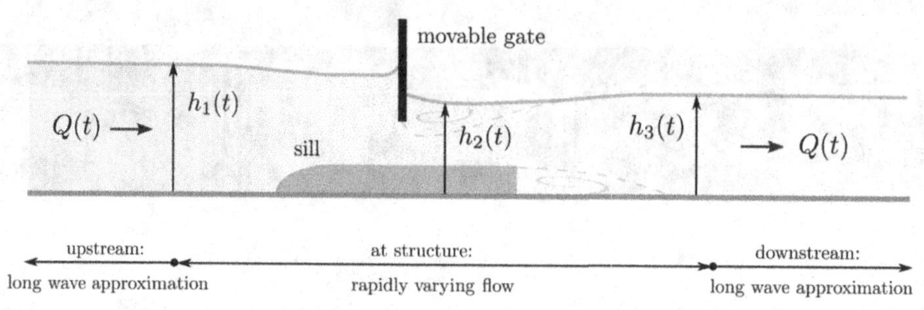

Fig. 9.1 Example control structure: movable gate above sill

The rapidly varying flow in control structures makes the calculation of the flow difficult, but on the other hand the short length scales involved allow certain simplifications in the modelling, which we will now consider.

A fluid particle passes the structure in a time of order ℓ/U, in which ℓ is the length scale of the structure, which is of the order of the flow depth d. For all situations of practical interest, this duration is small compared with the time scale of the variations in the boundary conditions. This implies that a fluid particle experiences virtually steady conditions while it passes the structure (see Box 2.2).

This allows the approximation of *quasi-steady flow*, in which case the flow can be determined with classical methods from steady-flow hydraulics, determined at each instant by the instantaneous boundary conditions. Specifically, we *neglect the local accelerations* ($\partial u_s/\partial t$ and $\partial u_n/\partial t$) and we *neglect storage* ($\partial h/\partial t$) inside the domain of the structure. The latter implies the approximation that *the discharge is spatially uniform* ($\partial Q/\partial s = 0$) over the length of the structure.

In view of the short length of the structure and the corresponding large accelerations, *boundary resistance is negligible*. Nevertheless, rapid decelerations will cause significant *expansion losses*.

9.1.2 Flow Patterns

In one way or another, control structures cause a local constriction of the cross section available for the flow. Accordingly, we can distinguish three principal zones:

- an inflow section, where the flow is accelerated
- the cross section with minimal area, the so-called throat or *vena contracta*, where the maximum flow velocities occur
- the outflow section, where the flow is decelerating.

This distinction is relevant because it is known from experience that the dynamics in accelerating flows and in decelerating flows are quite different.

Inflow Section

In accelerating flow, potential energy is converted into kinetic energy, which is an efficient process with negligible internal energy losses, for which reason Bernoulli's law (see below)

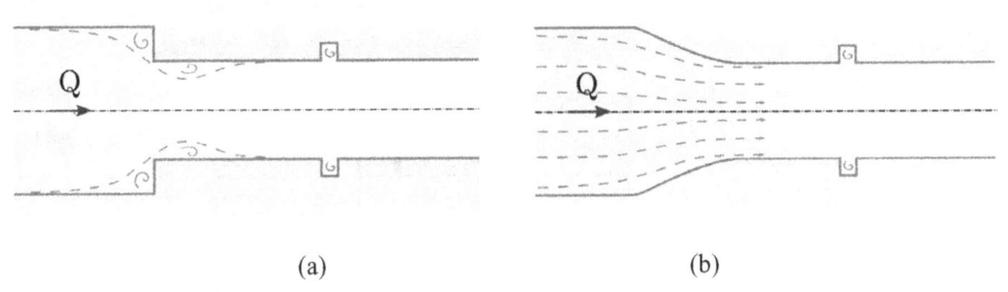

Fig. 9.2 Flow towards structure: (partially) obstructed (a), streamlined (b)

can be used in good approximation. This applies to the major part of the inflow region. Yet, within this region, there may be local zones of decelerating flow, because at angular profile variations, and at rounded changes with a too small radius of curvature, flow separation and contraction occur, followed by local expansions with their inherent losses. This is sketched in Figure 9.2a.

To prevent such local expansion losses, profile changes should be streamlined so that separation does not occur; see Figure 9.2b. This usually requires a longer structure. It may or may not be economical to accept the extra costs involved. In the case of a pumping station, the extra construction costs may be compensated by savings on fuel consumption. On the other hand, where there is ample head available to the flow, there is little point in making extra investments to reduce the losses.

Minimal Cross Section

Obviously, at the minimal cross section, the flow velocities are maximal. It is the most downstream location of the domain of applicability of the Bernoulli equation. Therefore, the flow characteristics at this location form an important input for the determination of the discharge.

Outflow Section

Downstream of the minimal cross section, where the flow velocities are maximal, the flow must spread so as to occupy the full cross section of the reach downstream of the structure. Here, kinetic energy is converted into potential energy, which is an inefficient process if it is accompanied by flow separation, in which case significant expansion losses occur.

Expansion losses can be reduced by making the profile expansion gradual, as in a diffuser. This gradual expansion can be in the vertical plane, e.g. in the form of a sloping fixed bed, or in plan. A simple example of the latter type can be found in streamlining the downstream ends of bridge piers or of the pillars between gates.

To minimize the expansion losses, the flow should not separate. This may be relevant for the design of evacuation sluices, particularly if the available head is low. As a rule of thumb, an expansion angle of not more than 1:7 is required for this, reducing the loss to about 15% of what it would be if the pillars ended abruptly. However, this implies a

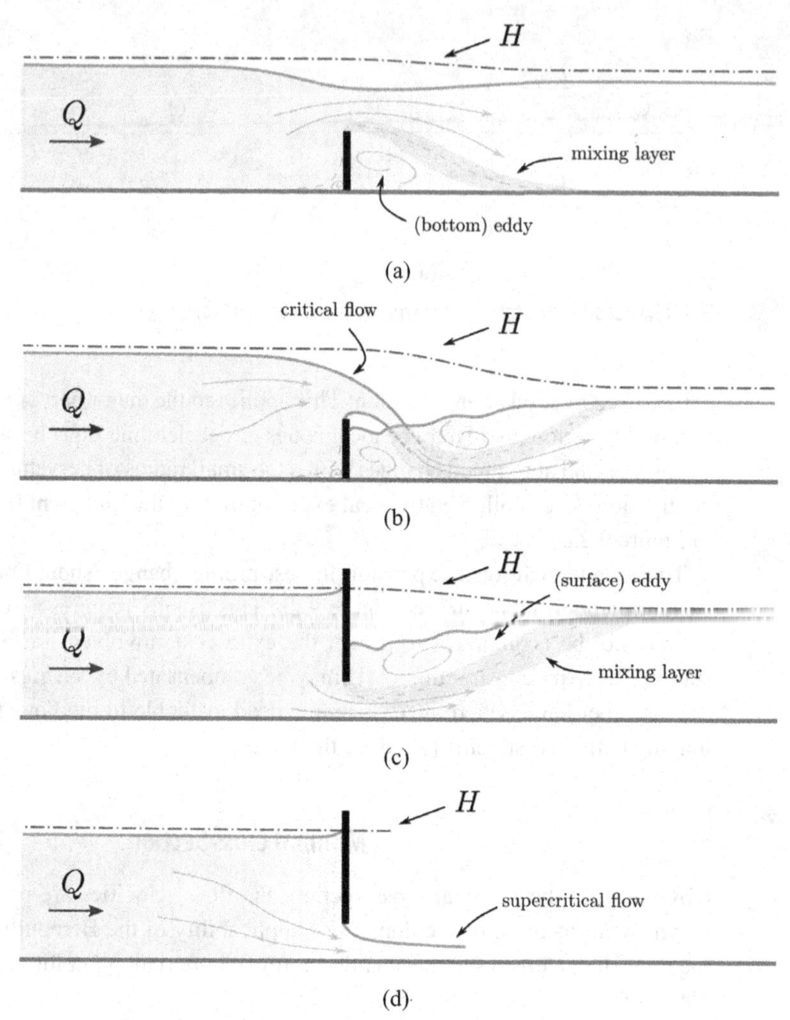

Fig. 9.3 Flow over short-crested weir and through underflow gate; hampered and free flow conditions

long structure. A divergence angle of about 1:4 reduces the loss to about 30% and may be a more acceptable compromise. Short extensions, such as a semi-circular cylindrical extension spanning the width of the pillar, can be effective on the upstream side, but when placed downstream they do not reduce the expansion loss.

The flow patterns on the downstream side can vary greatly, depending on the type of structure and the water levels. This is illustrated in Figure 9.3 for a sharp-crested weir, panels (a) and (b), and for an underflow gate, panels (c) and (d), respectively. In panels (a) and (c), the tail water level is high, hampering a free flow. In panels (b) and (d), on the other hand, there is free flow, independent of the downstream water level (as long as this is low enough).

In panels (a), (b) and (c), a narrow jet of water impinges into a thicker layer of water. The strong difference in flow velocity in the two regions causes the development of vortices at the interface, the so-called *mixing layer*, indicated in the figure as a shaded area. It forms the transition between the ongoing flow and a mass of water that is more or less stagnant except for a circulating motion, i.e. a large eddy.

The vortices generated in the mixing layer cause an exchange of momentum, which on the one hand drives the large-scale eddy, and on the other hand slows down the rapid flow in the jet, so decelerating it and (consequently) causing it to expand until it occupies the cross section available at the downstream side. As a rule of thumb, the establishment of parallel flow over the entire depth takes a distance of some 10 times the depth. The formation of the vortices in the mixing layer goes at the expense of the energy of the ongoing flow, which therefore experiences a so-called expansion loss.

9.1.3 Bernoulli Equation

The most relevant theoretical result for the determination of the discharge through a control structure is the Bernoulli equation for a stream tube of finite cross section, expressing conservation of energy. It is applicable in the inflow region, where the flow accelerates over a short distance. For steady motion, this implies that the rate of energy transfer through a cross section (F) is constant along the stream tube: $\mathrm{d}F/\mathrm{d}s = 0$. This transfer consists of work done by the pressure, and advection of potential and kinetic energy:

$$F = \int_{A_c} \left(p + \rho g z + \tfrac{1}{2}\rho u_s^2\right) u_s \,\mathrm{d}A \tag{9.1}$$

In a cross section where the streamlines are straight and parallel, the sum $p+\rho g z$ is constant and can be taken outside the integral, in which case Eq. (9.1) can be rewritten as

$$F = \left(p + \rho g z + \beta \tfrac{1}{2}\rho U^2\right) U A_c = \left(z + \frac{p}{\rho g} + \beta \frac{U^2}{2g}\right) \rho g Q = \rho g Q H \tag{9.2}$$

in which H is the so-called *energy head*, and the coefficient β is defined as the cross-sectional average value of $(u_s/U)^3$. It corrects for the non-uniformity of the particle velocity over the cross section, and is ignored in the following.

On the basis of mass conservation, $\rho g Q$ is constant along a given stream tube, and on the basis of energy conservation, F is constant; therefore, so is H: the equation of Bernoulli for a stream tube.

In a cross section with straight and parallel streamlines, the energy head H can be presented geometrically as an energy level by adding the velocity head $U^2/2g$ to the elevation of the free surface (setting the atmospheric pressure at zero). This same energy level then applies also in regions with curved streamlines, for which the pressure is not hydrostatic, and the particle velocities not uniform.

9.1.4 Relations between Water Level and Discharge

Relations between water level and discharge have a theoretical foundation, viz. the application of Bernoulli's law in the inflow region (accelerating flow) and the momentum balance in the outflow region (decelerating flow). Nevertheless, there is always the need for additional empirical information to account for unknown effects of the following:

- flow conditions upstream
- streamline curvature
- local contractions
- expansion losses
- boundary resistance.

Moreover, for convenience some additional approximations are often introduced in the theoretical relations, such as the use of the upstream water level in the expression for the discharge, instead of the upstream energy level (see below), which also requires some correction. For all these reasons, the definitive relations have to be determined through a process of calibration. The final result is expressed in an elementary mathematical form with one or more empirical coefficients. In some cases, coefficients are used for each of a number of factors such as those listed above. In other cases, they are lumped into one overall coefficient.

In the course of time, numerous formulas of theoretical origin, supplemented with empirical calibration data and coefficient values, have been collected for a large variety of structures in different handbooks (see e.g. Bos (1989) or ISO standards ICS-17.120.20, available online). In the following, we restrict ourselves to a few archetypal examples.

Flow under a Gate

A classical case for a first application of the above is the flow under a gate. We first take the case of a drowned outflow, shown in Figure 9.4. The streamwise uniformity of the energy level up to the location of the minimal flow cross section is shown by the horizontal chain line representing the energy level. Beyond that point, expansion losses occur, resulting in a lowering of the energy line.

Let h_1 be the water level just upstream of the gate, and h_2 the water level at the location of the minimum cross section. The Bernoulli equation applied to the flow between the cross sections 1 and 2 gives $H_1 = H_2$, or

$$h_1 + \frac{U_1^2}{2g} = h_2 + \frac{U_2^2}{2g} \tag{9.3}$$

in which U_2 is the flow velocity in the *vena contracta*. Assuming two-dimensional flow (no lateral variations) and a horizontal bed ($z = 0$), in absence of a sill, so that $h_1 = d_1$ and $h_2 = d_2$, this becomes

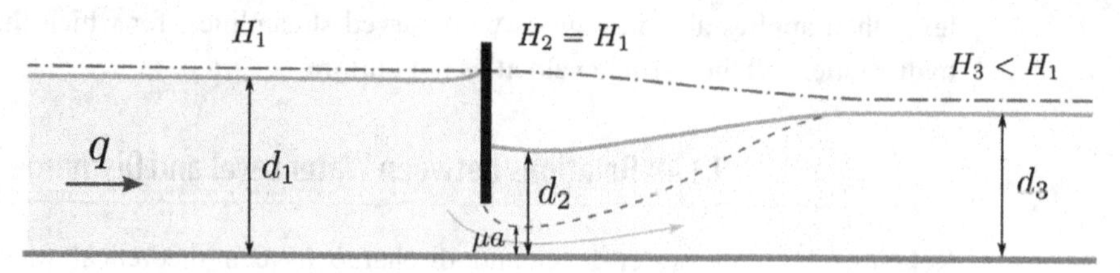

Fig. 9.4 Underflow gate: drowned outflow

$$d_1 + \frac{q^2}{2g{d_1}^2} = d_2 + \frac{q^2}{2g(\mu a)^2} \tag{9.4}$$

in which a is the gate opening and μ the coefficient of contraction.

For a given discharge, Eq. (9.4) provides a relation between the two depths against the gate on either side. However, in computations of the flows in the two reaches upstream and downstream of the gate, we need a link between the depths in the regions of established parallel flow, d_1 and d_3, respectively.

The Bernoulli equation cannot be applied beyond the minimal flow cross section because of the head loss occurring where the flow expands to the parallel flow further downstream. Instead, the value of d_3 can be computed from a momentum balance:

$$\tfrac{1}{2}\rho g d_2^2 + \rho\frac{q^2}{\mu a} = \tfrac{1}{2}\rho g d_3^2 + \rho\frac{q^2}{d_3} \tag{9.5}$$

In principle, d_2 can be eliminated between Eqs. (9.4) and (9.5). This provides an algebraic relation between the depths d_1 and d_3 for a given discharge q that cannot be solved explicitly. However, for dimensional reasons it is necessarily of the form

$$d_1 - d_3 = \xi\frac{U_3^2}{2g} \tag{9.6}$$

in which ξ is a head loss coefficient that is determined by the geometry and the Froude number of the flow; see for instance Box 9.1. Relations such as given in Eq. (9.6) are used to impose boundary conditions for the flows on either side of the underflow gate, resulting from its operation (Chapters 4 and 5).

In the case of free flow (Figure 9.3d), d_2 equals μa, in which case Eq. (9.4) yields

$$q = \sqrt{2g\frac{d_1^2 d_2^2}{d_1 + d_2}} \tag{9.7}$$

Using potential-flow theory, Kirchhoff calculated the value of the contraction coefficient for free flow (Figure 9.3d) in the case where the upstream depth is far greater than the gate opening, with the result (Lamb, 1932)

Box 9.1 Head loss coefficient of an underflow gate with small opening

Under certain circumstances we can make simplifying assumptions allowing the derivation of an explicit expression for the head loss coefficient. For weak flows ($Fr \ll 1$), the rise of the free-surface elevation in the region where the flow expands is relatively small. In view of the horizontal bottom, this can be expressed in terms of the depths: $\Delta d \equiv d_3 - d_2 \ll d_3$. Neglecting a term of $O(\Delta d)^2$ in Eq. (9.5), as well as the advection of momentum in the downstream reach (the second term in the right-hand side), which is allowed if $Fr_3^2 \ll 1$, this equation yields $\Delta d = q^2/(g\,\mu a\, d_3)$. Substituting this into Eq. (9.4) and neglecting the upstream velocity head, we arrive at Eq. (9.6), with ξ given by $\xi = (d_3/\mu a)^2 - 2d_3/\mu a$. For ξ to be positive, the relative gate opening should obey the condition $\mu a/d_3 < \frac{1}{2}$. Furthermore, the condition $\Delta d \ll d_3$ can be expressed as $\Delta d/d_3 = q^2/\left(g\,\mu a\, d_3^2\right) = \left(U_3^2/gd_3\right)(d_3/\mu a) = Fr_3^2\, d_3/\mu a \ll 1$. Altogether, the above results are valid for $Fr_3^2 \ll \mu a/d_3 < \frac{1}{2}$.

$$\mu = \frac{\pi}{\pi + 2} \cong 0.61 \tag{9.8}$$

In the beginning of this subsection, several items were listed that can violate the theoretical assumptions. However, in the present case of flow under a vertical gate their effects are negligible, so that the preceding equations can be used with only the contraction coefficient to be determined empirically or, if applicable, from Eq. (9.8).

Broad-Crested Weir

Another well-known situation in the context of discharge relations is the flow over a broad-crested weir, as sketched in Figure 9.5a for the case with subcritical flow both upstream and downstream. The weir is supposed to have sufficient length in the flow direction for the establishment of straight and parallel streamlines (the figure is not to scale), yet short enough for boundary resistance to be negligible. The streamwise uniformity of the energy level up to the downstream separation point is shown by the horizontal chain line. Beyond the point of separation, expansion losses occur, resulting in a lowering of the energy line.

Assuming two-dimensional flow (no lateral variations), the Bernoulli equation applied to the flow between the cross sections 1 and 2 gives

$$d_1 + \frac{q^2}{2gd_1^2} = a + d_2 + \frac{q^2}{2gd_2^2} \tag{9.9}$$

For given weir height a, this is a cubic algebraic equation in the three variables d_1, d_2 and q. If two of these are known, the other is determined. This can be used to determine the discharge for given (measured) values of d_1 and d_2.

Suppose now that we keep the upstream energy level constant but lower the downstream water level in Figure 9.5a somewhat and then hold it steady. This results in a negative

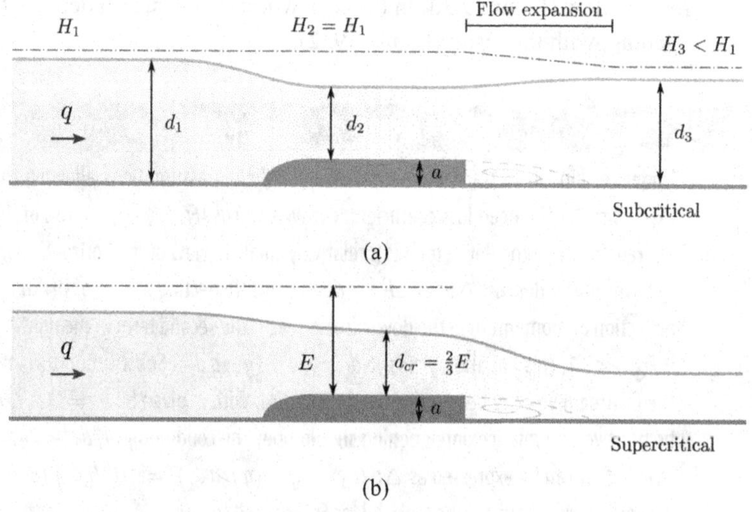

Fig. 9.5 Broad-crested weir: subcritical outflow (a) and supercritical outflow (b)

translatory wave travelling upstream, after which a new steady-flow situation is established with a lower water level and enhanced flow velocities over the weir.

If this lowering of the downstream water level is repeated time and again, a situation is reached in which a further lowering of the downstream water level no longer has any effect on the flow over the weir or further upstream. This is because with decreasing depth and increasing (counter) flow velocities over the weir, the local Froude number increases until it reaches the value 1 at some point, i.e., the flow has become critical. Disturbances from downstream then can no longer propagate upstream. This situation is referred to as free flow; see Figure 9.5b.

For free parallel flow over the weir, the local flow velocity is $U = U_{cr} = \sqrt{gd_{cr}}$, the velocity head is $U_{cr}^2/2g = \frac{1}{2}d_{cr}$, and the energy head above the crest, denoted as E, is $E = d_{cr} + U_{cr}^2/2g = \frac{3}{2}d_{cr}$, so that $d_{cr} = \frac{2}{3}E$. It follows that

$$q = d_{cr}\sqrt{gd_{cr}} = \tfrac{2}{3}E\sqrt{\tfrac{2}{3}gE} \tag{9.10}$$

This result is based on several theoretical assumptions that are only partly justified for flow over a weir. In particular the length may not be long enough for the establishment of parallel flow, or not short enough for bed resistance to be negligible. Moreover, in practice, the height of the upstream water level above the level of the crest of the weir (h_1) is often used in Eq. (9.10) instead of the energy level, because it can be directly measured. To account for this, and for the effects listed above, an empirical so-called discharge coefficient is used, m, say:

$$q = m\,\tfrac{2}{3}h_1\sqrt{\tfrac{2}{3}gh_1} \tag{9.11}$$

Sharp-Crested Weir

The preceding result, derived for flow over a broad-crested weir, is also used for sharp-crested weirs (Figure 9.3b). The curvature of the streamlines near the crest causes higher velocities and a larger discharge compared with a long-crested weir with the same upstream water level above the weir crest, resulting in m-values in excess of unity. Moreover, the nappe can be aerated or not. If not, entrainment of air bubbles by the jet impinging into the downstream water mass reduces the pressure inside the air pocket beneath it, and with it the size of that air pocket, until it totally vanishes. This reduced pressure beneath the nappe causes an even higher discharge. To achieve a stable and well-defined flow condition, needed for purposes of calibration or measurement, the nappe should be aerated so that there is atmospheric pressure not only above it but also below it.

An accurate calibration result for aerated sharp-crested weirs has been given by Rehbock (1929):

$$q = m\,\tfrac{2}{3}h_e\sqrt{\tfrac{2}{3}gh_e} \tag{9.12}$$

Here, $h_e = h_1 + 1.1$ mm and the discharge coefficient m is given by

$$m = 1.045 + 0.141\frac{h_e}{a} \tag{9.13}$$

in which a is the height of the weir crest above the upstream bed level. The water level h_1 should be measured at a distance of at least $2a$ upstream of the weir. The height of 1.1 mm, added to h_1 in a dimensionally inhomogeneous manner, accounts for effects of surface tension. Needless to say, it is relevant in application to small-scale flumes only.

In order to obtain the total discharge Q, the discharge per unit width (q) should be integrated laterally across the flow over the weir. In practice, this amounts to a multiplication of q with the total flow width B. This introduces an additional error, e.g. due to sidewall resistance and/or lateral contraction, which is compensated by adopting another discharge coefficient, m', say. Applied to Eq. (9.11), this yields

$$Q = m' \tfrac{2}{3} B h_1 \sqrt{\tfrac{2}{3} g h_1} \tag{9.14}$$

In cases where the width of the weir crest (measured normal to the approaching flow) is less than the upstream flow width, increasing contraction occurs with increasing water depth over the weir, resulting in systematic variations of m' with the water level. In order to reduce this variation, the weir cross section can be made trapezoidal, as sketched in Figure 9.6a for the so-called *Cipoletti weir*, which has sides sloping at 4:1. The corresponding value of m' in Eq. (9.14) is about 1.09. Note that in this case B is the length of the weir crest, not the width of the free surface in the weir.

In order to maintain sufficient accuracy in the measurement of small discharges, it is advisable to reduce the flow width as the discharge decreases. This has resulted in the V-shaped weir, the so-called *Thomson weir*, sketched in Figure 9.6b. The discharge over sharp-edged weirs of this type can be expressed as

$$Q = m' \left(\tan \tfrac{1}{2}\theta\right) h_1^2 \sqrt{g h_1} \tag{9.15}$$

The product $\left(\tan \tfrac{1}{2}\theta\right) h_1^2$ can be recognized as the area of the triangle enclosed between the weir profile and the upstream water level (Figure 9.6b). The discharge coefficient m' varies with the angle θ. Its value for the commonly used angle $\theta = 90°$ is approximately 0.45.

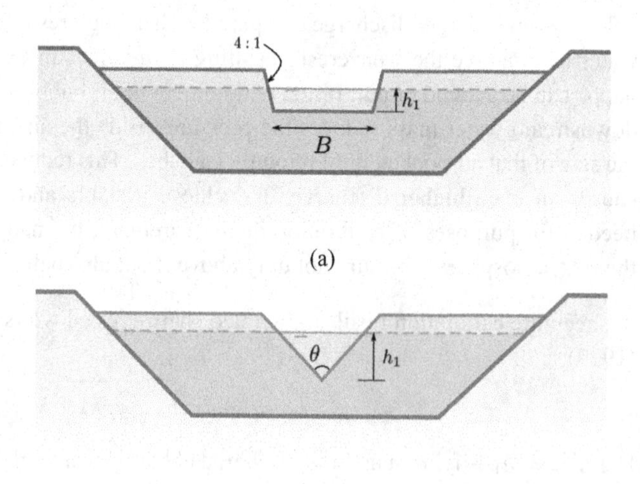

Fig. 9.6 Sharp-crested weirs: Cipoletti weir (a) and Thomson weir (b)

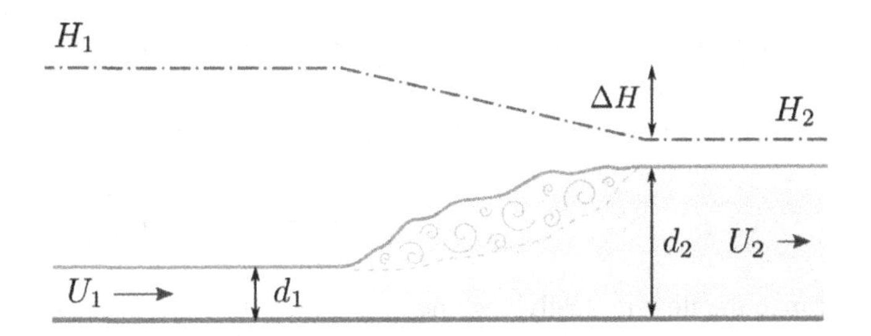

Fig. 9.7 Hydraulic jump

9.1.5 Hydraulic Jump

In the case of free flow under a gate (no influence from downstream), the gate causes a transition from the subcritical approaching flow to supercritical flow immediately downstream of it. Most commonly, the flow farther downstream is subcritical, with a higher water level. The transition from supercritical to subcritical flow takes place rather abruptly in a hydraulic jump, as sketched in Figure 9.7. Disregarding the complex dynamics of the highly turbulent internal flow, the overall properties of the jump can be calculated in a straightforward manner on the basis of conservation of mass and momentum.

Consider a control volume encompassing the jump, with vertical sides upstream and downstream at a sufficient distance from the jump, such that the local streamlines are straight and parallel and the vertical pressure distribution is hydrostatic. We assume steady, two-dimensional flow (no lateral variations) and denote the upstream and the downstream depth as d_1 and d_2, respectively, and likewise for the mean flow velocity.

Conservation of mass (or volume, because of incompressibility) implies that the flow rate per unit width (q) is constant across the jump, or

$$q = U_1 d_1 = U_2 d_2 \tag{9.16}$$

We assume that the bed is horizontal, so that the streamwise component of the weight of the fluid inside the control volume is zero. Furthermore, the bed resistance is neglected. This is allowed because the jump is short compared with the length scale of significant influence of bed resistance. It follows that in this approximation the streamwise momentum is conserved across the jump, or

$$\tfrac{1}{2}\rho g d_1^2 + \rho U_1^2 d_1 = \tfrac{1}{2}\rho g d_2^2 + \rho U_2^2 d_2 \tag{9.17}$$

Elimination of the flow velocities from Eqs. (9.16) and (9.17) yields the following relation between the discharge and the depths at both sides of the jump:

$$\frac{q^2}{g} = d_1\, d_2\, \frac{d_1 + d_2}{2} \tag{9.18}$$

The depths (d_1, d_2) obeying this relation for given q are said to be each other's conjugate. This result is used in Section 9.2 in the context of gradually varying steady flow, for the determination of the location of the jump.

The rate of energy loss in the jump can be calculated from an energy balance. To do so, we use the expression for the rate of energy transfer through a fixed cross section, Eq. (9.2). For parallel flow, with $p + \rho gz = \rho gd$, its value per unit width can be written as

$$F = \rho gqH = \rho gq\left(d + \frac{U^2}{2g}\right) \tag{9.19}$$

Since the flow is steady, the energy balance for the control volume containing the jump is simply $F_1 - F_2 = G$, in which G denotes the rate of energy loss per unit width in the jump. Substituting Eqs. (9.16), (9.18) and (9.19) and doing a little algebra yields

$$G = \rho gq\frac{(d_2 - d_1)^3}{4d_1 d_2} \tag{9.20}$$

The fraction in the right-hand side represents the head loss, i.e., the energy loss per unit weight of fluid. It can be seen that the rate of energy loss is proportional to the third power of the jump height. The capacity of a hydraulic jump to dissipate energy is utilized in hydropower projects for the design of spillways and the transition to parallel flow downstream.

9.2 Gradually Varying Flow

Manipulation of control structures in an initially uniform flow leads to disturbances that propagate away from the structure. If for a sufficiently long time the setting of the structure is not changed, and in absence of disturbances entering the domain of interest from the downstream side or the upstream side, a state of *steady flow* sets in, which in the vicinity of the structure deviates strongly from the state of uniform flow and gradually varies with increasing distance from the structure. The resulting elongated profile of the free-surface elevation is called a *backwater curve*. The determination of backwater curves is the subject of the present section.

9.2.1 Governing Differential Equation

In steady flow, $\partial Q/\partial t = 0$ and $\partial h/\partial t = 0$. Using the latter in the continuity equation, Eq. (1.20), it follows that $\partial Q/\partial s = 0$. Therefore, Q is constant in time and in space, so that the surface elevation h remains as the only unknown, a function of s only: $h = h(s)$. The momentum balance, Eq. (1.21), then reduces to a first-order ordinary differential equation for $h(s)$, written as:

$$\frac{Q^2}{gA_c}\frac{\mathrm{d}A_c^{-1}}{\mathrm{d}s} + \frac{\mathrm{d}h}{\mathrm{d}s} + c_f\frac{Q^2}{gA_c^2 R} = 0 \tag{9.21}$$

Because we deal here with steady flow, the flow velocity does not change sign, allowing the replacement of $|Q|Q$ by Q^2.

As we saw in the context of flood waves in rivers, it is most natural to use the sloping bed level as a reference, and to measure h with respect to this. Therefore, we write $h = z_b + d$, and use the depth d as the new unknown. The slope of the free surface, which delivers the driving force in Eq. (9.21), can then be written as $\mathrm{d}h/\mathrm{d}s = \mathrm{d}d/\mathrm{d}s + \mathrm{d}z_b/\mathrm{d}s = \mathrm{d}d/\mathrm{d}s - i_b$.

The derivative in the first term of Eq. (9.21) must now be transformed into a derivative of the depth. To do this, we substitute $A_c = B_c\, d$ and neglect longitudinal variations of the free-surface width B_c, so that $\mathrm{d}A_c/\mathrm{d}s = B_c\, \mathrm{d}d/\mathrm{d}s$. With this, we obtain

$$\frac{Q^2}{gA_c}\frac{\mathrm{d}A_c^{-1}}{\mathrm{d}s} = -\frac{Q^2}{gA_c^3}\frac{\mathrm{d}A_c}{\mathrm{d}s} = -\frac{Q^2B_c}{gA_c^3}\frac{\mathrm{d}d}{\mathrm{d}s} = -\frac{U^2}{gd}\frac{\mathrm{d}d}{\mathrm{d}s} = -Fr^2\frac{\mathrm{d}d}{\mathrm{d}s} \tag{9.22}$$

Substitution of the above intermediate results into Eq. (9.21) yields

$$\frac{\mathrm{d}d}{\mathrm{d}s}\left(1 - Fr^2\right) = i_b - i_f \tag{9.23}$$

in which the friction slope i_f is given by

$$i_f = c_f\frac{Q^2}{gA_c^2R} = c_f\frac{Q^2P}{gA_c^3} \tag{9.24}$$

(we revert to the wetted perimeter $P = A_c/R$ for reasons of symmetry; see below). Assuming $Fr^2 \neq 1$ (flow not critical), the result can be written in compact form as

$$\frac{\mathrm{d}d}{\mathrm{d}s} = \frac{i_b - i_f}{1 - Fr^2} \tag{9.25}$$

Written explicitly in terms of the discharge Q, this becomes

$$\frac{\mathrm{d}d}{\mathrm{d}s} = \frac{i_b - c_fQ^2P/gA_c^3}{1 - Q^2B_c/gA_c^3} \tag{9.26}$$

For wide, shallow conveyance cross sections, the wetted perimeter P and the width B_c are almost equal; the same then holds for the two expressions in the numerator and the denominator starting with Q^2. It follows that the relative magnitude of i_b and c_f plays an important role in the behaviour of the solutions of this equation, as we will indeed see below.

9.2.2 Integral Curves

For a given discharge, the geometric profile parameters and the resistance coefficient are supposed to vary in a known, monotonic manner with the depth, so that Eq. (9.26) prescribes $\mathrm{d}d/\mathrm{d}s$ as a function of s and d. This can be visualized by plotting short line elements in a dense network of points in the (s, d)-plane with a direction given by the local value of $\mathrm{d}d/\mathrm{d}s$. Solutions of Eq. (9.26) can graphically be constructed or represented as the so-called integral curves, i.e. curves in the (s, d)-plane that are everywhere tangent to the local line element.

In uniform flow, $i_f = i_b$, and $\mathrm{d}d/\mathrm{d}s = 0$. The corresponding depth is often called the normal depth or the equilibrium depth. We designate it as d_u, and the corresponding Froude number as Fr_u.

The denominator in the right-hand side of Eq. (9.25) (or Eq. (9.26)) is zero for critical flow, with depth d_{cr} and $Fr = 1$. As $d \to d_{cr}$, then $\mathrm{d}d/\mathrm{d}s \to \pm\infty$.

For very large depths, as in a reservoir, the flow velocities are virtually zero, and so are their dynamic effects and the resistance. It follows that $\mathrm{d}d/\mathrm{d}s = i_b$ very nearly; i.e., the free surface is horizontal.

The above is illustrated in Figure 9.8 for the situation in which $d_u > d_{cr}$, and in Figure 9.9 for the situation in which $d_u < d_{cr}$, showing line element graphs of a prismatic conduit, in which the line elements have a direction that for constant d does not vary with s. (The modification for nonprismatic channels is obvious.) The uniform-flow depth is represented by a line with long dashes, and the critical depth by a line with short dashes.

The flow depth must be given at some point, which provides the boundary condition (the heavy dots in the Figures 9.8 and 9.9). From there, the corresponding integral curve of Eq. (9.26) can be constructed by drawing a smooth curve that is everywhere tangent to the local line element. This has been in done Figure 9.8 for three cases, for which $d > d_u$, $d_{cr} < d < d_u$, and $d < d_{cr}$, respectively. The boundary condition has been indicated by a heavy dot, and the direction of integration (see below) by an arrow. Figure 9.9 shows essentially the same for the case where $d_u < d_{cr}$.

The figures clearly illustrate that the state of uniform flow is approached asymptotically, since $\mathrm{d}d/\mathrm{d}s \to 0$ as $d \to d_u$. In contrast, the critical depth is approached increasingly rapidly, because as $d \to d_{cr}$, then $\mathrm{d}d/\mathrm{d}s \to \pm\infty$. In the latter case, the variations become so rapid that the assumption of gradually varied flow, with a hydrostatic pressure distribution, is no longer valid.

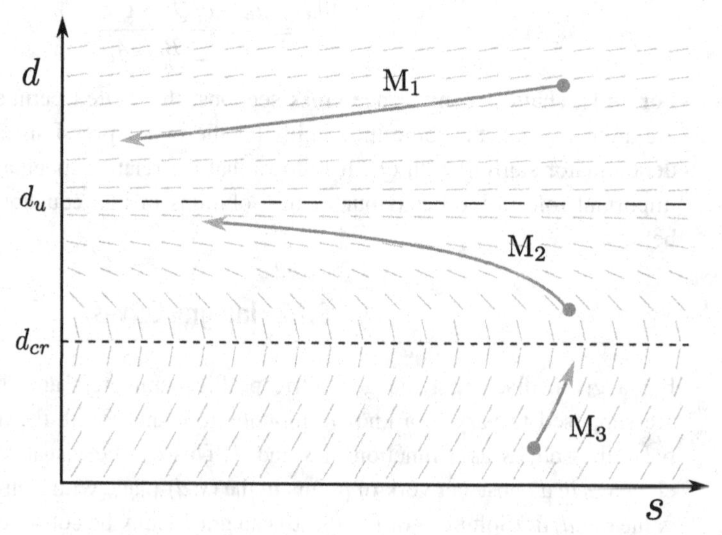

Fig. 9.8 Integral curves for mild slopes; dots indicate location of the boundary condition; arrows indicate direction of integration; flow is to the right

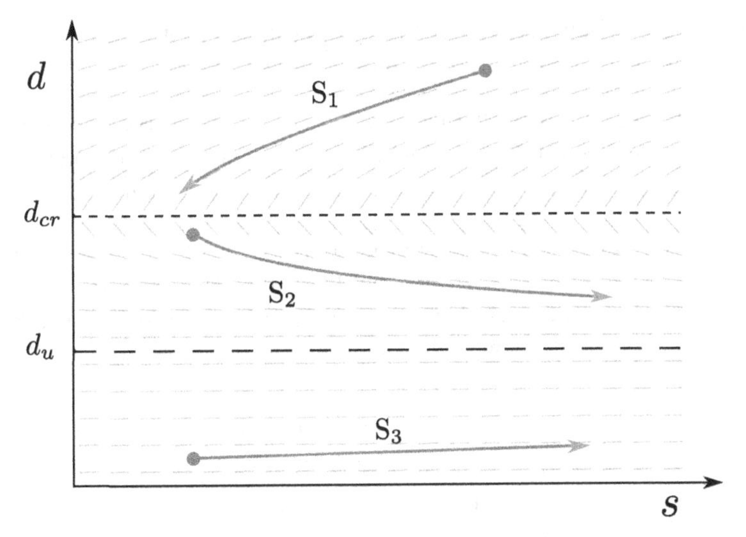

Fig. 9.9 Integral curves for steep slopes; dots indicate location of the boundary condition; arrows indicate direction of integration; flow is to the right

9.2.3 Classification of Backwater Curves

The preceding results show that the behaviour of the integral curve is quite different for the two cases shown, i.e. for d_u larger than d_{cr} and the other way around. This leads to a classification of bed slopes as a function of the type of flow in the case where it is uniform, as follows:

Steep slope (labelled 'S'): $d_u < d_{cr}$, the uniform flow is supercritical ($Fr_u > 1$).

Critical slope (labelled 'C'): $d_u = d_{cr}$, the uniform flow is critical ($Fr_u = 1$).

Mild slope (labelled 'M'): $d_u > d_{cr}$, the uniform flow is subcritical ($Fr_u < 1$).

Horizontal bed (labelled 'H'): the bed slope equals zero.

Adverse slope (labelled 'A'): the bed rises in the downstream direction.

Note that this classification does not depend on the actual state of flow. In contrast, the following classification refers to the actual depth d in relation to d_u and d_{cr}:

Type 1: refers to situations in which d exceeds both d_u and d_{cr}. The flow in this category is subcritical. A backwater curve of this type is labelled as M_1, C_1 or S_1 on slopes of type M, C or S, respectively. (For a horizontal bed or an adverse slope, uniform flow is not possible.) In type 1 curves, the numerator and the denominator in Eq. (9.25) are both positive, so that $\mathrm{d}d/\mathrm{d}s > 0$; i.e., the depth increases in the downstream direction.

Type 2: if the actual depth is between d_u and d_{cr}, the backwater curve is labelled M_2, etc. For this category, the numerator and the denominator in Eq. (9.25) are of opposite sign;

i.e. the depth decreases in the downstream direction. The flow is subcritical on mild slopes ($d_u > d_{cr}$) and supercritical on steep slopes ($d_u < d_{cr}$).

Type 3: if d is smaller than both d_u and d_{cr}, the curves are labelled as M_3, etc. The flow is supercritical. In these cases, the numerator and the denominator in Eq. (9.25) are both negative, so that $\mathrm{d}d/\mathrm{d}s > 0$; i.e., the depth increases in the downstream direction.

Uniform flow is not possible on a horizontal bed or on an adverse slope. In these cases, type 1 curves do not exist, and the curves are classified as type 2 or 3 if $d > d_{cr}$ or $d < d_{cr}$, respectively, corresponding to subcritical flow with depth decreasing downstream, or supercritical flow with depth increasing downstream, respectively.

9.2.4 Boundary Conditions

As we have seen in Chapter 5, subcritical flow is determined by a downstream boundary condition, and supercritical flow by a boundary condition at an upstream location. In the present context, this means a prescribed depth at one of such locations. In the examples shown in Figures 9.8 and 9.9, the (arbitrarily chosen) locations where the boundary conditions are prescribed are indicated by heavy dots. The integration proceeds in the direction of the arrow.

The transition from supercritical flow to subcritical flow can take place only in a hydraulic jump. The location is not fixed, but varies along the conduit with the flow conditions. We return to this in the examples presented below.

The transition from subcritical to supercritical flow always takes place at some kind of discontinuity in the conduit, a so-called control section, e.g. a structure or an abrupt steepening of the bed slope from mild to steep. This too will be illustrated in the examples below.

Control structures impose internal conditions, such as a relation between the depths at both sides of the structure for the given discharge, as we have seen in the section on rapidly varying flow.

9.2.5 Explicit Representation

The arguments presented above are valid for arbitrary cross sections. Both i_f and Fr are implicit functions of the depth, increasing monotonically with decreasing depth (for given discharge) in a manner that for arbitrary cross-sectional profiles can be accounted for only numerically. This generality goes at the expense of explicit relations. In order to obtain better insight into the problem and the possible solutions, we develop approximate but explicit expressions in terms of the flow depth. To this end, we replace the hydraulic radius (R) by the mean depth of the conveyance cross section (d). This is a good approximation for relatively wide and shallow flow cross sections, and it has no effect on the qualitative behaviour of the solutions.

For a compact notation, we use the cross-sectional average of the discharge per unit width: $q \equiv Q/B_c = Ud$. Together with the above approximation, Eq. (9.26) is transformed into

$$\frac{\mathrm{d}d}{\mathrm{d}s} = \frac{i_b - c_f q^2/gd^3}{1 - q^2/gd^3} \tag{9.27}$$

The denominator in the right-hand side is zero for critical flow; the corresponding critical depth obeys

$$d_{cr}^3 = q^2/g \tag{9.28}$$

The numerator is zero for uniform flow. The corresponding normal depth obeys

$$d_u^3 = \frac{c_f}{i_b}\frac{q^2}{g} = \frac{c_f}{i_b} d_{cr}^3 \quad (\text{for } i_b > 0) \tag{9.29}$$

with corresponding Froude number given by

$$Fr_u^2 = \frac{U_u^2}{gd_u} = \frac{q^2}{gd_u^3} = \frac{i_b}{c_f} \quad (\text{for } i_b > 0) \tag{9.30}$$

We see that in this approximation the criteria for steep, critical or mild slopes can be written as $i_b > c_f$, $i_b = c_f$ and $0 < i_b < c_f$, respectively.

Using the above expressions, and restricting ourselves to non horizontal beds ($i_b \neq 0$), Eq. (9.27) can be written as

$$\frac{\mathrm{d}d}{\mathrm{d}s} = i_b \frac{d^3 - d_u^3}{d^3 - d_{cr}^3} \tag{9.31}$$

This equation allows an analytical integration (named after Bresse), but we do not present it here because it does not give added insight. Instead, we derive an approximate expression for a characteristic length of the backwater curves for the special case in which the actual depth deviates by only a small amount from the uniform-flow depth. This allows a linearization of Eq. (9.31), as follows.

Writing $d = d_u + \Delta d$, in which $|\Delta d| \ll d$, we have

$$d^3 - d_u^3 = (d - d_u)\left(d^2 + dd_u + d_u^2\right) \approx 3d_u^2\,\Delta d \tag{9.32}$$

If also $d^3 - d_{cr}^3 \approx d_u^3 - d_{cr}^3$, Eq. (9.31) can be approximated as

$$\frac{\mathrm{d}d}{\mathrm{d}s} = \frac{\mathrm{d}\,(\Delta d)}{\mathrm{d}s} \approx i_b \frac{3d_u^2}{d_u^3 - d_{cr}^3}\,\Delta d = L^{-1}\,\Delta d \tag{9.33}$$

in which L is an adaptation length defined by

$$L \equiv \frac{d_u^3 - d_{cr}^3}{3i_b\, d_u^2} = \frac{1 - Fr_u^2}{3i_b} d_u = \frac{1 - i_b/c_f}{3i_b} d_u \tag{9.34}$$

Considering L as a constant on the s-interval considered, Eq. (9.33) can be integrated to

$$\Delta d = \exp\left(\frac{s - s_0}{L}\right)\,\Delta d_0 \tag{9.35}$$

The adaptation length is proportional to d_u, but inversely proportional to the bed slope i_b. Therefore, it is many times greater than the (uniform-flow) depth. As an example, consider the following situation: $i_b = 10^{-4}$, $c_f = 0.004$, $q = 4\,(\mathrm{m}^3/\mathrm{s})/\mathrm{m}$. It follows that

$d_u = (c_f q^2/g i_b)^{1/3} = 4\,\text{m}$ and $Fr_u^2 = i_b/c_f = 0.025$, so that $L = 13\,\text{km}$. This shows that the influence of a locally induced deviation from uniform flow extends over considerable distances.

An exact integration of Eq. (9.27) is possible for flow over a horizontal bed. In that case, the division by i_b in Eq. (9.27) is not permitted, but the equation reduces to

$$\frac{\mathrm{d}d}{\mathrm{d}s} = \frac{-c_f q^2/gd^3}{1 - q^2/gd^3} = -c_f \frac{d_{cr}^3}{d^3 - d_{cr}^3} \tag{9.36}$$

This can be integrated to

$$\tfrac{1}{4} d^4 - d_{cr}^3 d + c_f d_{cr}^3 s = \text{const} \tag{9.37}$$

in which the constant is determined by the boundary condition.

Example 9.1 Figure 9.10 gives an example of steady flow on a mildly sloping conduit in the presence of a partially opened vertical gate, such that the gate forces the subcritical flow from upstream to become supercritical at the downstream side. The gate is the only control structure in an otherwise undisturbed, prismatic reach, which has a free outflow at the downstream end.

Solution

Upstream of the gate, the water level is raised above the uniform-flow value (d_u) to a level that can be calculated with the method described in the preceding section on flow through control structures. Because the flow in the upstream reach is subcritical, the resulting depth at the gate is the boundary condition for this reach, so that an M_1-type backwater curve is established there.

Immediately downstream of the gate, where the flow is supercritical, an M_3-type backwater curve develops, with the minimum depth just past the gate as the upstream boundary condition. Going downstream, the depth in this reach approaches the critical depth with increasing rapidity until at some point the solution becomes invalid. Farther downstream, the flow is subcritical. Due to the free outflow the depth is critical at the downstream boundary, causing an M_2 backwater curve upstream of it.

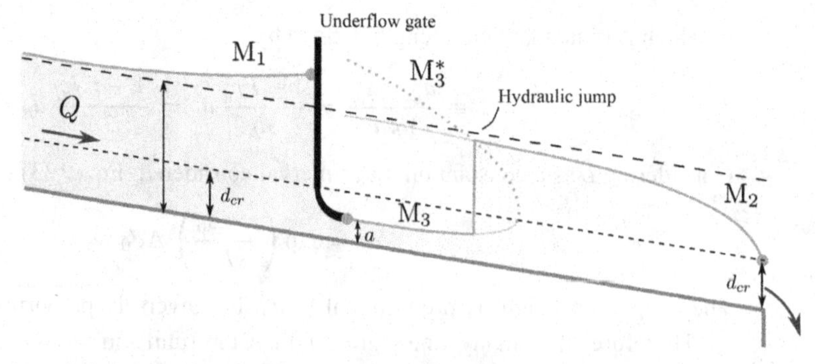

Fig. 9.10 Example 9.1: underflow gate with backwater curves and hydraulic jump

We see that over some distance downstream of the gate, two flow regimes with two different depths seem to overlap: a supercritical regime with an upstream boundary condition and a subcritical regime determined by the downstream boundary condition. The two branches are connected by a hydraulic jump, forming the transition from supercritical flow to subcritical flow.

In order to determine the location of the jump, we use the fact that the two depths at both sides of the jump form a conjugate pair obeying Eq. (9.18). For every point on the M_3-curve, the corresponding conjugate depth is plotted at the same cross section, together forming a smooth curve labelled M_3^*. The jump is located where this curve intersects the backwater curve determined by the downstream boundary condition (M_2).

Notice that with increasing distance upstream from the location of free outflow, the depth of the M_2-curve approaches the uniform-flow depth more and more. If the downstream control is sufficiently far away, the flow immediately downstream of the jump can be considered to be uniform.

Example 9.2 Consider a case where two very long prismatic channel reaches of equal slope (labelled I and III) are connected by a shorter reach of somewhat steeper slope (labelled II), although still mild in the formal classification; see Figure 9.11. Downstream controls, if present, are situated at large distances from the steeper reach, leading to a state of uniform flow immediately downstream of it.

Solution

The backwater curves in the respective reaches are of type M. Because of the change in bed slope, the depth of uniform flow changes at the two transitions, but the critical depth is the same everywhere because it depends on the discharge only, irrespective of the bed slope.

Because the slope is mild, and in absence of influence of a possible downstream control, the flow in the prismatic downstream reach III is necessarily uniform, as explained in the preceding example. The corresponding depth ($d = d_{u,III}$) provides the downstream boundary condition at point P_2 for the depths in reach II. Since reach II is steeper than reach III, its uniform-flow depth is smaller, so that an M_1 backwater curve develops in reach II, extending up to point P_1, at the transition between reach I and reach II. The value

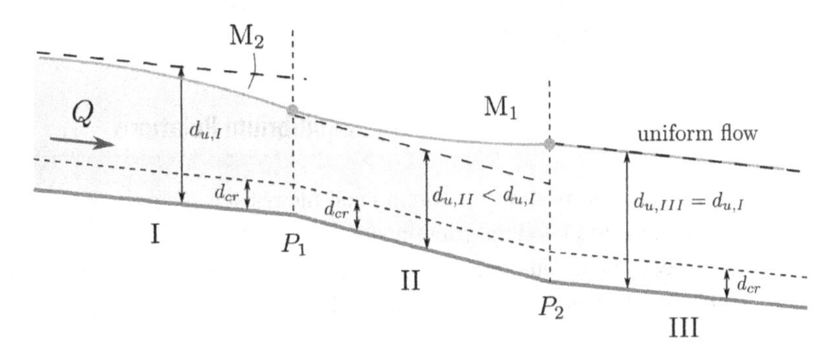

Fig. 9.11 Example 9.2: backwater curves caused by varying bed slope

of the M_1-depth at point P_1 forms the downstream boundary condition for reach I, and since it is smaller than $d_{u,I}$ (which equals $d_{u,III}$), an M_2-type backwater curve is present in reach I. Going upstream, the depth approaches the uniform-flow value asymptotically.

9.3 Uniform Flow

Flow resistance in long waves is usually expressed through a quadratic dependence of the boundary shear stress on the instantaneous cross-sectionally averaged flow velocity, as if the flow were steady and uniform. The value of the dimensionless resistance factor c_f was so far left unspecified. Its estimation is the subject of the present section.

Uniform flow has been the subject of numerous investigations through the centuries, spurred by the need to design and build water conveyance systems for irrigation or urban water supply. These investigations pertained to free-surface flows as well as pressurized flow in pipes. The purpose was mainly to determine the discharge under given forcing or, equivalently, the resistance to such flows for given discharge.

Up to the beginning of the twentieth century, the approach was purely empirical, resulting in a host of different expressions containing dimensional coefficients, which in general were implicit boundary roughness parameters. Of the numerous empirical expressions relating the mean flow velocity to the geometric parameters of the conduit, three have survived in engineering practice, namely those presently named after Chézy (1768), Manning (1889) and Strickler (1923).

The early twentieth century saw the advent of theories for turbulent boundary layers and flow resistance, in particular for pressurized flow in pipes. The results were later generalized to free-surface flow. They could in principle have superseded the older, empirical results, but the latter are still widely used in engineering practice, not least because of extensive empirical knowledge of the required coefficient values for varying types and sizes of boundary roughness.

The results of both approaches are briefly summarized in the following, beginning with results from turbulence theory. As pointed out by Rouse and Ince (1957) and Sturm (2001), the empirical expressions have gone through quite an evolution. Interesting though this may be, we will merely deal with these equations in their present form.

9.3.1 Equilibrium Relations

In steady, uniform flow, the forcing and the resistance are in balance; see Figure 9.12. The surface slope (i_s) and the friction slope (i_f) are both equal to the bed slope (i_b). Eq. (1.11) then reduces to an algebraic expression of the equilibrium between the downslope component of the fluid weight and the boundary resistance:

$$\tau_b = \rho g R i_b \tag{9.38}$$

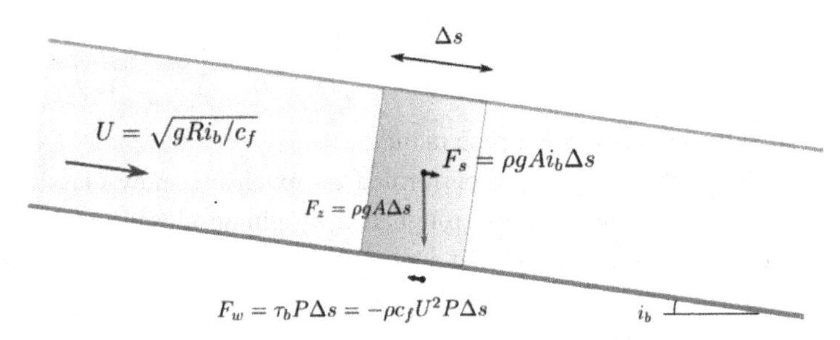

Fig. 9.12 Balance of forces in a uniform flow

Substituting $\tau_b = c_f \rho U^2$ yields the following well-known expression for the mean flow velocity U

$$U = \sqrt{gRi_b/c_f} \tag{9.39}$$

so that

$$Q = A_c\sqrt{gRi_b/c_f} \tag{9.40}$$

The remainder of this section deals with the determination of numerical values of the resistance coefficient c_f or its equivalents.

9.3.2 Resistance Relations

The Log Law

The theory of turbulent boundary layer flow results in a logarithmic variation of the mean particle velocity with distance from the boundary (see Chapter 10). For free-surface flows, the profile is given by Eq. (10.39). We will now elaborate on this result in order to establish an expression for c_f.

Integration of Eq. (10.39) over the depth from $z = z_0$ to $z = d$, with $z_0 \ll d$, yields the following expression for the depth-averaged velocity U:

$$U = \frac{u_*}{\kappa} \ln\left(\mathrm{e}^{-1}\frac{d}{z_0}\right) \tag{9.41}$$

in which u_* is the so-called shear velocity, defined as $u_* \equiv \sqrt{\tau_b/\rho}$; κ is a constant of about 0.4, named after Von Karman; and z_0 is the distance from the wall where the (extrapolated) logarithmic profile gives a zero velocity. Combining this with $c_f = \tau_b/\rho U^2$ yields the following expression for c_f in uniform free-surface flow:

$$\frac{U}{u_*} = \frac{1}{\sqrt{c_f}} = \frac{1}{\kappa}\ln\left(\mathrm{e}^{-1}\frac{d}{z_0}\right) \tag{9.42}$$

The logarithmic velocity profile given in Eq. (10.39) also applies to pipe flow (for which in fact it was originally derived). Repeating the steps above yields the following result for pipe flow:

$$\frac{U}{u_*} = \frac{1}{\sqrt{c_f}} = \frac{1}{\kappa} \ln \left(e^{-3/2} \frac{r_0}{z_0} \right) \tag{9.43}$$

in which r_0 is the inner pipe radius.

Nikuradse (1933) has performed an extensive, now classical set of experiments on pressurized flow in pipes, roughened by gluing closely packed sand grains of virtually uniform diameter k on the inner pipe wall. His result confirmed the validity of the logarithmic profile and yielded explicit expressions for z_0. For hydraulically rough boundary layer flow, he found $z_0 = k/30$.

The two expressions for c_f given above for uniform free-surface flow and for pipe flow, respectively, have later been generalized to arbitrary cross sections by expressing both in terms of the hydraulic radius R instead of the depth ($R = d$) and the pipe radius ($R = \frac{1}{2} r_0$). We then find $e^{-1} d/z_0 \approx 0.37R/z_0$ and $e^{-3/2} r_0/z_0 \approx 0.45R/z_0$. For given R/z_0, the corresponding values of c_f do not differ much because of the weak variation of the logarithmic function with its argument. Therefore, it is justified to use a kind of average between these two results for arbitrary cross sections, rounded to $0.4R$, resulting in

$$\frac{1}{\sqrt{c_f}} = \frac{1}{\kappa} \ln \left(0.4 \frac{R}{z_0} \right) \tag{9.44}$$

Substituting $\kappa = 0.4$ and $z_0 = k/30$, and switching from the natural logarithm to the logarithm with base 10, the following result is obtained, which is used in practice:

$$\frac{1}{\sqrt{c_f}} = 5.75 \log \left(\frac{12R}{k} \right) \tag{9.45}$$

The ratio k/R is called the *relative roughness*.

In Nikuradse's experiments, the roughness was brought about by gluing sand grains of virtually uniform diameter (k) in close packing on the inner pipe wall. In natural channels with alluvial beds, the size of the roughness elements is not uniform but is characterized by a distribution. An equivalent value of k can then be determined, such that substitution of its value in to Eq. (9.45) yields the same value of c_f as the one that is obtained from the measurements.

The resistance is mainly determined by the larger elements such as D_{84}, i.e. the value of the 'diameter' such that 84% is smaller by weight (for a Gaussian distribution, this limit corresponds to the mean plus one standard deviation). Sturm (2001) lists values of the ratio k/D_{84}, determined from field measurements by various authors, ranging from 2.4 to 3.5. The fact that k is several times larger than D_{84} is due to the fact that these relatively large elements are not closely packed, as were the grains in Nikuradse's experiments, but are much more widely spaced, which allows the development of a larger wake behind each, and therefore a higher resistance.

Williamson (1951) has reanalyzed the data obtained by Nikuradse, made some corrections to his calculations, and supplemented the data with results from measurements in large-diameter pipes (up to about 6 m diameter), which resulted in a more than six-fold increase in the range of relative roughness to very low values. He found that in the range of small relative roughness, the log law overestimates the value of the resistance coefficient. In contrast, he obtained a very good fit over the entire range of relative roughness with a Manning- or Strickler-type power-law expression (see below), as follows:

$$\lambda = 0.180 \left(\frac{k}{D}\right)^{0.330} \tag{9.46}$$

in which λ is the Darcy–Weisbach coefficient, equal to $8c_f$. Making this substitution, replacing D by $4R$, and approximating the exponent as the rational fraction 1/3, we obtain

$$c_f = 0.0142 \left(\frac{k}{R}\right)^{1/3} \tag{9.47}$$

The latter relationship, although based on measurements of pipe flow, is assumed to be approximately valid for arbitrary cross sections, including those of open channels, in the same vein as had been assumed for the log law given in Eq. (9.45). Substitution of Eq. (9.47) into Eq. (9.39) yields

$$U = 8.4 \left(\frac{R}{k}\right)^{1/6} \sqrt{gRi_b} \tag{9.48}$$

Chézy

The French engineer Chézy (1768) reasoned (correctly) that the resistance to uniform, turbulent flow is proportional to the square of the velocity and to the wetted perimeter of the (conveyance) cross section (P), whereas the driving force is proportional to the product of the area of the conveyance cross section and the bed slope ($A_c i_b$). With these two being in equilibrium in uniform flow, U should be proportional to $\sqrt{A_c i_b/P}$ or to $\sqrt{Ri_b}$. This proportionality was later written as an equality through insertion of a proportionality coefficient:

$$U = C\sqrt{Ri_b} \tag{9.49}$$

This is the so-called Chézy equation, and C is the so-called Chézy coefficient. Comparison of Eq. (9.49) with Eq. (9.39) shows that C and c_f are related through

$$C = \sqrt{g/c_f} \tag{9.50}$$

Obviously, a factor $\sqrt{g}$ is hidden in C, causing it to have the dimension of $L^{1/2}T^{-1}$ ($m^{1/2}/s$ in the metric system).

In applications, C may be given an *a priori* value from experience, such as $C = 50\,m^{1/2}/s$, or it can be used as a calibration factor. Quite often, it is expressed in terms of the relative roughness, based on the log law, by substituting Eq. (9.45) into Eq. (9.50):

$$C = 5.75\sqrt{g}\log\left(12\frac{R}{k}\right) \tag{9.51}$$

Needless to say, this is in fact a hybrid expression, in which a newer theoretically derived result is used to assign numerical values to an outdated coefficient.

Manning

As did many of his predecessors, the Irish engineer Manning (1889) (re)analyzed data on uniform flow in a wide variety of open channels and pipes, and concluded that the data were best represented by an expression that is presently written (in the metric metres/seconds system) as

$$U = \frac{1}{n} R^{2/3} i_b^{1/2} \tag{9.52}$$

Obviously, as was the case with the Chézy equation, this equation is not dimensionally homogeneous. The quantity n, usually called Manning's n, has the dimension $L^{-1/3}$ T. Published values bear the unit $m^{-1/3}$ s, although this is generally not stated; these n-values are commensurate with U being expressed in metres per second and R in metres. Using the feet/seconds system, with U expressed in feet per second and R in feet, the value of n should in principle be adapted, but in practice it is left unchanged. In order to compensate for this, the right-hand side of Eq. (9.52) is multiplied with the factor 1.486 if the feet/seconds system is used, this being the value of $(1\ \mathrm{m}/1\ \mathrm{ft})^{1/3}$. Actually, the number 1.486 suggests a quasi-accuracy that is neither achievable nor necessary. Using a factor 1.5, which differs less than 1% from the nominal value, would be more appropriate.

Similar to the Chézy coefficient, $1/n$ contains a hidden factor $\sqrt{g}$. The remaining dimension is $L^{1/6}$. Comparison with Eq. (9.48) shows that n is proportional to the 1/6-th power of the characteristic size of the boundary roughness elements. In fact, Eqs. (9.48) and (9.52) are equivalent if

$$n = \frac{k^{1/6}}{8.4\sqrt{g}} \tag{9.53}$$

However, it is not common to express n in terms of the size of the roughness elements. Instead, extensive tables have been developed (Ven Te Chow (1959), Henderson (1966), Sturm (2001)), listing empirical values of n for boundary roughnesses and other types of resistance elements of various sizes and kinds, including vegetation (Sturm lists more than 100 entries). This wealth of empirical information explains and justifies the continuing wide use of Manning's equation. To give an impression, Table 9.1 presents a brief listing of n-values. Comparable information linking equivalent Nikuradse's k-values to various kinds of boundary roughness or other resistance elements has not been developed to the same extent.

The connection of Manning's n with Chézy's C and c_f is

$$\frac{1}{n} R^{1/6} = C = \sqrt{g/c_f} \tag{9.54}$$

Table 9.1 Values of Manning's *n* for a variety of boundary materials; single *n*-values are the mean of a range of approximately $\pm 0.001\ \mathrm{m}^{-1/3}\mathrm{s}$; data from Henderson (1966)

Boundary material	Manning's n ($\mathrm{m}^{-1/3}\mathrm{s}$)
Glass, plastic, machined metal	0.010
Dressed timber, joints flush	0.011
Sawn timber, joints uneven	0.014
Cement plaster	0.011
Concrete, steel troweled	0.012
Concrete, timber forms, unfinished	0.014
Untreated gunite	0.015–0.017
Brickwork or dressed masonry	0.014
Rubble set in cement	0.017
Earth, smooth, no weeds	0.020
Earth, some stones and weeds	0.025
Natural river channels	
Clean and straight	0.025–0.030
Winding, with pools and shoals	0.033–0.040
Very weedy, winding and overgrown	0.075–0.150
Clean, straight alluvial channels	$0.031\, D_{75}{}^{1/6}$ (D_{75} in ft)

Strickler

Not being satisfied with existing formulae for flow resistance, the Swiss engineer Strickler (1923) set out to (re-?)analyze data on uniform flow in open channels and pipes. Apparently unaware of Manning's work, he arrived nevertheless at the same power law for the mean velocity, written by him as

$$U = KR^{2/3} i_b^{1/2} \tag{9.55}$$

with K given as

$$K = 4.75\sqrt{2g}\, D_m^{-1/6} \tag{9.56}$$

in which D_m is the mean diameter of the roughness elements (sand, gravel, shingle, cobbles, etc.). Clearly, Strickler's K is the reciprocal of Manning's n, and his expression for K reconfirms that the latter is proportional to the 1/6-th power of the roughness size. Substituting by approximation $k \approx 3D_{84} \approx 4D_m$, we obtain

$$U \approx 8.5 \left(\frac{R}{k}\right)^{1/6} \sqrt{gRi_b} \tag{9.57}$$

which is close to Williamson's result given by Eq. (9.48).

9.3.3 The Overall Resistance of a Channel

In the preceding summary of formulations of flow resistance, it has been assumed that the flow is uniform, with the tacit exception of the micro-scale flow around relatively small boundary roughness elements. Since these are relatively small, the resulting spatially averaged resistance can be modelled as a bed shear stress with a coefficient whose value could be expressed in terms of a relative roughness.

More generally, large-scale profile variations may be present, causing expansion losses and so adding to the overall resistance to the flow. Shoals and dunes that may develop in natural channels with mobile beds of alluvial material are a case in point. It is not feasible to take these into account individually; their effect can be lumped in a single flow-dependent resistance coefficient such as c_f, but relations such as Eq. (9.56) lose their significance. Additional resistance may be due to groynes, bridge piers etc. In these conditions, it is very difficult to assign proper values to the resistance coefficient *a priori*. Calibration in the target area is necessary. This must be done using the chosen profile schematization because c_f, A_c and R occur together in the single resistance term $c_f|Q|Q/(A_cR)$, so that errors in the geometric profile parameters are compensated by errors in the estimated value of c_f.

9.3.4 Applicability to Unsteady Flow

Last, we should discuss the applicability of the quasi-steady approximation for the determination of the flow resistance in unsteady flow. This rests on the assumption that the velocity distribution in the vertical is the same as it is for steady, uniform flow with the same cross-sectionally averaged velocity. In reality, it takes some time for this distribution to adapt to changes in the mean velocity.

This process is similar to the establishment of uniform, steady flow in the case of inflow from a reservoir or in the case of an abrupt change in bed roughness. As a rule of thumb, such adaptation length is about 40 times the depth, corresponding to an adaptation time, following the mean flow, of about $40\,d/U$. If this is small compared with the time scale of the imposed flow variations, the assumption of quasi-steady resistance is justified.

Taking some realistic values of U and d for tidal flow in channels, such as $U = 1$ m/s and $d = 10$ m, we find an adaptation time of about 400 s. This is small compared with the tidal period of more than 12 h, justifying the quasi-steady approximation for tides. Translatory waves ususaly occur in canals of less depth, but also cause lower velocities, so that the adaptation time is comparable to those for tides. Since their time scale is typically of the order of minutes (see Chapter 4), the quasi-steady approximation is less justified for this category.

Problems

9.1 Explain why the unsteady flow through, over or under a control structure can be treated as quasi-steady (under normal operating conditions).

9.2 The discharge in a laboratory flume is measured with a sharp-crested weir. The width of the flume is 0.6 m and the height of the weir above the bed is 0.5 m. The upstream water level is 0.2 m above the weir crest. Calculate the discharge.

9.3 The discharge in an irrigation canal is measured with a 90° Thomson weir. The upstream water level is 0.5 m above the lowest point of the weir crest. Calculate the discharge for free flow.

9.4 A Cipoletti weir has a crest width (B) of 1 m. The upstream water level in a situation of free flow is 0.4 m above the crest level. Calculate the discharge.

10 Transport Processes

This chapter deals with the transport of dissolved or suspended matter or heat in free-surface flows. This subject is of great importance in the context of problems of water quality, sedimentation, erosion etc. The present chapter is restricted to so-called *passive transports*, based on the assumption that the presence of the transported matter or heat in the water does not affect the flow. This restricts the treatment to relatively low concentrations, such that the bulk density and the bulk viscosity are affected to a negligible degree only. However, a warning is in order here. There are numerous situations in which this approximation does not hold. A well-known example is the effect of salinity variations on the fluid density, which in estuaries, for instance, is sufficiently strong so as to affect the flow significantly. Likewise, effects of temperature variations and high concentrations of fine suspended sediments occurring in practice are often not negligible. Nevertheless, as stated, these effects are ignored in the present introductory chapter. A detailed account, also treating flows induced by density differences, can be found in Fischer et al. (1979) and Rutherford (1994).

10.1 Introduction

We distinguish the following types of physical transport processes, in order of increasing length scale:

Molecular diffusion: the time-averaged value of fluctuating micro-transports due to random thermal molecular motions (Brownian motion in fluids)

Advection: transports following the fluid at continuum scale

Turbulent diffusion: the time-averaged value of fluctuating advective transports due to turbulent motion of finite fluid 'parcels'

Dispersion: mixing arising in the two-dimensional or one-dimensional modeling of three-dimensional transports due to integration over the depth or the cross section of a water course.

The larger the scale considered, the higher the transport rates are. We start the modelling of transport processes with the formulation of a generic balance equation, not restricted to a particular type of transport or of transported substance. Thereafter, the types of transport mentioned above are dealt with in succession.

10.2 Generic Balance Equation

In order to avoid the complexities of the derivation and solution of transport equations in curvilinear coordinates (s, n, b), we will temporarily use Cartesian coordinates (x, y, z).

Suppose a quantity 'X' (salt, sediment, heat etc.) is being transported in three dimensions, possibly in flowing water. Without paying attention to the type of physical transport being considered from among those listed above, we represent the total transport of X at a point by a vector $\vec{T}$ pointing in the direction of the transport and whose magnitude equals the amount of X being transported per unit time through a unit area normal to $\vec{T}$.

We apply this in the derivation of the balance for the amount of X inside a small rectangular control volume in (x, y, z)-space, with sides given by $x = x_1$, $x = x_2 = x_1 + \Delta x$, etc., whose surface areas are given by $\Delta y \Delta z$, etc., and with volume $\Delta V = \Delta x\, \Delta y\, \Delta z$; see Figure 10.1.

The amount of X transported per unit time into the control volume across the side given by $x = x_1$ equals $T_{x,1}\, \Delta y \Delta z$, in which $T_{x,1}$ is the x-component of $\vec{T}$ at $x = x_1$. Likewise, an amount $T_{x,2}\, \Delta y \Delta z$ leaves the control volume in unit time through the side for which $x = x_2$. The net inward transport per unit time through these two sides is therefore

$$(T_{x,1} - T_{x,2})\, \Delta y \Delta z \cong \left(-\frac{\partial T_x}{\partial x}\, \Delta x\right) \Delta y \Delta z = -\frac{\partial T_x}{\partial x}\, \Delta V \tag{10.1}$$

Similar expressions apply to the other four sides, so that the total transport per unit time into the control volume through the six sides is approximately given by

$$-\left(\frac{\partial T_x}{\partial x} + \frac{\partial T_y}{\partial y} + \frac{\partial T_z}{\partial z}\right) \Delta V \tag{10.2}$$

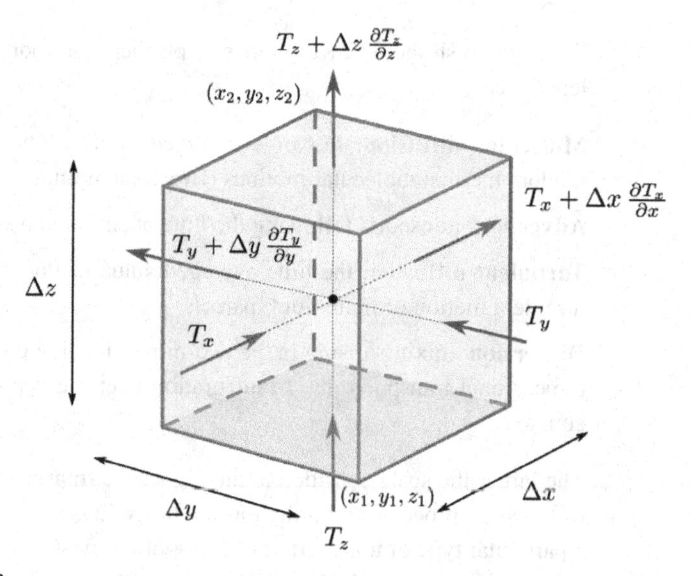

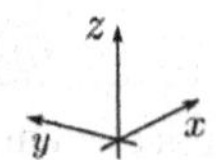

Fig. 10.1 Cartesian control volume

The quantity between parentheses is the so-called *divergence* of $\vec{T}$, or $\nabla \cdot \vec{T}$, in which ∇ is the differential operator defined by

$$\nabla (\cdot) \equiv \frac{\partial \cdot}{\partial x} \vec{e}_x + \frac{\partial \cdot}{\partial y} \vec{e}_y + \frac{\partial \cdot}{\partial z} \vec{e}_z \tag{10.3}$$

in which $\vec{e}_x$ is the unit vector in the direction of x-positive, etc. We see that *the divergence of a transport vector equals the net outflow per unit time and per unit volume through a closed surface*. In fact, this may be taken as the definition of the divergence of a (transport) vector.

In addition to transport, production or loss of X may take place, represented through so-called *source terms*. We denote the net production of X per unit time and per unit volume as P.

Finally, we consider the *storage* of X, i.e. the amount of X inside the control volume. The amount of X per unit volume is called the volume concentration, here denoted as c. (Wave propagation is not considered in this chapter, so there will be no confusion with the wave propagation speed c used in preceding chapters.)

Collecting the above results, we obtain the following balance equation:

$$\frac{\partial c}{\partial t} \Delta V = -\left(\nabla \cdot \vec{T}\right) \Delta V + P \Delta V \tag{10.4}$$

or

$$\frac{\partial c}{\partial t} + \nabla \cdot \vec{T} = P \tag{10.5}$$

This is the archetype of all balance equations, regardless of the nature of the transport process or the transported quantity. *For a conserved quantity, production and dissipation are zero*, i.e., $P = 0$, in which case the balance equation takes the form of a *conservation equation*:

$$\frac{\partial c}{\partial t} + \nabla \cdot \vec{T} = 0 \tag{10.6}$$

10.3 Molecular Diffusion

In common turbulent flows, the role of molecular diffusion is utterly negligible compared with effects of turbulence. Yet we start with considerations on molecular diffusion because the formulation of this process is relatively straightforward and is a model for the more complicated processes of turbulent diffusion and dispersion, which will be treated further on.

We could take any diffusive medium and any diffused quantity, but to emphasize that we deal with molecular diffusion only we will occasionally refer specifically to the conduction of heat in a solid medium, the physical process that stood at the basis of the theory of diffusion.

10.3.1 Fick's Law of Diffusion

Molecular diffusion (or conduction, in the case of heat transfer) is the net transport resulting from the thermal micro-motions of the molecules of which the medium consists, averaged over a time interval that is just long enough to smoothen the molecular variations (here we introduce the continuum approximation).

In an *isotropic* medium, the transport by molecular diffusion takes place in the direction of strongest decrease of the concentration c of the transported quantity; i.e. it is directed opposite to the gradient of c, i.e. ∇c. Because the length scale of the underlying micro-motions is quite small compared with the length scale of the spatial variations of the mean concentration (such as the flow depth), the diffusive transport rate can be assumed to be proportional to that gradient. The proportionality coefficient, called the (coefficient of) molecular diffusivity, has the dimension (length)^2/(time) (in SI units: m^2/s); it is here denoted as ϵ:

$$\vec{T} = -\epsilon\,\nabla c = -\epsilon\left(\frac{\partial c}{\partial x}\vec{e}_x + \frac{\partial c}{\partial y}\vec{e}_y + \frac{\partial c}{\partial z}\vec{e}_z\right) \tag{10.7}$$

The conservation equation for a conservative quantity, subject to molecular diffusion, can be found by substituting Eq. (10.7) into Eq. (10.6), with the result

$$\frac{\partial c}{\partial t} = \nabla\cdot(\epsilon\nabla c) \tag{10.8}$$

In the case of a homogeneous medium, ϵ is uniform in space, or $\nabla\epsilon = 0$, in which case we obtain

$$\frac{\partial c}{\partial t} = \epsilon\nabla^2 c = \epsilon\left(\frac{\partial^2 c}{\partial x^2} + \frac{\partial^2 c}{\partial y^2} + \frac{\partial^2 c}{\partial z^2}\right) \tag{10.9}$$

This is the standard *diffusion equation* in three dimensions. It is a linear, so-called parabolic partial differential equation with a constant coefficient. The operator ∇^2 is the so-called Laplace operator.

10.3.2 One-Dimensional Diffusion

In order to obtain insight into the meaning of the diffusion equation, and to get acquainted with the nature of its possible solutions, we restrict ourselves at first to one-dimensional diffusion, for which the transport takes place in the direction of x-positive or x-negative only. One can think of heat conduction along a thin metal rod. In that case, the concentration c is the amount of heat per unit length, integrated over the cross section.

For the one-dimensional case, Eq. (10.9) reduces to

$$\frac{\partial c}{\partial t} = \epsilon\frac{\partial^2 c}{\partial x^2} \tag{10.10}$$

In order to see in a qualitative sense how the concentration varies due to diffusion, we consider a short interval along the x-axis containing a maximum of c. This implies that the diffusive transport is directed outward at both end points of this interval, causing the local concentration to decrease. This can also be seen from Eq. (10.10) because at and near a local maximum, $\partial^2 c/\partial x^2 < 0$, which according to Eq. (10.10) implies a decrease in time of the local concentration. Conversely, where the concentration has a minimum, the local concentration increases. We see that diffusion reduces the spatial variations.

Solution in an Infinite Domain

To illustrate the above, we consider the archetypal example of the one-dimensional spreading of an initially concentrated amount. We present it in terms of the conduction of heat along a rod following an initial localized heating, but the result is also valid for diffusion of substances.

At time $t = 0$, the temperature is uniform everywhere, while at that instant, heat with an amount M is injected instantaneously at a point $x = x_0$. This is the initial condition. Let c be the amount of heat per unit length in excess of the original value. The two boundary conditions are that $c(x,t) \rightarrow 0$ as $x \rightarrow \pm\infty$. (This problem is mathematically the same as the flood-wave problem of Section 8.4.2.) The solution of Eq. (10.10) for this case is presented here without derivation (its validity can be verified through back-substitution):

$$c\,(x,t) = \frac{M}{\sqrt{2\pi}\,\sigma\,(t)} \exp\left(-\frac{(x-x_0)^2}{2\sigma^2(t)}\right) \equiv M\,\Phi\,(x - x_0; \sigma\,(t)) \tag{10.11}$$

in which

$$\sigma\,(t) = \sqrt{2\epsilon t} \tag{10.12}$$

It follows from these equations that at each instant the concentration varies along the x-axis as a Gaussian (bell-shaped) curve, centered at $x = x_0$. See Figure 10.2. We have introduced the symbol Φ to represent this Gaussian function in a shorthand fashion. As time goes on,

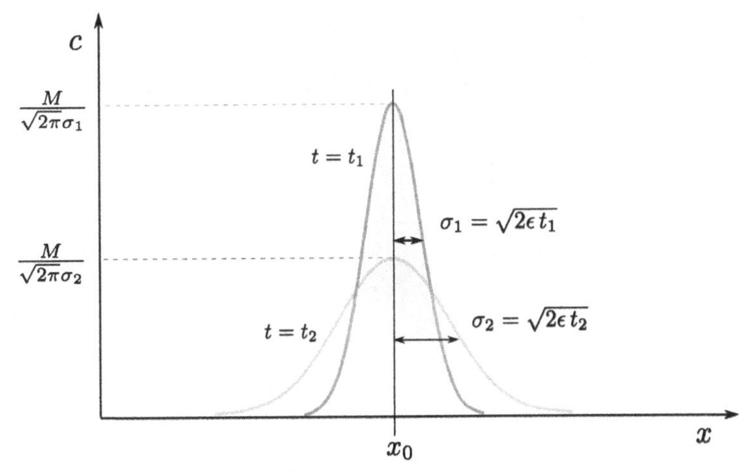

Fig. 10.2 Solution of the one-dimensional diffusion equation in an infinite domain: instantaneous injection of an amount M at the point $x = x_0$ at time $t = 0$

the peak value decreases at a rate proportional to $1/\sqrt{\epsilon t}$, while the 'standard deviation' σ, here defined as the distance of the inflection point of $c(x, t)$ to the location of the peak value, increases in proportion to $\sqrt{\epsilon t}$. Such variations are typical for all diffusion processes. Note that the product of the peak value and the width, which is a measure of the total heat content, is constant in time, in accordance with the fact that we deal with a conservative quantity.

The solution expressed in Eq. (10.11) applies to an instantaneous point injection in an infinitely long one-dimensional medium, for which the initial concentration was zero everywhere. An injection spread out in time and/or in space can be subdivided into a number of narrow local pulses of short duration. To each of these, the Gaussian solution can be applied. The overall solution is then found as the sum of these partial solutions. This is allowed because the diffusion equation (10.10) is linear. The same procedure can be followed to determine the evolution following non-uniform initial conditions (in absence of injection), by summation of a series of point sources; see Figure 10.3.

Solution in a Semi-Infinite Domain

In the case of diffusion in a finite one-dimensional domain, such as a rod, the transport should be zero at the end points. This requires that $\partial c/\partial x = 0$ at the end points. We will first see how this can be achieved for the case of instantaneous injection of an amount M at a point $x = x_0$ of a *semi-infinite domain* extending from $x = 0$ to ∞.

Imagine first that the domain extends from $-\infty$ to $+\infty$, as above, and apply the Gauss solution given by Eq. (10.11) to that extended domain. In that case, there would be an outward diffusive transport at $x = 0$ because $\partial c/\partial x > 0$ at $x = 0$; see Figure 10.4. Now fold the tail of this Gauss solution that extends into the imaginary domain $x < 0$ around $x = 0$ back into the physical domain $x > 0$. The x-derivative of this folded tail at $x = 0$ is exactly the opposite of the x-derivative of the original solution, so that the sum of these solutions has zero x-derivative there, so fulfilling the boundary condition.

As can be seen in Figure 10.4, the folded left tail of the original Gauss solution, centered at $x = x_0$, is the same as the right tail of an imaginary Gauss solution centered at $x = -x_0$, i.e. the mirror image of the original solution with respect to the impermeable boundary at

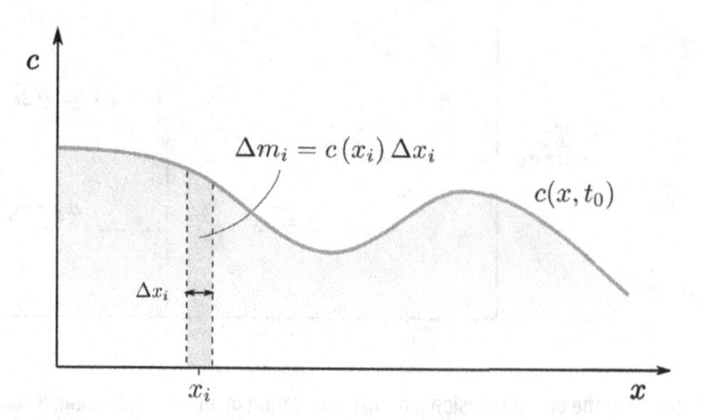

Fig. 10.3 Initial concentration distribution as a series of point sources

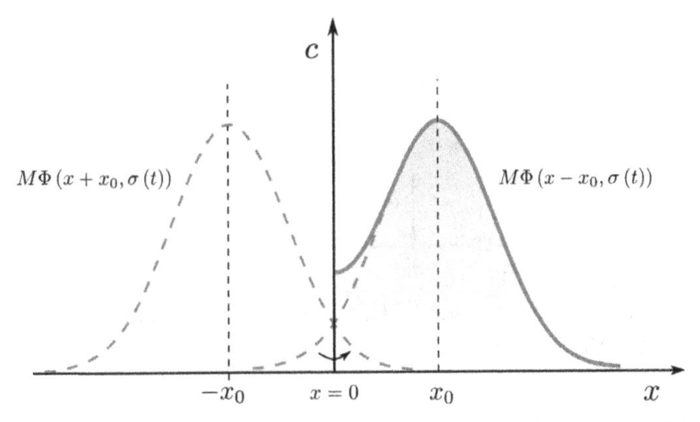

Fig. 10.4 Solution of the one-dimensional diffusion equation in the semi-infinite domain $[0; \infty)$: superposition of respective point sources at $x = x_0$ and $x = -x_0$

$x = 0$. Therefore, we can express the total solution analytically, in short-hand fashion using the Φ-notation, as

$$c(x,t) = M(\Phi(x - x_0; \sigma) + \Phi(x + x_0; \sigma)) \qquad \text{for } x \geq 0 \tag{10.13}$$

(Note that $\Phi(x - x_0; \sigma)$ is centered at $x = x_0$, i.e. where its argument $(x - x_0)$ is zero. Likewise, $\Phi(x + x_0; \sigma)$ is centered where $x + x_0 = 0$, i.e. at $x = -x_0$.) For injection at $x = 0$, the solution is

$$c(x,t) = 2M\,\Phi(x; \sigma) \qquad \text{for } x \geq 0 \tag{10.14}$$

In this case, exactly half of a Gaussian function is folded back into the semi-infinite physical domain. The solution in the physical domain is the same as it is for the injection of the double amount in an infinite domain.

Solution in a Finite Domain

Next, we consider the instantaneous release of an amount M at a point $x = x_0$ of a *finite domain*, extending from $x = 0$ to $x = \ell$, say. In this case, the tails of the original distribution extend across both boundaries; thus mirroring at both boundaries is necessary. Moreover, the tails of the mirrored functions in their turn extend across both boundaries, requiring that these be mirrored too, and so on; see Figure 10.5. For the special case of injection in $x = 0$, this can be expressed mathematically as follows:

$$c(x,t) = 2M\left(\Phi(x; \sigma) + \Phi(x - 2\ell; \sigma) + \Phi(x + 2\ell; \sigma) + \Phi(x - 4\ell; \sigma) + \cdots\right) \tag{10.15}$$

One can think of this succession of terms as representing the effect of repeated inward folding of the tails of the original Gauss function at the boundaries, and adding the resulting concentrations at each point.

Shortly after the injection, the original Gauss curve $\Phi(x; \sigma)$, with its maximum at $x = 0$, is still highly peaked, in the sense that $\sigma(t) \ll \ell$. In other words, it extends over only a relatively small distance into the physical domain $0 \leq x \leq \ell$, and the influence of the

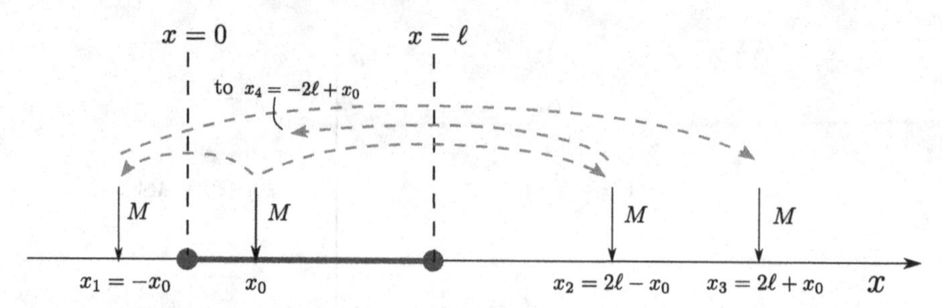

Fig. 10.5 Solution of the one-dimensional diffusion equation in a finite domain: superposition of sources for injection of an amount M at the point $x = x_0$

boundary at $x = \ell$ is hardly felt, so that mirroring (or folding) there is not yet necessary; the first term between the parentheses in the right-hand side of Eq. (10.15) suffices initially.

As time goes on, the Gauss curve collapses and spreads, with $\sigma(t)$ increasing in proportion to $\sqrt{t}$. In other words, the tails become longer so that after some time, mirroring at $x = \ell$ becomes necessary, with an imaginary source at $x = 2\ell$ (the second Φ-term in Eq. (10.15)). After an even longer time, remirroring at $x = 0$ becomes necessary as well (the third Φ-term, centered in $x = -2\ell$). In other words, as time goes on we have to take more and more terms of Eq. (10.15) into account. At the same time, the total solution becomes more and more nearly uniform on the entire interval $0 \leq x \leq \ell$.

In fact, using Eq. (10.15), it can be verified that the distribution varies by less than 3% between the two ends when $\sigma(t) = \ell$, or after a time $t = \ell^2/2\epsilon$. This applies to the case of injection at one of the two end points. If the injection takes place halfway between the two end points, such nearly uniform lateral distribution is reached when $\sigma(t) = \ell/2$, or after a time $t = \ell^2/8\epsilon$. Ultimately, the distribution tends to a uniform value given by $c_\infty = M/\ell$.

10.3.3 The Random Walk Model

The fact that an initially concentrated amount diffuses as a collapsing Gaussian function, with standard deviation proportional to $\sqrt{t}$, can also be shown and understood by consideration of the underlying random molecular motions. To do this, we use the model of the so-called *random walk*, here presented for the one-dimensional case.

Consider a particle that makes a sequence of a large number of N stochastically independent random displacements along the x-axis, written as $\Delta\underline{x}_i$, $i = 1, 2, ..., N$, starting at $x = 0$ (the underscore denotes a random variable). The particle location after N steps is given by

$$\underline{x}_N = \sum_{i=1}^{N} \Delta\underline{x}_i \tag{10.16}$$

Since $\underline{x}_N$ is the sum of a large number of stochastically independent contributions, the central limit theorem of probability theory applies to it, implying that $\underline{x}_N$ is Gaussian distributed in the limit as $N \to \infty$.

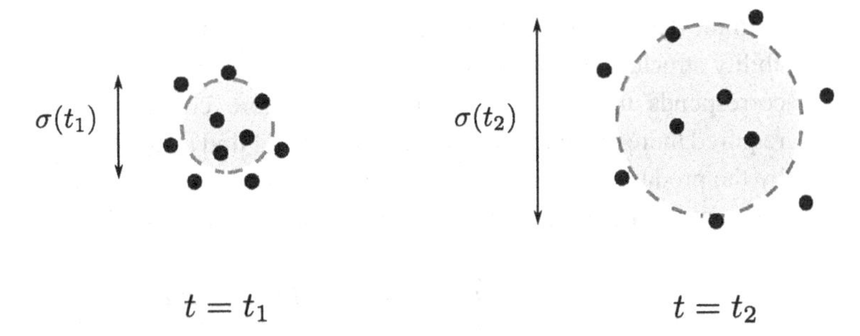

Fig. 10.6 Diffusion: discrete versus continuum approach

To see the link of the above with the diffusion process, imagine that a very large number of particles is released at $x = 0$, each subsequently making a random walk as described above. It then follows that these particles will be distributed along the x-axis according to the Gaussian density function, in accordance with Eq. (10.11). (See Figure 10.6 for a schematic representation in two dimensions.)

The random walk model can be elaborated further in order to express the resulting diffusivity in terms of the characteristics of the underlying micro-scale motions. For simplicity, we assume that all steps have the same length, given by $\Delta\ell$, and that only the direction in which each step is made is random, with a 50% chance of the particle going to the right and a 50% chance of it going to the left, independent of the preceding steps. With these definitions, the mean value of the displacement $\Delta \underline{x}_i$ is zero for all i, and the standard deviation is $\sigma_{\Delta \underline{x}_i} = \Delta\ell$ for all i.

The mean value of the sum $\underline{x}_N$ equals the sum of the mean values of its contributions, i.e. zero, which also follows directly from symmetry. Furthermore, since the displacements are stochastically independent, the variance of their sum is the sum of their variances, or

$$\sigma^2_{\underline{x}_N} = N(\Delta\ell)^2 \tag{10.17}$$

(This quantity is to remain finite as $N \to \infty$, so $\Delta\ell \sim 1/\sqrt{N}$.) We now introduce time and assume that successive steps are made at equal time intervals Δt, with the same average velocity U, so $\Delta\ell = U\Delta t$. Let this expression replace one of the factors $\Delta\ell$ in Eq. (10.17) to obtain

$$\sigma^2_{\underline{x}_N} = N(\Delta\ell)\,(U\Delta t) = (U\Delta\ell)\; N\Delta t = (U\Delta\ell)\; t \tag{10.18}$$

in which $t = N\Delta t$ is the elapsed time since the injection of the particles at $x = 0$. We reproduce here the important result that the particles spread in proportion to $\sqrt{t}$, as in the diffusion model. Furthermore, we can now express the diffusivity in terms of the characteristic velocity and displacement of the underlying random motions by comparing Eqs. (10.18) and (10.12), with the result

$$\epsilon_{\text{rw}} = \tfrac{1}{2} U\Delta\ell \tag{10.19}$$

in which the subscript 'rw' on ϵ indicates that it is the diffusivity resulting from the random walk model.

In the derivation of Eq. (10.19), we have made simplifying assumptions about the probability structure, the time interval between successive steps and the step size. Eq. (10.19) corresponds to these particular choices. More complicated assumptions would have required more algebra, but in the end we would find again that the diffusivity is determined by the product of a characteristic velocity and a characteristic displacement of the underlying micro-motions. We will find a similar result for turbulent diffusion (Section 10.5).

10.3.4 Two-Dimensional Diffusion

For diffusive transport in a two-dimensional (x, y) isotropic and homogeneous medium, Eq. (10.9) reduces to

$$\frac{\partial c}{\partial t} = \epsilon \left(\frac{\partial^2 c}{\partial x^2} + \frac{\partial^2 c}{\partial y^2} \right) \tag{10.20}$$

Here, the meaning of c is that of an amount per unit area (for heat conduction in a thin plate, it is the amount of heat per unit area integrated over the plate thickness). In the case of an injection of an amount M at time $t = 0$ at a point $x = 0, y = 0$ of an infinite domain, with similar initial and boundary conditions as above for the one-dimensional case, the solution is the two-dimensional Gaussian density function (a bell shape), which can be written as

$$c(x, y, t) = \frac{M}{2\pi\sigma^2} \exp\left(-\frac{x^2 + y^2}{2\sigma^2} \right) = M\,\Phi\,(x; \sigma)\;\Phi\,(y; \sigma) \tag{10.21}$$

in which $\sigma = \sqrt{2\epsilon t}$. For this isotropic case, the concentration distribution is axially symmetric, with circular iso-concentration contours ($x^2 + y^2$ = constant), centered on the point of injection. The radial standard deviation increases in proportion to $\sqrt{t}$, so that the injected amount is spread over an area that increases in proportion to t. The peak of this bell-shaped concentration distribution decreases proportional to $1/t$, so the volume of the bell is constant, as it should be.

The solution for two-dimensional diffusion from a point injection in a finite domain can be constructed similar to the one-dimensional case, i.e. by using the Gauss solution for diffusion in an infinite domain as the building block, mirrored at the boundaries so as to fulfill the boundary condition of zero transport. Non-uniform initial conditions and injections spread out in time and/or space can likewise be dealt with similar to the one-dimensional case.

10.4 Advection and Molecular Diffusion

Our next subject is transport by a combination of advection and molecular diffusion. We repeat that molecular diffusion is irrelevant in civil engineering practice, where turbulent diffusion and dispersion are dominant, but we treat it here nevertheless because of the straightforward formulation and the qualitative similarity to the processes just mentioned.

Advection is by definition the transport along with the mass of a moving fluid. Therefore, the advection per unit area, of the quantity being considered with volume concentration c, is given by $c\vec{u}$, in which $\vec{u}$ is the fluid particle velocity. Together with the transport due to molecular diffusion $(-\epsilon\nabla c)$, the total transport is

$$\vec{T} = c\vec{u} - \epsilon\nabla c \tag{10.22}$$

Substituting this into the generic conservation equation, Eq. (10.6), we find

$$\frac{\partial c}{\partial t} + \nabla \cdot (c\vec{u}) - \nabla \cdot (\epsilon\nabla c) = 0 \tag{10.23}$$

For an incompressible $(\nabla \cdot \vec{u} = 0)$ and homogeneous $(\nabla\epsilon = 0)$ fluid, it reduces to

$$\frac{\partial c}{\partial t} + \vec{u} \cdot \nabla c - \epsilon\nabla^2 c = 0 \tag{10.24}$$

This is the standard *advection–diffusion equation*. The sum of the first two terms is the total derivative of c, following the fluid. In other words, it is the material derivative. Using the corresponding operator $D/Dt = \partial/\partial t + \vec{u} \cdot \nabla$, Eq. (10.24) can be written as

$$\frac{Dc}{Dt} = \epsilon\nabla^2 c \tag{10.25}$$

This has a structure that is analogous to that of the standard diffusion equation without advection, given in Eq. (10.9). The difference is that Eq. (10.25) describes the effect of diffusion in a frame of reference moving with the fluid.

For one-dimensional transport in x-space, Eq. (10.24) reduces to

$$\frac{\partial c}{\partial t} + u_x\frac{\partial c}{\partial x} - \epsilon\frac{\partial^2 c}{\partial x^2} = 0. \tag{10.26}$$

(Note the correspondence of this equation with the advection–diffusion equation (8.17) that appeared in the context of flood waves in rivers; Chapter 8.) The solution of Eq. (10.26) for the case of an instantaneous release of an amount M at time $t = 0$ at a point $x = 0$, with the boundary conditions $c \to 0$ as $x \to \pm\infty$, and assuming that u_x and ϵ are constant, is

$$c\,(x,t) = M\Phi\left(x - u_x t; \sqrt{2\epsilon t}\right) \tag{10.27}$$

At each instant, this represents a Gauss function in x with its maximum where $x = u_x t$; see Figure 10.7. The effect of advection is to *displace* the total amount, whereas the effect of diffusion is to *spread* this as it moves. In a fixed position, c does not vary in time as a Gauss function because σ increases in time, which causes asymmetry. The rise in c occurs at a higher rate than its fall, as we have seen for flood waves in rivers (Chapter 8). Also, the maximum value of c at a fixed point occurs somewhat ahead of $t = x/u_x$. However, the asymmetry decreases with increasing distance from the point of injection.

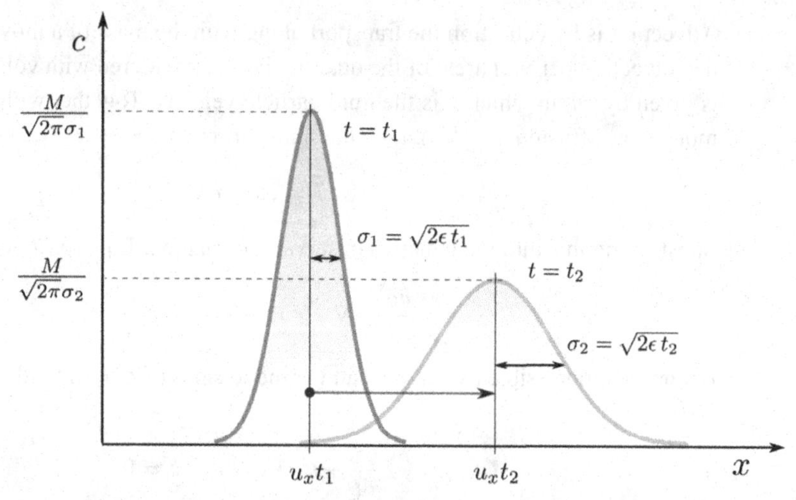

Fig. 10.7 Solution of the one-dimensional advection–diffusion equation: injection of an amount M at $x = 0$ at time $t = 0$

10.5 Turbulent Diffusion

Fluid turbulence is characterized by irregularly fluctuating, eddying motions of 'parcels' of fluid. The particle velocity varies randomly in magnitude and direction. The same is true for the concentration of any substance, or heat, that is carried along with the fluid.

We have seen above that molecular diffusion is the time-averaged effect of random underlying molecular motions. In like manner, turbulent diffusion is the time-averaged effect of random turbulent displacements of fluid parcels.

10.5.1 Reynolds Averaging

It is not possible to describe the irregular turbulent motions in a deterministic manner. However, a description in terms of averages is feasible, and at the same time sufficient for practical applications. This method was introduced by Reynolds (1895). It forms the basis of the modelling of virtually all kinds of turbulent flows.

The key element in Reynolds' approach is to separate the instantaneous value of the fluctuating quantities from their average values. The averages are taken over a time interval long enough to cancel the effects of the turbulent fluctuations on the outcome, yet short enough that we can describe relevant variations on a longer time scale, e.g. those due to unsteady boundary conditions. (On this longer time scale, turbulence-averaged quantities may be unsteady, but we will not indicate such time dependence because it is irrelevant in the present context.) The resulting averages are indicated with an overbar and the fluctuations around them by a prime, as in

$$c(t) = \overline{c} + c'(t) \tag{10.28}$$

It follows that $\overline{c'(t)} = 0$.

While the mean of each individual turbulent fluctuating quantity is zero, this does not in general apply to the mean of the product of two fluctuating quantities. To see this, consider the example of the vertical transport of a dissolved substance like salt, for the case of a continuously stratified flow with –on average– saltier water near the bottom than higher in the fluid column, due to gravity ($\partial\bar{c}/\partial z < 0$).

Neglecting molecular diffusion, the instantaneous upward transport of salt per unit area can be written as $T_z = cu_z$. Taking the average, we obtain

$$\overline{T_z} = \overline{c\,u_z} = \overline{(\bar{c} + c')\left(\bar{u}_z + u_z'\right)} = \overline{\bar{c}\,\bar{u}_z} + \overline{\bar{c}\,u_z'} + \overline{c'\,\bar{u}_z} + \overline{c'u_z'} \tag{10.29}$$

Since $\overline{u'}_z = 0$ and $\overline{c'} = 0$, this implies that

$$\overline{T_z} = \overline{cu_z} = \bar{c}\;\bar{u}_z + \overline{c'u_z'} \tag{10.30}$$

The transport by the averaged motion is given by $\bar{c}\,\bar{u}_z$. We see that the turbulence adds to this in the amount $\overline{c'u_z'}$. Even though the averages of c' and of u_z' are zero, this does not hold for their product if they are correlated.

Since the time-averaged salt concentration decreases upward due to gravity, an upward moving fluid parcel ($u_z' > 0$) will advect salt upward with a concentration that *on average* is higher than the local mean value, so $c' > 0$, while a downward velocity fluctuation ($u_z' < 0$) *on average* advects salt downward with a lower-than-average concentration ($c' < 0$). It follows that $\overline{c'u_z'} > 0$, as a consequence of the fact that $\partial\bar{c}/\partial z < 0$. We see that *the turbulent fluctuations cause a mean transport in the direction of decreasing values of the mean concentration, as in a diffusion process.* This is an important result.

10.5.2 Closure Hypothesis

In order to be able to calculate the net turbulent transport, additional equations are required for the mean product of fluctuating quantities. However, these introduce new unknowns. Somehow, this sequence must be cut off by a so-called *closure hypothesis*, expressing a link between the relevant turbulence properties and the mean motion. Such hypothesis was introduced by Prandtl (1925) with the concept of the *mixing length.* The idea behind it is that local properties are advected with the turbulent eddies over some distance, and so cause deviations from the average state in the new locality in the case of a mean gradient. These deviations (fluctuations) are estimated in order of magnitude as the change of the average concentration ($\bar{c}\,(z)$) over a distance equal to the characteristic displacement of the eddies, called the mixing length ℓ_t. The velocity fluctuations are estimated as a characteristic particle velocity of the turbulent eddies (u_t):

$$c' \sim \ell_t\,\frac{\partial\bar{c}}{\partial z}, \qquad u' \sim u_t \quad \text{and} \qquad \overline{c'u_z'} \sim u_t\,\ell_t\,\frac{\partial\bar{c}}{\partial z} \tag{10.31}$$

In this formulation, the average transport due to the turbulence is proportional to the gradient of the mean concentration, as it is in molecular diffusion. Thus, *the average turbulent transport can be described as a diffusion process.* The proportionality factor is called the *turbulence diffusivity*, written as ϵ_t:

$$\overline{c'u'_z} = -\epsilon_t \frac{\partial \bar{c}}{\partial z} \tag{10.32}$$

with which the total averaged transport (neglecting molecular diffusion) becomes

$$\overline{T}_z = \overline{cu}_z = \bar{c}\,\bar{u}_z - \epsilon_t \frac{\partial \bar{c}}{\partial z} \tag{10.33}$$

The turbulence diffusivity is a new unknown, in order of magnitude given by $\epsilon_t \sim u_t\,\ell_t$. (Notice the similarity with the diffusivity resulting from the random walk model.) For each turbulent-flow problem, ℓ_t and u_t have to be estimated or calculated.

10.6 Vertical Diffusion in Free-Surface Flows

10.6.1 Turbulence Diffusivity

The so-called turbulence diffusivity ϵ_t is a flow property (not a fluid property, as ϵ), therefore not known beforehand. Its value can be estimated only by rough approximation, using the notions of mixing length ℓ_t and eddy velocity u_t according to the order-of-magnitude relationship $\epsilon_t \sim \ell_t u_t$.

Free-surface flows are boundary-layer flows. The interaction with the bed is crucial for the generation of flow turbulence and determines its properties. Near the bed, the mixing length is constrained vertically by the proximity of the bed, but less so with increasing height above it, denoted as z (we take $z = 0$ at the bed). By approximation, we will assume that it increases in proportion: $\ell_t \sim z$. However, this does not hold over the entire depth (d) because the free surface damps the eddying motions vertically, reducing them to nearly zero because gravity counteracts the tendency of eddies to rise above the mean free surface. This effect is likewise represented as a proportional increase of ℓ_t with the distance below the free surface. Taken together, we assume

$$\ell_t \sim z\left(1 - \frac{z}{d}\right) \tag{10.34}$$

This is a parabolic variation of ℓ_t with z. Because the turbulence arises as a result of bed shear stress (τ_b), we assume (confirmed by measurements) that the characteristic eddy particle velocity (u_t) is of the order of magnitude of the so-called shear velocity u_*, defined as $u_* \equiv \sqrt{\tau_b/\rho}$.

Altogether, we express these order-of-magnitude relations as an algebraic equality for ϵ_t, with a proportionality coefficient κ (named after Von Karman):

$$\epsilon_t = \kappa u_* z\left(1 - \frac{z}{d}\right) \tag{10.35}$$

The value of κ has empirically been determined as about 0.4. The parabolic variation of ϵ_t over the vertical implies that its maximum value occurs at mid-depth, with the value $\frac{1}{4}\kappa u_* d$, or about $0.1\, u_* d$. The depth-averaged value is $\frac{1}{6}\kappa u_* d$, or about $0.07\, u_* d$.

10.6.2 Vertical Distribution of Horizontal Velocity

The preceding result is now applied to determine the vertical distribution of the mean streamwise particle velocity in a horizontally uniform flow. The approach is based on the modelling of the turbulent vertical mixing of streamwise momentum in combination with an equilibrium condition.

An upward displacement of a fluid parcel, with velocity u_z, implies an upward transport of streamwise momentum (ρu_x per unit volume) at a rate given by $u_z(\rho u_x)$ or $\rho u_x u_z$ per unit horizontal area. Since $\overline{u}_z = 0$ in a horizontally uniform flow, the average of this is $\rho \overline{u'_x u'_z}$. Because $\overline{u}_z$ decreases downward, due to bed resistance, $\rho \overline{u'_x u'_z} < 0$, implying a net downward transport of forward momentum, equivalent to a shear stress, the so-called Reynolds shear stress τ_{xz}. In line with Eqs. (10.31) and (10.32), with c replaced by ρu_x, i.e. the x-momentum per unit volume, it is modelled as a diffusion process:

$$\tau_{xz} = -\rho \overline{u'_x u'_z} = \rho \nu_t \frac{\partial \overline{u}_x}{\partial z} \tag{10.36}$$

Here, ν_t is the so-called *eddy viscosity*, i.e. the turbulence diffusivity for momentum exchange. It is modelled in the same manner as ϵ_t in the preceding section. For simplicity, their values are here set equal to each other: $\nu_t = \epsilon_t$, given in Eq. (10.35).

The condition of equilibrium in a steady, uniform free-surface flow implies that the mean shear stress τ_{xz} varies linearly from the value at the bed, τ_b, to zero at the free surface (see Figure 10.8). Neglecting the viscous contribution, we obtain

$$\tau_{xz} = \tau_b \left(1 - \frac{z}{d}\right) = \rho \nu_t \frac{\partial \overline{u}_x}{\partial z} = \rho \kappa u_* z \left(1 - \frac{z}{d}\right) \frac{\partial \overline{u}_x}{\partial z} \tag{10.37}$$

Substituting $\tau_b = \rho u_*^2$ and dividing by the common factor $\rho u_* \,(1 - z/d)$ yields

$$\frac{\partial \overline{u}_x}{\partial z} = \frac{u_*}{\kappa} \frac{1}{z} \tag{10.38}$$

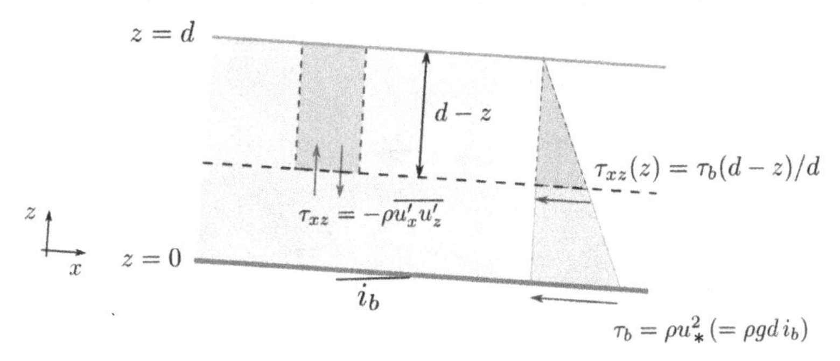

Fig. 10.8 Vertical variation of the mean shear stress (τ_{xz}) in a uniform free-surface flow (flow direction is from left to right)

Integration yields the well-known logarithmic profile for the mean horizontal particle velocity:

$$\overline{u}_x(z) = \frac{u_*}{\kappa} \ln \frac{z}{z_0} \tag{10.39}$$

The details of the flow very near the bed are not resolved by this model. Their effect on the mean velocity profile is expressed through the height z_0 at which theoretically, i.e. according to the (extrapolated) logarithmic profile, the mean velocity vanishes. The logarithmic velocity distribution has been elaborated in Chapter 9 for the derivation of a theoretical expression for the resistance coefficient c_f.

10.7 Horizontal Transport in Free-Surface Flows

10.7.1 Introduction

A substance or heat that is released in flowing water is advected downstream and at the same time spread or mixed by turbulence. Rivers and canals usually have a width (B) that is far greater than the depth, while the width in turn is usually small compared with the downstream distances that are of interest in practice. Therefore, we can distinguish three zones with respect to the advection and mixing from a point source (Figure 10.9):

The near field (I), extending downstream from the source over roughly fifty times the depth, in which the mixing is *three-dimensional* (3D), though dominated by the vertical transport; at the downstream end of this zone, complete mixing over the vertical has taken place.

The intermediate field (II), an intermediate region that is (very) long compared with the width, in which the released quantity, already more or less uniformly distributed over the depth as a result of vertical diffusion in zone I, is gradually spread laterally by turbulent mixing while it is advected downstream, as in a plume of smoke issuing from a smokestack, until it is virtually uniformly distributed over the entire cross section. The transport processes in this region are described with depth-integrated, *two-dimensional* models in the longitudinal and lateral coordinates (2DH).

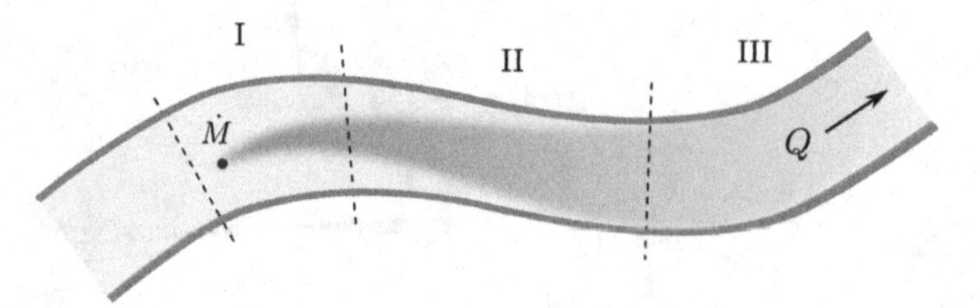

Fig. 10.9 Horizontal mixing of a substance released from a point source in a river (in reality the plume width would increase at a slower rate than indicated)

The far field (III), an end zone in which the released quantity is virtually uniformly distributed over the cross section as a result of the vertical and lateral transports in zones I and II. Therefore, in zone III only longitudinal advection and mixing take place, to be described with depth- and width-integrated, *one-dimensional* models in the longitudinal coordinate (1DH).

In the following, we ignore the near field, because it is relatively short, and deal exclusively with zones II and III, which in rivers may easily extend over tenths to hundreds of kilometres, respectively. Furthermore, from here on we deal exclusively with turbulence-averaged quantities without expressing this all the time. In line with this, averaged turbulent quantities are written without an overbar. Instead, a single overbar will indicate a depth average (of turbulence-averaged quantities), and a double overbar a cross-sectional average, as in

$$\overline{c} \equiv \frac{1}{d}\int_d c\,\mathrm{d}z \quad \text{and} \quad \overline{\overline{c}} \equiv \frac{1}{A}\int_A c\,\mathrm{d}A \tag{10.40}$$

A local deviation from these averages is denoted by a prime.

10.7.2 Two-Dimensional Horizontal Transport

We integrate the amount of transported matter per unit volume, and its horizontal transport per unit area at a point of the vertical, $\vec{T}$, over the depth. This gives

$$\int_d c\,\mathrm{d}z = \overline{c}\,d \tag{10.41}$$

and

$$\int_d \vec{T}\,\mathrm{d}z = \overline{\vec{T}}d \tag{10.42}$$

With these definitions, the conservation equation for the amount of transported matter in a control volume in the form of a vertical prism with unit horizontal area, extending over the entire depth, reads

$$\frac{\partial}{\partial t}(\overline{c}d) + \nabla\cdot\left(\overline{\vec{T}}d\right) = 0 \tag{10.43}$$

Written in component form, this is

$$\frac{\partial}{\partial t}(\overline{c}d) + \frac{\partial}{\partial x}\left(\overline{T}_x d\right) + \frac{\partial}{\partial y}\left(\overline{T}_y d\right) = 0 \tag{10.44}$$

So far, we have made no restriction on the orientation of the (x, y)-axes in relation to the orientation of the conduit or the flow field. However, we deal with long, relatively narrow conduits for which it is meaningful to distinguish longitudinal transports and lateral transports. To this end, we let the x-axis from here on locally point downstream and the y-axis normal to it. With this convention, u_x is the streamwise component of the turbulence-averaged particle velocity at a point, $\overline{u}_x$ its depth average value, and $\overline{\overline{u}}_x$ its cross-sectionally averaged value, equal to $U = Q/A$. Furthermore, $\overline{T}_x d$ is the longitudinal transport per unit width, and $\overline{T}_y d$ is the lateral transport per unit length.

Longitudinal Transport

The longitudinal transport at a point (x, y) consists in part of *advection* with the local mean longitudinal velocity at that point, $c\,u_x$. Because the mean velocity at a point (u_x) varies vertically, the released matter that at some instant is present in a vertical is advected downstream with different velocities, thus being spread longitudinally (Figure 10.10). It is subsequently continually mixed over the vertical by turbulent diffusion. The net effect is a longitudinal spreading of the depth-integrated amount of transported substance, called *longitudinal dispersion*.

Note that this process does not need to be described explicitly in three-dimensional modelling of the transport because it is automatically represented therein. It arises in the process of depth-integration of the transport, so reducing the dimensionality of the formulation from three (x, y, z) to two (x, y). It comes in addition to the advection with the depth-averaged velocity $\overline{u}_x$, and should therefore be accounted for as a separate contribution in depth-integrated models.

Expressed mathematically, and omitting the turbulent diffusion for the time being, we have

$$\overline{T}_x\, d = \int_0^d c u_x \,\mathrm{d}z = \overline{c u_x}\, d = \left(\overline{c}\,\overline{u}_x + \overline{c' u'_x}\right) d = \overline{c}\, q_x + \overline{c' u'_x}\, d \tag{10.45}$$

in which $q_x = \overline{u}_x d$ is the discharge of water per unit width.

The last term in (10.45) represents the transport due to longitudinal dispersion. (Remember that a prime here means a deviation of a point value from the depth-averaged value, not a turbulent fluctuation.) Mathematically, it can be expressed as a longitudinal diffusion process, similar to turbulent diffusion, as in

$$\overline{c' u'_x} = -K_x \frac{\partial \overline{c}}{\partial x} \tag{10.46}$$

in which K_x is the coefficient of longitudinal dispersion. Its value has been derived by Elder (1959) from the parabolic distribution of the turbulent eddy viscosity given in Eq. (10.35), and the corresponding logarithmic velocity profile given in Eq. (10.39), with the result

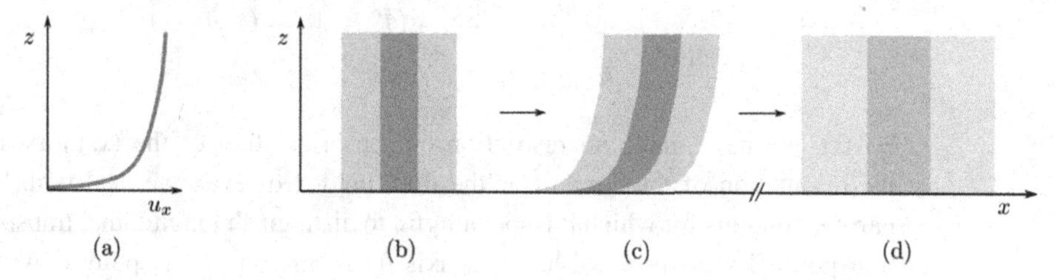

Fig. 10.10 Longitudinal dispersion: vertical velocity profile (a), initial concentration distribution (b), advection by local mean velocity (c) and vertical mixing by turbulent diffusion (d)

$$K_x \cong 6\, u_* d \tag{10.47}$$

This is a factor of about 10^2 larger than the depth-averaged turbulent diffusivity, which is about $0.07\, u_* d$. The effect of turbulent diffusion is qualitatively similar to that of longitudinal dispersion, but we need not take it into account separately because it is relatively small in a quantitative sense. We could say that it is included in the value of K_x given by Eq. (10.47).

Using Eq. (10.46), the total longitudinal transport per unit width and unit time, given in Eq. (10.45), can be written as

$$\overline{T}_x\, d = q_x\, \overline{c} - K_x\, d\, \frac{\partial \overline{c}}{\partial x} \tag{10.48}$$

Lateral Transport

In principle, the same considerations as those given above apply to the lateral transport, which therefore is expressed as

$$\overline{T}_y\, d = q_y\, \overline{c} - K_y\, d\, \frac{\partial \overline{c}}{\partial y} \tag{10.49}$$

Yet there are important differences because the x-axis is directed downstream and the y-axis is transverse to that. The flow velocities in the two directions are therefore quite different in magnitude. In fact, in steady flow, when there is no net storage taking place (no net flow towards or away from the river bank), q_y must be zero. Nevertheless, a net lateral advection can occur locally in the vertical because of secondary, spiralling currents, i.e. nonzero local average lateral velocities with zero depth average.

Secondary currents occur particularly in river bends, where the upper layers have an outward velocity and the lower layers an inward velocity. Therefore, the transported matter that at some instant is present in a vertical is being spread laterally. This effect, which manifests itself when considering the depth-integrated quantities, is called *lateral dispersion*. It comes in addition to the advection with the depth-averaged lateral velocity $\overline{u}_y$ (which is quite small or even zero), and should therefore be accounted for as a separate contribution.

Like the longitudinal dispersion, the lateral dispersion is expressed mathematically as a diffusion process, as in Eq. (10.49), in which K_y is the coefficient of lateral dispersion. Its value depends on the specific circumstances generating the secondary currents, such as river bends or lateral depth variations. Therefore, there is no generally valid expression for its value, unlike the coefficient of longitudinal dispersion. A very rough estimate for natural water courses with lateral depth variations and bends is (see Rutherford, 1994)

$$K_{y,\text{disp}} \cong 0.6\, u_* d \pm 50\% \tag{10.50}$$

This includes the effect of turbulent diffusion, which is a factor of about 10 smaller.

In the more or less academic special case of a laterally uniform flow, i.e. no lateral variations of the turbulence-averaged velocities or the depth, there are no secondary currents and therefore there is no lateral dispersion. The lateral transport in this case consists purely of *lateral turbulent diffusion*. The corresponding diffusivity is of the same order of magnitude as that of the vertical turbulence diffusivity, but about twice as large because the lateral turbulent displacements are not forced to (almost) zero at the bed and at the free surface: $K_{y,\text{diff}} \cong 0.15\, u_* d$ (Rutherford, 1994).

Two-Dimensional Advection–Diffusion Equation

Collecting the preceding results, the conservation equation (10.44) takes the following form:

$$\frac{\partial}{\partial t}(d\bar{c}) + \frac{\partial}{\partial x}\left(q_x\bar{c} - K_x d\frac{\partial \bar{c}}{\partial x}\right) + \frac{\partial}{\partial y}\left(q_y\bar{c} - K_y d\frac{\partial \bar{c}}{\partial y}\right) = 0 \qquad (10.51)$$

The depth-integrated continuity equation, expressing conservation of the volume of water, is

$$\frac{\partial d}{\partial t} + \frac{\partial q_x}{\partial x} + \frac{\partial q_y}{\partial y} = 0 \qquad (10.52)$$

We substitute this into the expanded Eq. (10.51), neglect the depth-integrated lateral velocity, i.e. q_y, and write $q_x = Ud$ instead of $\bar{u}_x d$. This yields

$$\frac{\partial \bar{c}}{\partial t} + U\frac{\partial \bar{c}}{\partial x} - \frac{1}{d}\frac{\partial}{\partial x}\left(K_x d\frac{\partial \bar{c}}{\partial x}\right) - \frac{1}{d}\frac{\partial}{\partial y}\left(K_y d\frac{\partial \bar{c}}{\partial y}\right) = 0 \qquad (10.53)$$

This has the appearance of the advection–diffusion equation, expressing the effects of longitudinal transport with the depth-averaged velocity (second term), supplemented with two terms representing the effects of longitudinal and lateral dispersion due to velocity variations over the vertical. The sum of the first two terms is the total derivative of $\bar{c}$ for an observer moving downstream with the depth-average velocity U.

In general, the parameters U, d, K_x and K_y vary with (x, y, t), such that numerical integration of Eq. (10.53) is required to find a solution for $\bar{c}\,(x, y, t)$. For simplicity, we restrict ourselves here to steady, horizontally uniform flows, in which U, d, K_x and K_y are constants, in which case Eq. (10.53) reduces to

$$\frac{\partial \bar{c}}{\partial t} + U\frac{\partial \bar{c}}{\partial x} - K_x\frac{\partial^2 \bar{c}}{\partial x^2} - K_y\frac{\partial^2 \bar{c}}{\partial y^2} = 0 \qquad (10.54)$$

This is the two-dimensional, non-isotropic counterpart of Eq. (10.24). Elementary solutions of this advection–diffusion equation for the concentration resulting from a point source take the form of Gauss functions in x and in y, which gradually spread in proportion to $\sqrt{t}$ and collapse in proportion to $1/t$ while being advected downstream with velocity U. We present two examples.

Case 1: Instantaneous Release at a Point

Suppose that at time $t = 0$ an amount M of some substance is released at a point $x = 0$, $y = 0$ in a horizontally uniform flow of infinite extent, and instantaneously distributed uniformly over the depth. This implies that zone II starts at the point of release. The concentration is initially zero everywhere, and it remains so at infinity for all times. The solution is analogous to that of Eq. (10.20) given in Eq. (10.21):

$$c(x,y,t) = \frac{M/d}{2\pi\sigma_x\sigma_y}\exp\left(-\frac{(x-Ut)^2}{2\sigma_x^2} - \frac{y^2}{2\sigma_y^2}\right) \tag{10.55}$$

or

$$c(x,y,t) = \frac{M}{d}\,\Phi\left(x-Ut;\sigma_x\right)\,\Phi\left(y;\sigma_y\right) \tag{10.56}$$

in which

$$\sigma_x = \sqrt{2K_x t} \quad \text{and} \quad \sigma_y = \sqrt{2K_y t} \tag{10.57}$$

The solution in Eq. (10.55) represents the concentration in an elongated cloud of the transported substance that is advected downstream with velocity U. It has the shape of a Gaussian hat with ellipsoidal iso-concentration contours ($\sigma_x > \sigma_y$ because $K_x > K_y$). At any instant, the concentration varies longitudinally (x) and laterally (y) as a Gauss curve, with its peak at $x = Ut, y = 0$, spreading in these directions in proportion to $\sqrt{2K_x t}$ and $\sqrt{2K_y t}$, respectively, due to dispersion.

If the flow is bounded at one impermeable bank, e.g. at $y = 0$, we have the boundary condition $\overline{T}_y = 0$ at $y = 0$, or $\partial\overline{c}/\partial y = 0$ at $y = 0$. The solution given in Eqs. (10.55) or (10.56) no longer suffices because it does not fulfill this boundary condition. As before, a valid solution can be obtained by mirroring this solution at $y = 0$ and adding the results. For a point source at a distance y_0 from the bank, we obtain

$$c(x,y,t) = \frac{M}{d}\,\Phi\left(x-Ut;\sigma_x\right)\left(\Phi\left(y-y_0;\sigma_y\right) + \Phi\left(y+y_0;\sigma_y\right)\right) \tag{10.58}$$

In the case of two parallel banks, the solution that is valid in the infinite domain must repeatedly be mirrored at both banks, as explained in Section 10.3. As time goes on, the distribution occupies virtually the entire width B with a nearly uniform lateral concentration distribution. When the lateral distribution has become more or less uniform, let us say when σ_y has reached the value B, the longitudinal dispersion has given rise to a σ_x-value of $B\sqrt{K_x/K_y}$, or about $3B$ in the case of longitudinal and lateral dispersion (K_x and K_y given by Eqs. (10.47) and (10.50), respectively). This can be considered as the initial value of σ_x for the subsequent purely longitudinal advection and dispersion in zone III, treated below.

The lateral spreading is a relatively slow process. Accordingly, it takes a long time, and therefore a large distance, to establish a laterally uniform distribution following a point release. Box 10.1 gives a quantified estimate.

Box 10.1 **The lateral spreading rate**

If the release takes place at one of the two banks, the state of near-uniformity is reached to within a few percent when $\sigma_y \cong B$, or after a duration of about $B^2/2K_y$ (see Section 10.3), at a distance $\ell_B = UB^2/2K_y$ from the source. Substituting $K_y = K_{y,disp} \cong 0.6u_*d$, for lateral dispersion, and $u_* = \sqrt{c_f}\,U$, we find $\ell_B = (B^2/d)\,/\,(1.2\sqrt{c_f})$. Assuming $c_f = 0.004$ and $B/d = 100$, which are not unrealistic values, we find $\ell_B \cong 1300\,B$. These numbers show the extreme slowness of the lateral spreading.

Case 2: Continuous Release at a Point

Our second example deals with the continuous release at a fixed point $x = 0, y = 0$, at a constant rate $\dot{M}$, in a horizontally uniform flow. In this case, an elongated plume develops downstream of the source, similar to a plume of smoke issuing from a smokestack into the atmosphere. Because of the elongated shape, the longitudinal variations of the mean concentration in the plume are far smaller than the lateral ones, for which reason we can neglect the longitudinal dispersion in Eq. (10.54). Assuming moreover a steady state, the latter equation reduces to

$$U\frac{\partial \overline{c}}{\partial x} = K_y \frac{\partial^2 \overline{c}}{\partial y^2} \tag{10.59}$$

This is just the one-dimensional diffusion equation, Eq. (10.10), if we replace the time t by x/U. With the 'initial' condition of a constant $\dot{M}$ in $x = 0, y = 0$, and with the boundary conditions that $\overline{c} \to 0$ as $y \to \pm\infty$, valid for an infinite extent of the flow, the solution is analogous to that given in Eq. (10.11), if we replace t by x/U, x by y and ϵ by K_y:

$$c(x,y) = \frac{C}{\sqrt{2\pi}\,\sigma_y} \exp\left(-\frac{y^2}{2\sigma_y{}^2}\right) = C\,\Phi\left(y;\sigma_y\right) \qquad \text{for} \quad x > 0 \tag{10.60}$$

in which $\sigma_y = \sqrt{2K_y x/U}$ and C is a constant. The value of C follows from the condition that the total transport through each cross section equals the rate of release $\dot{M}$, as must be the case for a conserved quantity and assumed steady conditions. Neglecting (again) longitudinal diffusion and dispersion, we obtain

$$\int_{-\infty}^{+\infty} (Ud\,\overline{c})\,\mathrm{d}y = Ud \int_{-\infty}^{+\infty} \overline{c}\,\mathrm{d}y = Ud\,C = \dot{M} \tag{10.61}$$

so that $C = \dot{M}/(Ud) = \dot{M}/q$, in which $q = Ud$, the discharge per unit width. Substituting this into Eq. (10.60), we obtain

$$c(x,y) = \frac{\dot{M}/q}{\sqrt{2\pi}\,\sigma_y} \exp\left(-\frac{y^2}{2\sigma_y{}^2}\right) \qquad \text{for } x > 0 \tag{10.62}$$

This represents a plume centered at $y = 0$, with a lateral concentration distribution given by the standard bell-shaped Gauss function, with standard deviation σ_y that increases downstream in proportion to $\sqrt{x}$.

In the case of the presence of one or two banks, this solution must be mirrored to satisfy the condition of zero transport through the banks, as above. The distribution widens with increasing distance downstream. If the release takes place at one of the two banks, it is virtually laterally uniform where $\sigma_y \cong B$, or where $x = x_B \cong UB^2/2K_y$. Downstream from this point, the distribution is more or less uniform in each cross section as well as longitudinally, with the limit value

$$\bar{c}_\infty = \frac{\dot{M}}{qB} = \frac{\dot{M}}{Q} \qquad \text{for } x > x_B \tag{10.63}$$

This relation is sometimes used for a discharge measurement, by purposefully releasing some substance at a constant rate $\dot{M}$ and measuring the (cross-sectionally averaged) concentration at a sufficient distance downstream of the source. The value of Q can then be calculated from Eq. (10.63).

10.7.3 One-Dimensional Horizontal Transport

In zone III, the concentration distribution is assumed to be virtually uniform in each cross section, so that a one-dimensional model can be used to calculate its longitudinal variations. In order to emphasize that we follow the water course in a one-dimensional model, we denote the longitudinal coordinate by s (as in the preceding chapters) instead of the Cartesian coordinate x. The longitudinal transport rate will be denoted as T, without subscript. Neglecting molecular and turbulent diffusion, the total longitudinal transport takes place by advection, as expressed in

$$\int_A uc \, \mathrm{d}A = A\overline{\overline{T}} = A\left(\overline{\overline{u}}\,\overline{\overline{c}} + \overline{\overline{u'c'}}\right) = Q\overline{\overline{c}} + A\overline{\overline{u'c'}} \tag{10.64}$$

Remember that a double overbar indicates a cross-sectional average, whereas a prime refers to a local deviation from the cross-sectional average. The last term of Eq. (10.64) represents the effect of *longitudinal dispersion*, i.e. the effect of cross-sectional variations of the local longitudinal advection. Again, it is mathematically represented as a diffusion process:

$$\overline{\overline{u'c'}} = -K\frac{\partial \overline{\overline{c}}}{\partial s} \tag{10.65}$$

in which K is the so-called *coefficient of one-dimensional dispersion*.

The difference between the present coefficient of longitudinal dispersion and that given in Eq. (10.47), in the context of two-dimensional transport (zone II), should be noted. In the latter case, only depth integration took place, so that only the variations of the flow velocity in the vertical played a role. In the present case (zone III), the integration took place over the entire cross section, so that also lateral variations of streamwise velocities are relevant. These are mainly a consequence of lateral variations of the depth, and can be significant. It follows that the value of K is much larger than that of K_x, given in Eq. (10.47).

Since the value of K depends on numerous geometrical characteristics, which in general are not known *a priori*, it is impossible to estimate its value with any precision. On the basis of some general considerations, Fischer et al. (1979) present an expression for K in the form

$$K = \gamma \frac{U^2 B_c^2}{u_* d} = \frac{\gamma}{\sqrt{c_f}} \frac{Q B_c}{d^2} \tag{10.66}$$

in which we have written U for $\overline{\overline{u}}$, and γ is a proportionality coefficient whose value is of the order of 10^{-2}. However, this is no more than a rough indication. Comparisons with observations show deviations from this expression of a factor of 4 both upward and downward (1/4). For better estimates, local calibration or the use of two-dimensional or even three-dimensional numerical models is required. (In the latter case, dispersion need not be represented explicitly.)

We now write the one-dimensional balance equation for the transported matter that is instantaneously present between two cross sections a unit distance apart:

$$\frac{\partial A \overline{\overline{c}}}{\partial t} + \frac{\partial A \overline{\overline{T}}}{\partial s} = 0 \tag{10.67}$$

or

$$\frac{\partial A \overline{\overline{c}}}{\partial t} + \frac{\partial Q \overline{\overline{c}}}{\partial s} - \frac{\partial}{\partial s}\left(KA \frac{\partial \overline{\overline{c}}}{\partial s} \right) = 0 \tag{10.68}$$

Expansion of the derivatives, assuming a prismatic conduit, and substitution of the continuity equation (Eq. (1.3)), given by

$$\frac{\partial A}{\partial t} + \frac{\partial Q}{\partial s} = 0 \tag{10.69}$$

yields

$$\frac{\partial \overline{\overline{c}}}{\partial t} + U \frac{\partial \overline{\overline{c}}}{\partial s} - K \frac{\partial^2 \overline{\overline{c}}}{\partial s^2} = 0 \tag{10.70}$$

Once more, we find the one-dimensional advection–diffusion equation. Its solutions are similar to those we have seen above.

Case 3: Instantaneous Release in a Cross Section

Assume that at $t = 0$ an amount of material M is released in a river at $s = 0$, which is instantaneously distributed uniformly in the cross section. Initially, the concentration is zero everywhere. The solution is again a Gaussian density function, analogous to that in Eq. (10.27), in which the proportionality factor now follows from the condition that the amount of transported (conservative) matter that is present in the entire river is constant in time, equal to the released amount M:

$$\int_{-\infty}^{+\infty} A\, \overline{\overline{c}}(s,t)\, \mathrm{d}s = A \int_{-\infty}^{+\infty} \overline{\overline{c}}(s,t)\, \mathrm{d}s = M \tag{10.71}$$

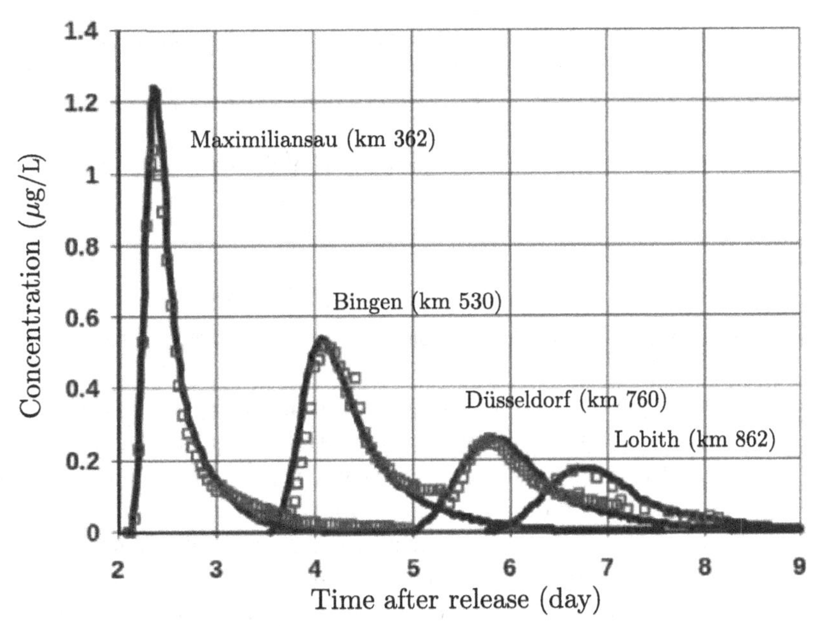

Fig. 10.11 Calibration of a one-dimensional horizontal transport model of the river Rhine: model result (solid) and measurements (squares); from Van Mazijk (1996)

The solution can then be written as

$$\bar{\bar{c}}(s,t) = \frac{M/A}{\sqrt{2\pi}\,\sigma_s(t)} \exp\left(-\frac{(s-Ut)^2}{2\sigma_s{}^2(t)}\right) = M\,\Phi\,(s-Ut;\sigma_s(t)) \tag{10.72}$$

in which $\sigma_s(t) = \sqrt{2Kt}$. The solution is –again– a Gaussian distribution that gradually lowers and broadens as it is advected downstream. Figure 10.11 gives a set of rhodamine concentration distributions measured in the river Rhine following an abrupt injection of the tracer rhodamine in a cross section; see Box 10.2. The curves in Figure 10.11 have been calculated from Eq. (10.72), with an empirical correction to the mean velocity in the conveyance channel to account for the effect of groynes, and a best-fit estimate of the coefficient γ in Eq. (10.66). Therefore, the results are more like a calibration than an independent validation. Nevertheless, it is gratifying that the theory simulates the observed pattern quite well.

Box 10.2 One-dimensional transport model of the river Rhine

In an effort to verify the validity of the theoretical result expressed in Eq. (10.72), a major experiment was performed in the river Rhine as an international cooperative project (Van Mazijk, 1996). A tracer substance, rhodamine, was abruptly released at a cross section near Basel, in Switzerland, after which the concentrations were measured at a number of stations downstream, covering a distance of nearly 700 km. Figure 10.11 shows the results. The peak passes the most downstream measurement station (Lobith), at a distance of some 688 km from the source, approximately 6.8 days after the release, corresponding to an average velocity of nearly 1.2 m/s. The expected downstream decrease in peak concentration, the widening of the distribution, and the asymmetry of the variation in time at each fixed point can be clearly seen.

The total amount of tracer material passing any fixed cross section can be expressed as

$$\int_0^{+\infty} AU\overline{\overline{c}}(s,t)\,\mathrm{d}t = Q\int_0^{+\infty} \overline{\overline{c}}(s,t)\,\mathrm{d}t = M \tag{10.73}$$

This relation can be used for the purpose of discharge measurement, similar to Eq. (10.63) in the case of a continuous release.

Problems

10.1 What is the physical meaning of the mathematical quantity 'divergence of a vector'?

10.2 What is the difference between a balance equation and a conservation equation?

10.3 Describe in your own words the meaning of the notions of molecular diffusion, turbulent diffusion and dispersion.

10.4 Why, and under what circumstances, can turbulent fluctuations cause a net transport, even though their averages are zero?

10.5 Which characteristics of turbulence determine the value of the turbulence diffusivity?

10.6 Why is there no need to represent dispersion explicitly in 3D-transport models?

11 Numerical Computation of Solutions

This chapter provides an introduction to numerical methods by presenting the numerical counterparts of some analytical models introduced earlier in the book. We will discuss a particular type of solution method for a selection of problems, explain its most important properties and present the resulting numerical code down to the level of the individual program statements. The use of these program scripts will be demonstrated by means of computational examples. In the wake of each method treated we will also summarize some other methods, providing the reader with an overview of alternative solution strategies.

11.1 Introduction

The preceding chapters dealt with various types of unsteady flow, resulting in analytical solutions of the governing mathematical equations. These solutions generally provide relations between the characteristics of a water system and its forcing, on the one hand, and the system's response, on the other. In terms of readiness and the overall insight they provide, analytical approaches are still unmatched. However, for practical applications involving complex geometries, arbitrary forcings, and possibly nonlinear effects, they are also somewhat limited due to the assumptions and simplifications on which they rely.

Since the 1950s computer models have therefore gradually, and decisively, replaced the former analytical approaches for computing solutions to real-world problems. The use of advanced numerical tools to compute flow problems in today's engineering practice is ubiquitous and has witnessed a tremendous increase in possibilities. Yet however sophisticated these models may be, they are based on the same theoretical concepts as their analytical predecessors, the study of which is still relevant to understand the results provided by modern computer tools.

In this chapter we will take a first step into the realm of numerical computing by working out some simple examples from previous chapters using the Python programming language. For this purpose some basic knowledge of Python will be needed (see for instance Hetland (2005) or Langtangen (2009)), but this may also be learned as we proceed along the examples in this chapter.

11.2 Canal-Basin System

Our first numerical endeavor concerns a simple model of a small tidal basin that is connected to the sea by a narrow gap or entrance channel. The imposed tidal water level at sea (h_s) leads to a time-varying water level in the basin (h_b) that does not depend on the position within the basin. The resulting discrete system has been discussed extensively in Chapter 6, providing analytical solutions of the linearized equations. Using a numerical model we may refrain from such linearization and include nonlinear resistance and possible variations of the basin area and channel cross section with the water level as well. Because of its simplicity, the resulting model perfectly illustrates some basic concepts of numerical modelling.

11.2.1 Model Equations

For convenience we restate here the corresponding mathematical problem. The water level variation in the basin and the net inward discharge in the entrance, denoted by Q, are related by the continuity equation, (see also Eq. (6.1))

$$A_b \frac{dh_b}{dt} = Q \tag{11.1}$$

where A_b is the area of the free surface in the basin connected to the sea, possibly depending on the water level in the basin.

The discharge in the entrance is found by solving the following momentum equation (see Eq. (6.11)),

$$\ell \frac{dQ}{dt} = gA_c \left(h_s - h_b\right) - \chi \frac{|Q|Q}{A_c} \tag{11.2}$$

where h_s is the water level at sea, ℓ is the length of the entrance (possibly zero), A_c is the conveyance area of the entrance, and χ is a generalized resistance coefficient (including expansion losses and bed friction). Both A_c and χ may depend on the water levels h_b and/or h_s. The water level at sea is supposed to be known from measurements or tidal predictions. Bathymetric data are required to determine A_b and A_c as functions of the respective local water levels.

To complete the problem formulation, the water level in the bay and the discharge in the entrance must be prescribed at the initial time. Problems of this type are commonly referred to as *initial value problems*. Starting from the initial conditions, their solution is obtained by marching *forward* in time by evaluating the rate of change of the state variables.

Our task is now to compute the water level h_b and the discharge Q for some time interval of interest $I = [t_0; t_N]$, where t_0 and t_N are the start time and end time of our simulation, respectively.

11.2.2 Discretization

In order to solve the above problem by means of a computer model, the ordinary differential equations (11.1) and (11.2) must be transformed into *algebraic equations*. To that end, instead of treating the dependent variables h_b and Q as continuous functions of time t, we will represent them as ordered sequences of discrete values. Consequently, the model equations have to be approximated in some way or another, since the variables are no longer continuous but have been reduced to finite sets of numbers. These two steps are essentially intertwined and together define a particular numerical method.

We set out by representing the time interval of interest I as a sequence of $N+1$ discrete time levels: $I = [t_0, t_1, \ldots, t_{N-1}, t_N]$, where N is the number of time steps. This is called a *partitioning* of the time domain I. Individual time levels are indicated with t_n where n is an index corresponding to the order in the sequence. Such a sequence will be denoted by $[t_n]$. Within reasonable limits, to be discussed later on, the individual time levels may be chosen arbitrarily, but it may be convenient to use equal time step lengths or to use the time instances from the water level measurements. The partitioning also defines the time step sizes $\Delta t_n \equiv t_{n+1} - t_n$; see Figure 11.1.

Next, the water levels at sea corresponding to the time levels t_n are defined by the sequence $[h_s^n]$, whose elements are specified using the measurements or predictions. In a similar way, we define the sequences $[h_b^n]$ and $[Q^n]$ representing h_b and Q, respectively. These can be specified only for $n = 0$ as yet (using the initial conditions), with the remaining elements awaiting computation.

The time partitioning and definitions of the various discrete sequences complete the first step of the solution process.

11.2.3 Semi-implicit Method

Since initial value problems are solved by marching forward in time, the sequences $[h_b^n]$ and $[Q^n]$ must be constructed by computing their elements sequentially. That is, starting from the initial values for $n = 0$ we compute the elements indexed $n = 1$, which are then used to compute the elements $n = 2$, and so on. Each next element is thus expressed as a function of the preceding ones, which is called a *recurrence relation*. The second step of the discretization involves finding a general formulation to advance the solution from an

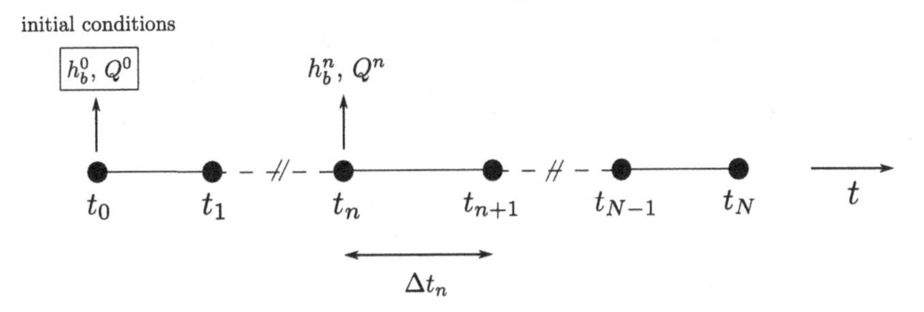

Fig. 11.1 Partitioning of the time domain

arbitrary time level t_n to the next time level t_{n+1}. We will do this in two consecutive steps, using the continuity equation and momentum equation, respectively.

Continuity Equation

Expressing the continuity Eq. (11.1) at time $t = t_n$ gives the following *semi-discrete* equation,

$$A_b^n \left.\frac{dh_b}{dt}\right|_{t_n} = Q^n \tag{11.3}$$

Supposing that h_b^n and Q^n are known, the right-hand side and the coefficient A_b^n have known values. The time derivative of the water level, on the other hand, must be estimated from the elements of the sequence $[h_b^n]$. Using for instance the *forward Euler method* it can be approximated as follows:

$$\left.\frac{dh_b}{dt}\right|_{t_n} \approx \frac{h_b^{n+1} - h_b^n}{t_{n+1} - t_n} \tag{11.4}$$

Substitution into Eq. (11.3) leads to the following recurrence relation:

$$h_b^{n+1} := h_b^n + \Delta t_n \frac{Q^n}{A_b^n} \tag{11.5}$$

This assignment advances the discrete solution for h_b from time level t_n to time level t_{n+1} directly, because the right-hand side of Eq. (11.5) contains only quantities already known at the previous time level t_n. Such a relation, expressing the solution at the new time level entirely in terms of the solution from the previous time level, is called *explicit.*

Momentum Equation

In order to find the discharge Q^{n+1} a similar procedure using the forward Euler method could be applied to the momentum balance equation. However, combined with the forward Euler method already applied to discretize the continuity equation, this leads to a recurrence relation that will amplify small deviations in the solution. Small deviations are always present, due to for instance rounding errors, and their unlimited growth will ultimately dominate the sequence, which renders the solution useless. This is referred to as *instability*, an important topic that will be discussed shortly. To avoid it, the *backward Euler method* may be used to approximate the time derivative of the discharge, giving the following approximation:

$$\left.\frac{dQ}{dt}\right|_{t_n} \approx \frac{Q^n - Q^{n-1}}{t_n - t_{n-1}} \tag{11.6}$$

Using this expression to approximate Eq. (11.2) at time $t = t_{n+1}$ leads to the following recurrence relation:

$$Q^{n+1} : = Q^n + \frac{\Delta t_n}{\ell} \left(g A_c^{n+1} \left(h_s^{n+1} - h_b^{n+1} \right) - \chi^{n+1} \frac{|Q^{n+1}| Q^{n+1}}{A_c^{n+1}} \right) \tag{11.7}$$

The right-hand side contains several quantities that are unknown at the previous time level t_n and the expression is therefore said to be *implicit*. Equations of this type are usually harder to solve than their explicit counterparts since they require an inversion rather than a plain substitution of already known values.

Fortunately, solution of Eq. (11.7) is not too complicated because h_b^{n+1} and the geometrical parameters A_c^{n+1} and χ^{n+1} are known after Eq. (11.5). What remains is a single nonlinear equation (due to the resistance term) having Q^{n+1} as unknown. It can be solved iteratively using the Newton–Raphson method or the Picard method. The quadratic resistance term in Eq. (11.7) is then linearized as, respectively,

$$\left|Q^{n+1}\right| Q^{n+1} \approx |Q^*| \left(2Q^{n+1} - Q^* \right) \quad \text{or} \quad |Q^*| \, Q^{n+1} \tag{11.8}$$

in which Q^* is an estimated value for the discharge at time t_{n+1}. Initially, Q^* is set to Q^n after which Eq. (11.7) can be solved to obtain a first estimate of the solution Q^{n+1}. Replacing Q^* with this value and repeating the procedure will improve the estimate Q^{n+1}, and so forth. In most cases, it suffices to perform just one step, in which case we simply set $Q^* = Q^n$ and compute Q^{n+1} directly.

Due to its alternate use of explicit and implicit steps the resulting method is referred to as *semi-implicit*. It is the simplest member of the class of so-called *symplectic methods* in which a particular sequence of explicit and implicit steps is performed that preserves the dynamical structure of the underlying mass–spring–damper system; see Sanz-Serna and Calvo (1994).

11.2.4 Some Other Solution Methods

The backward Euler method may be used in the discrete continuity equation as well (by replacing Q^n in Eq. (11.5) with Q^{n+1}) giving the *fully implicit method*. The solution strategy becomes slightly more complicated since now an expression for h_b^{n+1} in terms of Q^{n+1} (rather than its numerical value) must be substituted into the discrete momentum equation (11.7) to obtain an expression for Q^{n+1}. Back substitution of Q^{n+1} into the discrete continuity equation then gives the solution for h_b^{n+1}.

Another possibility is using a weighted average of the forward and backward Euler methods to approximate the time derivatives in the continuity and momentum equations. This can be expressed in terms of a weighting parameter θ, where $\theta = 0$ corresponds to the forward Euler method and $\theta = 1$ to the backward Euler method. Taking the average of the forward and backward Euler methods ($\theta = \frac{1}{2}$) gives the so-called *Crank–Nicolson method*, which is the most accurate scheme belonging to this class of methods.

For completeness we finally mention *Runge–Kutta methods* where both the continuity equation and the momentum equation are solved using explicit values of the water level and the discharge. By combining a limited number of explicit stages the approximation of the time derivatives is systematically improved; see for instance Sanz-Serna and Calvo (1994).

11.2.5 Properties of the Semi-implicit Method

It has already been mentioned a few times that the various approximations of the time derivatives involve errors. Application of numerical models without prior knowledge of the nature of these errors is not advisable, since it can easily lead to unreliable results that may lead to wrong conclusions or bad engineering decisions. Before coding the semi-implicit tidal basin model into a Python script, we will therefore briefly examine some important properties of discrete solutions obtained by Eqs. (11.5) and (11.7).

Accuracy

To determine the accuracy of our model approximation we assume that the analytical solution for h_b is a smooth function of time; its value at time $t = t_{n+1}$ can then be developed from the function value and its derivatives at time $t = t_n$ using a Taylor series expansion, as follows:

$$h_b\left(t_{n+1}\right) = h_b\left(t_n\right) + \Delta t_n \left.\frac{\mathrm{d}h_b}{\mathrm{d}t}\right|_{t_n} + \tfrac{1}{2}\Delta t_n^2 \left.\frac{\mathrm{d}^2 h_b}{\mathrm{d}t^2}\right|_{t_n} + \tfrac{1}{6}\Delta t_n^3 \left.\frac{\mathrm{d}^3 h_b}{\mathrm{d}t^3}\right|_{t_n} + \cdots \tag{11.9}$$

For Δt_n sufficiently small, the second-order term dominates the sequence of higher-order terms in the right-hand side. Therefore, a constant K exists such that this sequence is smaller than $K\Delta t_n^2$, provided Δt_n is sufficiently small. Supposing that the exact solutions for h_b and Q are known at time t_n, the following inequality holds:

$$h_b^{n+1} - h_b\left(t_{n+1}\right) \leq K\Delta t_n^2 \tag{11.10}$$

which follows after subtraction of Eq. (11.9) from Eq. (11.5) and using Eq. (11.3). The left-hand side of this inequality represents the difference between the numerical approximation h_b^{n+1} and the exact solution $h_b\left(t_{n+1}\right)$ and is called the *local error*. It occurs every time we apply the recurrence relation given by Eq. (11.5). Eq. (11.10) cannot be used to estimate the magnitude of the local error directly since the constant K is not known beforehand. It merely states that the local error is proportional to Δt_n^2.

This result may seem disappointing; however, it also implies that the local error can be made arbitrarily small by decreasing the time step size. In the limit $\Delta t_n \downarrow 0$ the local error vanishes completely; the discrete approximation is therefore *consistent* with the continuous form of the time derivative. The order of consistency is one in this case, meaning that a linear function is integrated exactly. By similar analyses it can be proven that the backward Euler method and Crank–Nicolson method are also consistent with consistency orders of one and two, respectively.

Stability

We now turn to the topic of stability, touched on earlier. A numerical method is said to be stable if small changes in the initial data cause only small changes in the computed result.

This guarantees that disturbances in the solution that may arise during a simulation (think of limited accuracy or finite precision of computer arithmetic) will not grow during the next sequence of time steps.

To examine the behaviour of such disturbances, imagine that the solution for the discharge at time t_n is perturbed with a small deviation δQ^n. Carrying out the assignment defined in Eq. (11.5), the water level at time t_{n+1} will then deviate by an amount $\delta h_b^{n+1} = \Delta t_n \delta Q^n / A_b^n$ from the regular solution. Performing next the discharge update according to Eq. (11.7), ignoring the resistance term, the discharge deviation at time t_{n+1} equals

$$\delta Q^{n+1} = \left(1 - \Delta t_n^2 \frac{g}{\ell} \frac{A_c^{n+1}}{A_b^n}\right) \delta Q^n \tag{11.11}$$

During the next recurrence, the new disturbances δh_b^{n+1} and δQ^{n+1} will influence the solution at time level $n + 2$, and so on. If these disturbances grow in magnitude during each time step, they will ultimately overrule the regular solution, rendering the computed result useless. In our example this occurs if in Eq. (11.11) the term between parentheses has a modulus exceeding one. Since this term includes the time step size, apart from the geometrical parameters and gravity, avoiding instabilities implies a bound on the time step size, depending on the system properties.

A more profound analysis, involving the eigenvalues of the discrete system formed by Eqs. (11.5) and (11.7), leads to the following necessary and sufficient stability criterion for the semi-implicit method (ignoring resistance):

$$\Delta t_n \leq \frac{2}{\omega_0^n} \tag{11.12}$$

where $\omega_0^n = \sqrt{gA_c^n / \ell A_b^n}$ is the eigenfrequency of the canal–basin system at time level n. The semi-implicit method is thus *conditionally stable*; the solution will be stable if and only if the above condition on the time step size is met.

Inclusion of the resistance term in the analysis results in a smaller allowable maximum time step size if the system remains dominated by inertia. If resistance dominates (see Chapter 6), the discrete solution will always be stable, independent of the time step size, which is referred to as *unconditional stability*. Without proof, we state that the semi-implicit method for the tidal basin problem has stable solutions if $\Delta t_n \leq \sqrt{2}/\omega_0^n$, regardless of the system parameters.

Convergence

An important consequence of stability is that local errors will not grow during the recurrency. The accumulated error during a simulation, the so-called *global error*, is therefore bounded by the sum of the local errors. Since the number of time steps N is inversely proportional to the time step size, it follows that the global error of a stable method is one order lower than the local error. The global error of the semi-implicit method is therefore proportional to Δt.

We finally state what is perhaps the most important result of this section: *the solution of a stable and consistent method converges to the exact solution if the time step size goes to zero*. Although the local and global errors are not known *a priori*, we can make them arbitrarily small by decreasing the time step size. In practice we will reduce the time step size until the computed solution does not change anymore, indicating that we are close to the exact solution. A comforting thought, indeed.

11.2.6 Python Implementation

Our computer code has to perform the following main tasks. First, the prescribed tidal water levels and the geometric data of the basin and its entrance have to be processed. To this end the tidal levels (h_s) are stored in a file containing two columns giving the discrete time levels and the corresponding tidal water levels, respectively. The bathymetry of the system (A_b, A_c and R) must be prescribed for a range of water levels. They will be interpolated for intermediate water levels. Second, the sequences representing the water level in the basin $[h_b^n]$ and the discharge $[Q^n]$ have to be computed using the recurrence relation. Finally, output of the model must be provided for further analyses and processing.

A Python script carrying out these tasks is given in Box 11.1. It is based on the semi-implicit method. The scripts can be executed by running Python in the directory containing the file `basin.py` and typing `import basin.py`. We will now examine the script in detail.

Input and Initialization

In line 2 the module `pylab` is loaded containing array-computing methods and plotting tools, which are not available in plain Python. Make sure this module has been downloaded and installed! In line 5 the gravitational acceleration `g` is specified. Line 8 reads the input file (`tidal-levels.dat`) and splits it into separate character strings which, after executing line 9, yields an array of numbers containing the prescribed tidal water levels and corresponding time instants.

The free-surface area of the basin as a function of the water level is specified in lines 12 and 13 by prescribing it for a number of intermediate water levels. This information is passed to a function `Ab(n)`, defined in line 16, which calculates the free-surface area for a water level h_b^n by means of interpolation. The function `interp`, used in this function, is available from `pylab`. The geometry of the entrance channel (conveyance cross section, hydraulic radius, and length) is specified in lines 19–21, respectively. In the example script these geometrical parameters are assumed constant, but, when necessary, they can be made functions of the water level as was done with the basin area. The dimensionless bed-friction coefficient (`cf`) is specified in line 22, from which the overall resistance coefficient (`chi`) is computed in line 23. This concludes the input section.

The computation proceeds by initializing some parameters first. The time levels, which are stored at even index numbers in the array `data`, are assigned to a new array `t`, for convenience, in line 26. The index specifier `[0::2]` refers to all elements in `data` from index 0 (first array element) up to the last one, with a step size of 2. The number of time

Box 11.1 Python script Basin.py

```
# import modules
from pylab import *

# physical constant(s)
g = 9.81                                                # gravitation              [m/s2]

# read water level records
data = open('tidal-levels.dat').read().split()
data = array([float(p) for p in data])

# wet surface area basin
level  = array([-2.0,-1.0, 0.0, 1.0, 2.0, 3.0])  # water level               [m]
area   = array([  80,  90, 100, 115, 120, 125])  # basin area                [km2]

# function := water level [m] -> basin area [m2]
def Ab(n): return interp(hb[n], level, area)*1.E6

# channel geometry
Ac = 3000                                               # conveyance area          [m2]
R  = 6.                                                 # hydraulic radius         [m]
L  = 600                                                # length                   [m]
cf = 4.E-3                                              # bed friction coefficient [-]
chi = .5 + cf*L/R

# time partitioning
t = data[0::2]                                          # discrete time levels     [s]
N = size(t) - 1                                         # number of time steps     [-]

# tidal water levels
hs = data[1::2]                                         # tidal level at sea       [m]

# initial conditions
hb = zeros(N+1); hb[0]=hs[0]
Q  = zeros(N+1); Q[0]=0

# time stepping
for n in range(N):
    dt = t[n+1] - t[n]                                  # time step size           [s]
    # continuity equation
    hb[n+1] = hb[n] + dt*Q[n]/Ab(n)                     # water level update       [m]
    # momentum equation
    kappa = chi*abs(Q[n])/Ac                            # resistance factor        [m/s]
    dQ = g*Ac*(hs[n+1]-hb[n+1])-kappa*Q[n]              # intermediate term
    Q[n+1] = Q[n] + dQ/(L/dt + 2*kappa)                 # discharge update         [m3/s]

# plot water levels
subplot(2, 1, 1)
plot(t/3600, hs, '-k',linewidth=2); plot(t/3600, hb, '--k',linewidth=2)
legend(['$h_s$','$h_b$'], fontsize=20)
xlabel('time [hrs]', fontsize=14)
ylabel('water level [m wrt datum]', fontsize=20)

# plot discharge
subplot(2, 1, 2)
plot(t/3600, Q, ':k',linewidth=4)
xlabel('time [hrs]', fontsize=14)
ylabel('$Q$ [m$ 3$/s]', fontsize=20)

# output to screen
show()
```

steps (`N`) equals the size of array `t` minus one, which is determined in line 27. The tidal water level at sea, which is stored at uneven index numbers in the array `data`, is assigned to a new variable `hs` in line 30. The water level in the basin and the discharge are initialized as arrays of zeros `hb` and `Q` in lines 33 and 34, respectively. The first array element (with index 0) prescribes the corresponding initial condition. In our example, the initial water level in the basin is equal to the initial water level at sea, while the initial discharge is set to zero.

Time Stepping

Next, the lists `hb` and `Q` are filled by applying the recurrence relations repetitively in a loop defined in line 37, iterating over time indices `n` from `0` to `N-1`. After determining the time step size `dt` (line 38) the new surface level element `hb[n+1]` is computed using Eq. (11.5), which is implemented literally in line 40.

Solving the momentum equation is somewhat more elaborate. First, the resistance term is linearized using a resistance factor $\kappa = \chi |Q^n|/A_c^{n+1}$, which is calculated in line 42. For ease of implementation, Eq. (11.7) for the momentum update is then rewritten as follows:

$$(\ell/\Delta t_n + 2\kappa)\,\Delta Q^n = gA_c\left(h_s^{n+1} - h_b^{n+1}\right) - \kappa Q^n \tag{11.13}$$

where ΔQ^n is the discharge increment from time level n to $n+1$. Newton–Raphson linearization has been used to linearize the resistance term; see also Eq. (11.8). The right-hand side of the above expression is calculated in line 43, using a temporary variable `dQ`, after which the new discharge `Q[n+1]` is obtained in line 44.

Output of Results

What remains is to plot the results. By means of `subplot` (lines 47 and 54) a plot window is created containing two figures, stacked vertically. The arguments of the `plot` function (line 48 for the water levels and line 55 for the discharge) are the values on the horizontal and vertical axes, respectively, and format specifiers for line style and markers. To distinguish between the water levels at sea and in the basin, a legend is added to the corresponding figure, specified in line 49. Horizontal and vertical axes are defined in lines 50 and 51 (water levels) and lines 58 and 59 (discharge), respectively. Note that Latex-style editing is used to represent mathematical symbols. All of this is rather basic; more advanced formatting examples can be found in the `pylab` user manual. The figures are finally plotted to screen using the `show()` command in line 60, which also concludes the tidal basin script.

11.2.7 Verification

Before using a numerical model in practice it should be tested to eliminate algorithmic errors and programming flaws. We will therefore briefly examine some results of our tidal basin model to see whether they make sense.

Example 11.1 We use the tidal basin from Example 6.2 ($A_b = 100\,\text{km}^2$, $A_c = 3000\,\text{m}^2$, $R = 6\,\text{m}$, $\ell = 600\,\text{m}$). Natural oscillation of the unforced canal–basin system is considered with initial conditions $h_b^0 = 0.1\,\text{m}$ and $Q^0 = 0$. In absence of resistance and for constant system parameters (A_b, A_c, and ℓ), the surface elevation in the basin and the discharge in the canal will vary harmonically, with an eigenfrequency $\omega_0 = \sqrt{gA_c/\ell A_b} = 7.0 \times 10^{-4}$ rad/s, and amplitudes $\hat{h}_b = h_b^0 = 0.10\,\text{m}$ and $\hat{Q} = \omega_0 A_b \hat{h}_b = 7004\,\text{m}^3/\text{s}$.

Choosing the time step size involves the stability criterion given in Eq. (11.12). The maximum allowable time step for stability equals $2/\omega_0 \approx 3000\,\text{s}$. This would lead to about three time steps per wave period. The simulation would be stable, but it would also lack resolution. Therefore a smaller time step size of 600 s is used to resolve the natural oscillation more accurately.

Result

Figure 11.2 shows that the computed solutions for the water level and discharge match the respective analytical solutions closely. Importantly, the computed solution maintains its amplitude, without damping. Not every numerical method has this desirable property. The fully implicit method for instance would yield a damped oscillation; see also Problem 11.3.

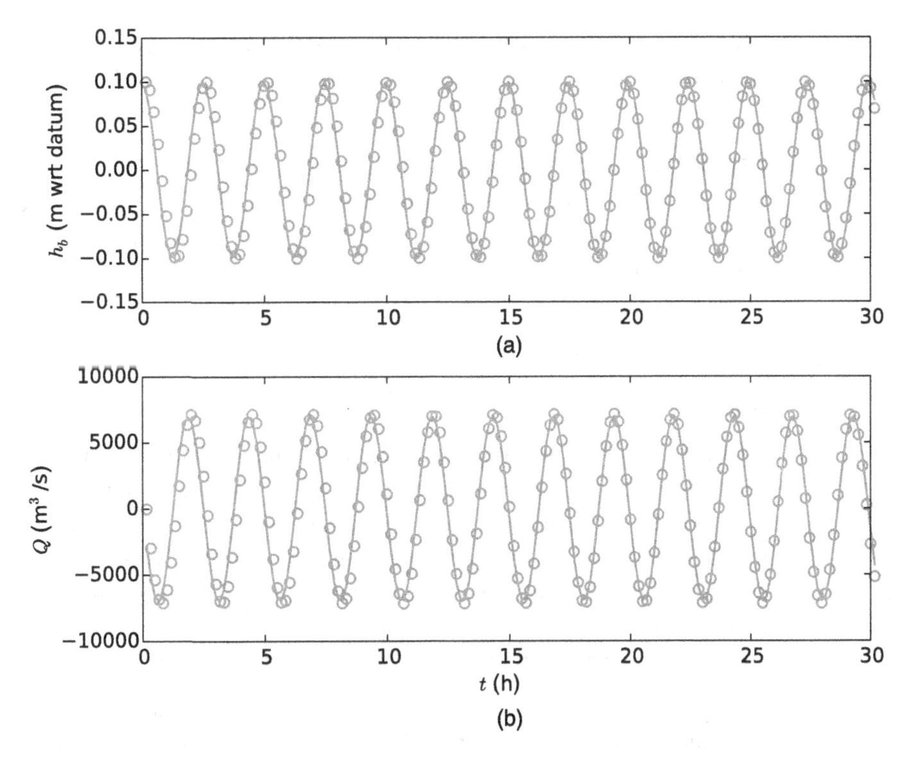

Fig. 11.2 Natural oscillation of a canal–basin system (Example 11.1); (a) exact (solid line) and computed (circles) surface elevation; (b) exact (solid line) and computed (circles) discharge

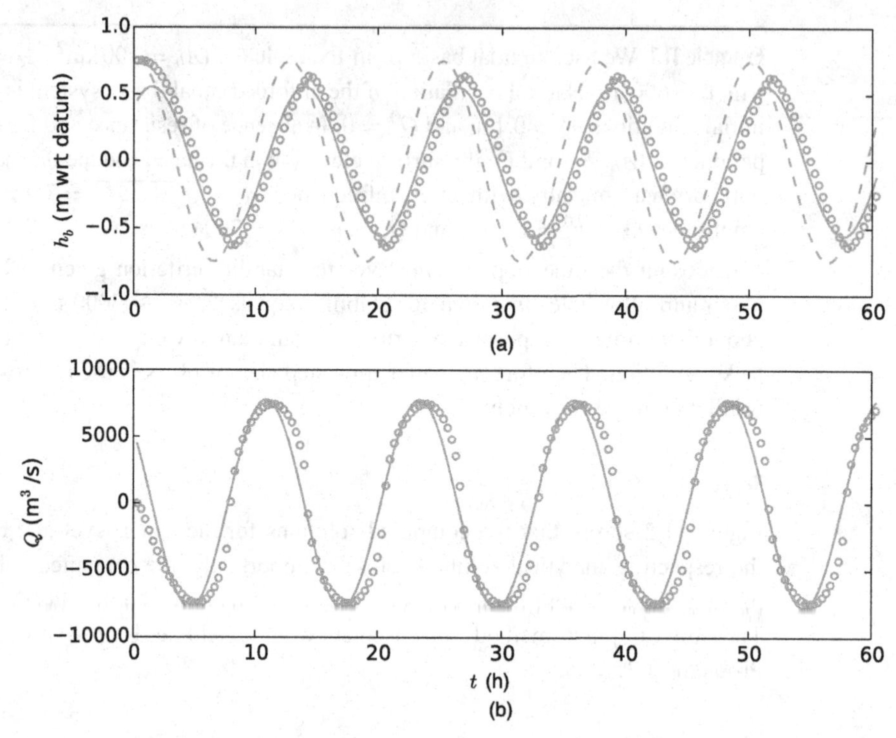

Fig. 11.3 Tidal basin (Example 11.2); (a) water level at sea (dashed line), linear (solid line) and computed (circles) solution of the water level in the basin; (b) linear (solid line) and computed (circles) solution of the discharge

Example 11.2 We add resistance to the system ($c_f = 0.004$) and force it with a sinusoidal M_2-tide with a surface level amplitude $\hat{h}_s$ of 0.75 m, as in Example 6.2. The initial conditions for the simulation are $h_b^0 = \hat{h}_s$ and $Q^0 = 0$; the time step size Δt is set to 1200 s, which is less than $\sqrt{2}/\omega_0$, satisfying the time step constraint. Anticipating a much larger period of oscillation (M_2-tide) than in the previous example, there seems no need (to be verified) to reduce the time step size in this case.

Result

Figure 11.3 shows the computed water level and discharge, together with the respective periodic solutions of the linearized problem. Resistance is dominant in this system and therefore the solution adapts quickly to a periodic state. The general characteristics of the computed periodic solution largely agree with the analytical solution of the linearized problem; the differences are mainly caused by the nonlinear resistance term; see also Box 11.2.

The observed absence of artificial damping is especially useful when modelling natural oscillation or near-resonant forcing of a canal–basin system. If the natural period of oscillation is much smaller than the period of the (tidal) forcing, the unconditionally stable fully implicit method (allowing a larger time step size) is perhaps more practical.

Box 11.2 **Nonlinear effects in a tidal basin**

Example 11.2 shows that around slack tide the discharge changes more rapidly than predicted by linear theory. Not only does the resistance term in the nonlinear model vanish at this time instant, but also its time derivative. This causes a larger water level amplitude in the basin, compared with the solution of the linearized model. On the other hand, around maximum ebb and flood the nonlinear resistance is larger than its linearized counterpart. Overall, the computed discharge tends to oscillate around the harmonic solution with a frequency twice that of the tidal forcing, clearly visible in Figure 11.3. This is the manifestation of a higher harmonic caused by the nonlinearity; see also the discussion in Section 7.8.

The above findings give confidence in the Python program. More advanced testing, by for instance comparing the model results with a wider range of analytical solutions, could provide a more quantitative assessment of the numerical error.

11.3 Semi-Implicit Method for Long Waves

In this section we will discuss numerical solution methods for one-dimensional models of shallow water flows in channels. In such systems the problem variables, being the water level h and the discharge Q, may vary both in time and in space. Besides discretization of the problem variables in the time interval of interest, as was done for the numerical model of the tidal basin, this requires a discretization in the spatial domain of interest. Apart from that, the ensuing solution algorithm will be quite similar to the previous methods for the tidal basin in that a recurrence relation is obtained in which the water level and discharge are computed sequentially.

11.3.1 Model Equations

For convenience, we repeat here the set of one-dimensional equations governing shallow water flow in open channels.

Variations of the water level (h) in time are related to variations of the discharge (Q) in space by the continuity equation (see also Eq. (1.20)),

$$B\frac{\partial h}{\partial t} + \frac{\partial Q}{\partial s} = 0 \tag{11.14}$$

where B is the storage width of the channel. If the channel is non-uniform, it will vary along the streamwise coordinate s. If the cross section is not rectangular, it will also vary in time through its dependency on h. Variations of the discharge in time are governed by the momentum balance equation (see also Eq. (1.21)),

$$\frac{\partial Q}{\partial t}+\frac{\partial}{\partial s}\left(\frac{Q^2}{A_c}\right)+gA_c\frac{\partial h}{\partial s}+c_f\frac{|Q|Q}{A_cR}=0 \tag{11.15}$$

where A_c and R are the conveyance area and hydraulic radius, respectively, and c_f is the dimensionless resistance coefficient. These may all depend on the spatial coordinate s, for instance where the cross-sectional profile or the bed roughness changes, and, through h, they may depend on time as well.

In analytical models the dependency of cross-sectional parameters on the local water level is usually ignored, while only highly schematized spatial configurations can be considered; see for example Section 7.6. Also, the nonlinear resistance term in Eq. (11.15) must be linearized to enable analytical treatment of the equations. Using a numerical model, however, there is no need for such simplifications; rather, inclusion of these nonlinearities is quite straightforward.

To solve Eqs. (11.14) and (11.15), initial conditions prescribing the state (h, Q) of the water system at some initial time must be specified, together with boundary conditions describing the interaction of the system with its environment. Problems of this type are called *initial boundary value problems*. They must be solved by marching forward in time, starting from the initial condition.

Consider therefore the spatial domain $S = [s_0; s_M]$, where s_0 and s_M are the coordinates of the respective end points of a channel reach of interest. The task lying ahead is to compute the water level h and discharge Q in the space–time domain $S \times I$, where $I = [t_0; t_N]$ is the time interval of interest, with t_0 and t_N being the start time and end time of the simulation.

11.3.2 Discretization

To solve Eqs. (11.14) and (11.15) approximately using computer arithmetic, they must be transformed into algebraic equations. To that end the dependent variables h and Q must be represented as discrete sets of numbers. Besides discretizing the variables in the time domain, introduced in Section 11.2 for the tidal basin model, this involves a discretization in space.

Spatial Domain

The spatial domain can be partitioned using the following sequence of points: $S = [s_0, s_1, \ldots, s_{M-1}, s_M]$. This defines M sections, numbered $m = 1, 2, \ldots, M$, having lengths $\Delta s_m \equiv s_m - s_{m-1}$. All sections may have the same length, though not necessarily, in which case the partitioning is called uniform. Furthermore, in each section m we define a corresponding discharge Q_m, the conveyance area A_m and hydraulic radius R_m. (For compact notation, and since there is no risk of confusion, A will be used from here on to denote the area of the conveyance cross section.)

At the transitions between sections and at the domain boundaries we next define additional computational nodes, numbered $m = 0, 1, \ldots, M$, holding the discrete water levels h_m. (Note that there is one more water level node than there are sections.) We also define the local storage widths at these nodes, and denote them by B_m. This arrangement,

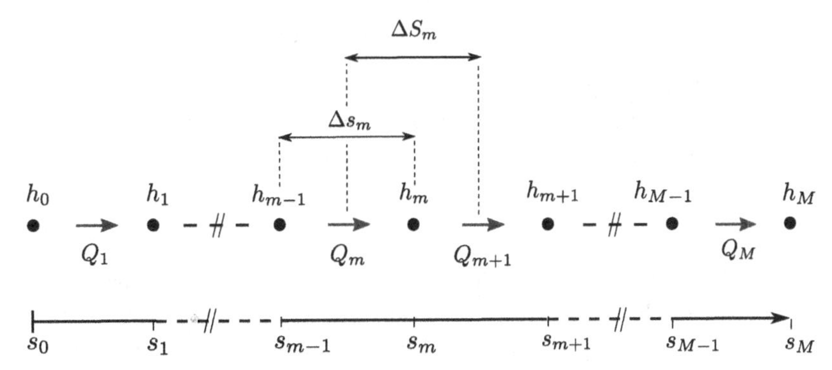

Fig. 11.4 Arrangement of variables in a staggered grid

where water levels and discharges reside in alternate locations, is called a *staggered grid*. It is depicted schematically in Figure 11.4.

Time Domain

The preceding provides a complete description of the channel geometry and the flow state in the channel at a particular time instant. To describe the problem variables in time, the time domain I is partitioned into N time intervals. We use a sequence of discrete time levels: $I = [t_0, t_1, \ldots, t_{N-1}, t_N]$, where t_0 and t_N are the start time and end time of the simulation, respectively.

We may now define the two-dimensional sequences $[h_m^n]$ and $[Q_m^n]$ of which the elements h_m^n and Q_m^n denote the water level at node m and the discharge in section m, respectively, at time level t_n. A similar convention can be used for the various parameters describing the geometry and the resistance of the channel. In a computer program these sequences are coded as two-dimensional arrays, covering the entire space–time domain or just a part of it (for instance the spatial values at time t_n and t_{n+1} only, which are overwritten alternately).

All elements having time index $n = 0$ are already specified by the initial conditions. Where water level boundary conditions are imposed, the corresponding elements $[h_0^n]$ (water level given at $s = s_0$) and/or h_M^n (water level given at $s = s_M$) are readily specified as well. All other elements of $[h_m^n]$ and $[Q_m^n]$ are still unknown and must be constructed by the time stepping algorithm.

11.3.3 Semi-Implicit Method

A general recurrence relation will be formulated to advance the solution from time level n, say, to time level $n + 1$. Starting from the initial conditions, the repeated application of such a relation will finally yield the solution over the entire time domain. In the following we will formulate a semi-implicit method in which the water level is updated first (using the continuity equation), followed by an update of the discharge (using the momentum equation). In this way the solution marches forward in time. This approach is conceptually similar to the method used for the tidal basin problem.

Continuity Equation

The first step involves the continuity equation Eq. (11.14). Approximating the time derivative of the water level at point $s = s_m$ and at time $t = t_n$ by the forward Euler method gives

$$B_m^n \frac{h_m^{n+1} - h_m^n}{\Delta t_n} \approx - \left. \frac{dQ}{ds} \right|_{s_m, t_n} \tag{11.16}$$

The right-hand side of this ordinary differential equation may be evaluated by comparing the discharges in the sections adjacent to water level node m, being section $m + 1$ and section m, respectively. The difference between the corresponding discharges divided by the streamwise distance between the midpoints of both sections is a measure of the spatial derivative of Q at the point $s = s_m$. This approach, commonly referred to as *central differencing*, leads to the following approximation:

$$\left. \frac{dQ}{ds} \right|_{s_m, t_n} \approx \frac{Q_{m+1}^n - Q_m^n}{\frac{1}{2}(\Delta s_m + \Delta s_{m+1})} \tag{11.17}$$

While this approximation of the spatial derivative of Q cannot be used at boundary nodes, its substitution into Eq. (11.16) leads to the following assignment at interior nodes:

$$h_m^{n+1} := h_m^n - \frac{\Delta t_n}{\Delta S_m} \frac{Q_{m+1}^n - Q_m^n}{B_m^n} \quad \text{for } m = 1, \ldots, M - 1 \tag{11.18}$$

using the definition $\Delta S_m \equiv \frac{1}{2}(\Delta s_m + \Delta s_{m+1})$, to denote the average section length around node m. The storage width B_m^n may depend on the associated water level h_m^n.

The right-hand side of the assignment in Eq. (11.18) is known explicitly at time level n. The expression is quite similar to Eq. (11.5), used to update the water level of the tidal basin, in that the water level increment equals the net inflow towards a particular control volume divided by its storage area.

Boundary Conditions

The assignment in Eq. (11.18) is defined at interior water level nodes only. To determine the water levels h_0^{n+1} and h_M^{n+1} at the left and right boundary, respectively, the boundary conditions must be applied.

At water level boundaries the imposed water level is simply assigned to the water level variable residing in the respective node, which is referred to as a *Dirichlet* boundary condition.

If the discharge is specified as a boundary condition, Eq. (11.18) is modified locally to give

$$h_0^{n+1} := h_0^n - \frac{\Delta t_n}{\Delta S_0} \frac{Q_1^n - Q_L^n}{B_0^n} \quad \text{or} \quad h_M^{n+1} := h_M^n - \frac{\Delta t_n}{\Delta S_M} \frac{Q_R^n - Q_M^n}{B_M^n} \tag{11.19}$$

in which Q_L^n and Q_R^n denote the imposed discharge at the left and right boundary, respectively, at time level t_n. They may be specified directly, leading to a so-called

Neumann boundary condition, but they may also depend on the local water level, in which case we have a so-called *Robin* boundary condition. An example of the latter is the weakly reflective boundary condition where the Riemann invariant along the ingoing characteristic is prescribed. Writing the Riemann invariants as $R^{\pm} = Q \pm Bc\,h$, and prescribing R^{+} at the left boundary and R^{-} at the right boundary, we obtain

$$Q_L^n = R_L^+ - (Bc)_L^n\, h_0^n \qquad \text{or} \qquad Q_R^n = R_R^- + (Bc)_R^n\, h_M^n \tag{11.20}$$

where the local Bc values are calculated using the geometric parameters of the section adjacent to the respective boundary node, at time t_n. The corresponding value of the local Riemann invariant is based on the flow state assumed in the domain outside the boundary.

Imposing Riemann boundary conditions avoids the artificial reflection of disturbances approaching the boundary from the interior of the domain. This would occur when the flow state in the canal does not match with the boundary condition. Robin-type boundary conditions may also be used to impose various other discharge relations treated in Chapter 5; for example those involving in/outflow head losses, control structures and the like.

Momentum Equation

Next, the discharge in the canal sections will be updated using the just obtained values of the discrete water levels at time level t_{n+1}. To simplify matters, we will temporarily ignore the advection of momentum and reintroduce it later on. Consideration of the momentum equation (11.15) in section m (denoted by S_m) at time $t = t_{n+1}$, and discretizing the time derivative using the backward Euler method, then leads to the following ordinary differential equation

$$\frac{Q_m^{n+1} - Q_m^n}{\Delta t_n} = -gA_m^{n+1} \left.\frac{\mathrm{d}h}{\mathrm{d}s}\right|_{S_m, t_{n+1}} - \frac{\chi_m^{n+1}}{\Delta s_m} \frac{|Q_m^{n+1}| Q_m^{n+1}}{A_m^{n+1}} \tag{11.21}$$

in which $\chi_m^{n+1} \equiv c_f \Delta s_m / R_m^{n+1}$ denotes the dimensionless bed resistance coefficient of section m, a definition similar to that for the tidal basin system used in Eq. (11.2). The water level gradient appearing in the right-hand side of Eq. (11.21) can be evaluated by central differences, using the water level nodes at both ends of each section

$$\left.\frac{\mathrm{d}h}{\mathrm{d}s}\right|_{S_m, t_{n+1}} \approx \frac{h_m^{n+1} - h_{m-1}^{n+1}}{\Delta s_m} \tag{11.22}$$

This approximation results in the following assignment to update the discharge,

$$Q_m^{n+1} :\, = Q_m^n - \frac{\Delta t_n}{\Delta s_m} \left(gA_m^{n+1} \left(h_m^{n+1} - h_{m-1}^{n+1} \right) + \chi_m^{n+1} \frac{|Q_m^{n+1}|\, Q_m^{n+1}}{A_m^{n+1}} \right) \quad \text{for all } m \tag{11.23}$$

This expression is defined for all sections; boundary conditions do not have to be specified at this stage of the algorithm. The assignment is implicit since the right-hand side of

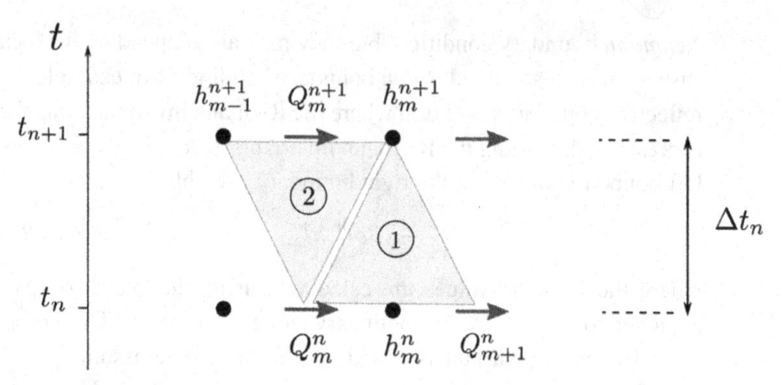

Fig. 11.5 Semi-implicit method: continuity equation (1), momentum equation (2)

Eq. (11.23) contains several quantities at the new time level $n+1$. The coefficient χ_m^{n+1} and cross-sectional area A_m^{n+1} may depend on the water level in a section and are readily evaluated once the new water levels are known.

What results is a set of M uncoupled equations, having unknowns Q_m^{n+1}, that can be solved sectionwise. The nonlinear resistance term can be linearized using the Picard method or the Newton–Raphson method, according to Eq. (11.8). The similarity of all the above with the discrete momentum equation of the tidal basin model, Eq. (11.7), is evident.

The overall algorithm, with its alternate use of explicit and implicit steps, is called semi-implicit. It is depicted schematically in Figure 11.5. Together with the staggered arrangement of the Q and h variables, the semi-implicit time stepping guarantees stability of the method; see also Section 11.3.5.

Inclusion of Momentum Advection

So far, the advection term has been ignored in the discrete momentum equation. This approximation is valid only for low waves propagating in (nearly) stagnant water. For high waves and/or significant ambient flow this assumption is no longer justified and we have to include momentum advection in the formulation.

The advection term in the momentum equation can be discretized by differencing the momentum flux $F = Q^2/A$ over a section, leading to an approximation similar to that in Eq. (11.22) for the water level gradient

$$\frac{\partial}{\partial s}\left(\frac{Q^2}{A}\right)\bigg|_{S_m} \approx \frac{\tilde{F}_m - \tilde{F}_{m-1}}{\Delta s_m} \tag{11.24}$$

in which $\tilde{F}_m$ and $\tilde{F}_{m-1}$ denote the momentum fluxes at nodes m and $m-1$, respectively. The discharge and cross-sectional area, which determine the momentum flux, are defined sectionwise, rendering the flux $\tilde{F}$ double-valued at nodes. A unique value may be obtained by averaging the fluxes in the sections adjacent to the considered node. However, the nodal flux will then depend on the flow state downstream of a node. This will cause a transfer of information in the upstream direction, which will lead to numerical artefacts.

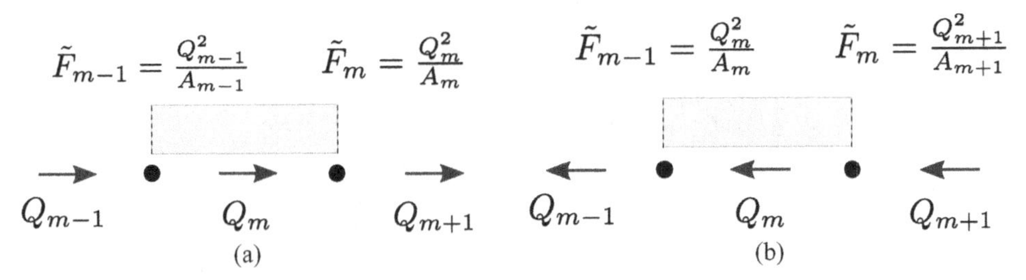

Fig. 11.6 Upwinding of the momentum flux: positive discharge (a), negative discharge (b)

A better approach is therefore to compute the nodal momentum flux using the section situated on the upstream side of the node only. This results in the following formulation for the nodal momentum flux:

$$\tilde{F}_m = \begin{cases} Q_m^2/A_m & \text{if } \; Q_m > 0 \; \text{ and } \; Q_{m+1} > 0 \\ Q_{m+1}^2/A_{m+1} & \text{if } \; Q_m < 0 \; \text{ and } \; Q_{m+1} < 0 \\ 0 & \text{otherwise} \end{cases} \tag{11.25}$$

This expression is then used in Eq. (11.24) to calculate the gradient of the momentum flux over section m. The result is multiplied with Δt_n and subtracted from the right-hand side of Eq. (11.23). The overall procedure is referred to as *upwind differencing*. It is depicted schematically in Figure 11.6.

The momentum flux gradient may be evaluated explicitly at time level n or implicitly at time level $n + 1$. The latter is consistent with the backward Euler method already used to discretize the momentum equation in time. However, a set of coupled equations will then result since the advection term depends on the unknown discharges in neighbouring sections. Solution of such a system involves the inversion of a tri-diagonal matrix, which is more elaborate than the sectionwise calculation of the discharge used so far. Implicit treatment of the advection term is further complicated by its nonlinearity, requiring Newton–Raphson or Picard linearization. It is therefore more practical to evaluate the advection term explicity. Fortunately, this will not compromise the performance of the semi-implicit method.

11.3.4 Some Other Solution Methods

The shallow water equations can be discretized in a number of ways; the semi-implicit method is just one of many possible approaches. Without pretending to be complete, a brief summary of some other well-known methods is given below. More comprehensive treatments can be found in for instance Vreugdenhil (1994), Abbott and Basco (1997) or Wesseling (2001).

Leapfrog Method

A slight modification of the semi-implicit method is obtained by staggering the variables Q and h in time as well, giving the *leapfrog method*. Using indices as in Figure 11.7,

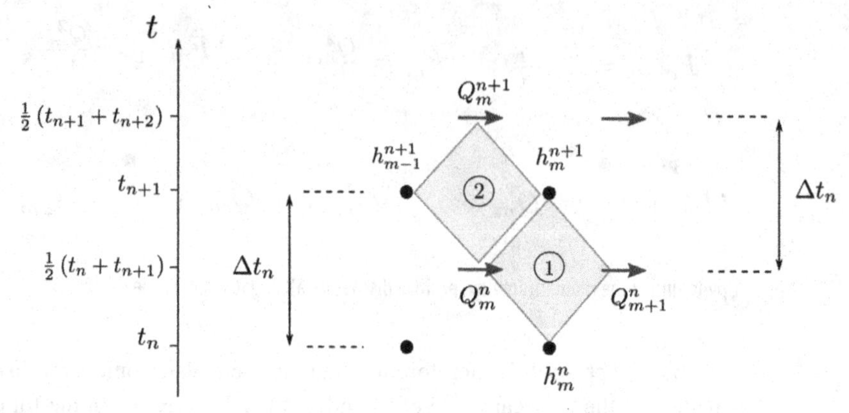

Fig. 11.7 Leapfrog method: continuity equation (1), momentum equation (2)

the method involves the same sequence of steps as the semi-implicit method, the only difference being that the discharge Q_m^n is attributed to time $t_n + \frac{1}{2}\Delta t_n$ instead of t_n.

Algorithmically the two methods are equivalent, except for the implementation of the initial and boundary conditions. The initial discharge Q_m^0 must now specify the discharge at time $t = t_0 + \frac{1}{2}\Delta t_0$, while the Neumann boundary condition for the continuity equation to obtain h^{n+1} must be specified at time $t = \frac{1}{2}(t_n + t_{n+1})$. This is a slight complication for which reason the semi-implicit method is often preferred. Formally, the leapfrog scheme is more accurate in time than its semi-implicit counterpart.

Implicit θ-Method

The semi-implicit method can be turned into a fully implicit method by using the backward Euler method in the discrete continuity equation as well. The right-hand side of Eq. (11.18) will then contain the unknown discharges Q_m^{n+1} and Q_{m+1}^{n+1}. Replacing these terms using the expression given in Eq. (11.23) gives a system of equations for the nodal water levels h_m^{n+1}. For linear waves in a uniform canal, ignoring resistance and momentum advection, this leads to the following implicit equation for the water level at an interior node m,

$$-\sigma^2 h_{m-1}^{n+1} + \left(1 + 2\sigma^2\right) h_m^{n+1} - \sigma^2 h_{m+1}^{n+1} = h_m^n - \Delta t \frac{Q_{m+1}^n - Q_m^n}{B\,\Delta s} \tag{11.26}$$

in which $\sigma \equiv c\Delta t/\Delta s$ is the so-called *Courant number*. It plays an important role in numerical analyses. Supplemented with the boundary conditions, a set of $M+1$ equations results with $M+1$ unknown nodal water levels. The terms in the right-hand side of Eq. (11.26) are computed explicitly. The combination of all left-hand sides constitutes a tri-diagonal matrix. It can be solved with relatively little effort using Gauss elimination. Substitution of the solution for the nodal water levels into the discrete momentum equation, which is a sectionwise operation, finally gives the solution for the discharge.

Using a weighted average of the forward and backward Euler method in both the continuity equation and the momentum equation gives the θ-method. The right-hand sides of Eqs. (11.18) and (11.23) are then formulated at time level $n+\theta$, where $0 \leq \theta \leq 1$. The resulting set of equations is implicit for $\theta > 0$ and has the same structure as Eq. (11.26). In practice it makes sense to only consider $\theta \geq \frac{1}{2}$ since the method is otherwise unstable for wave problems. For $\theta = \frac{1}{2}$ the Crank–Nicolson method is obtained, which of all these methods has the highest order of accuracy.

Preissmann Method

It is also possible to avoid any staggering of variables. By defining the water level and discharge variables at common spatial locations and at common time instances, a *collocated* grid is obtained. It is depicted in Figure 11.8.

To obtain a stable method for the shallow water equations, the continuity and momentum equations are discretized sectionwise, in a coupled fashion. For section m the discrete equations are (for linear waves in a uniform domain, ignoring resistance and momentum advection)

$$B\frac{h_{m+1}^{n+1}+h_m^{n+1}-h_{m+1}^n-h_m^n}{2\Delta t}=-\frac{Q_{m+1}^{n+1}-Q_m^{n+1}+Q_{m+1}^n-Q_m^n}{2\Delta s} \tag{11.27}$$

$$\frac{Q_{m+1}^{n+1}+Q_m^{n+1}-Q_{m+1}^n-Q_m^n}{2\Delta t}=-gA\frac{h_{m+1}^{n+1}-h_m^{n+1}+h_{m+1}^n-h_m^n}{2\Delta s} \tag{11.28}$$

Note that the time derivative has been averaged between nodes m and $m+1$ and that the spatial derivative has been averaged between time levels n and $n+1$, giving a nice skew-symmetry[1] of the formulation. Variations are possible using the general θ-method to discretize in time and upwind or downwind differencing in space. This class of algorithms is referred to as the *Preissmann method.*

For a domain of M sections, the method gives $2M$ equations, while the number of unknown Q and h variables equals $2\,(M+1)$ (since there is one more node than there are sections). Two of the unknowns are supplied by the boundary conditions after which a system of $2M$ equations with $2M$ unknowns is obtained. It can be solved by inversion of a matrix having a banded structure containing seven diagonals. The solution process, using Gauss elimination, will therefore be quite efficient.

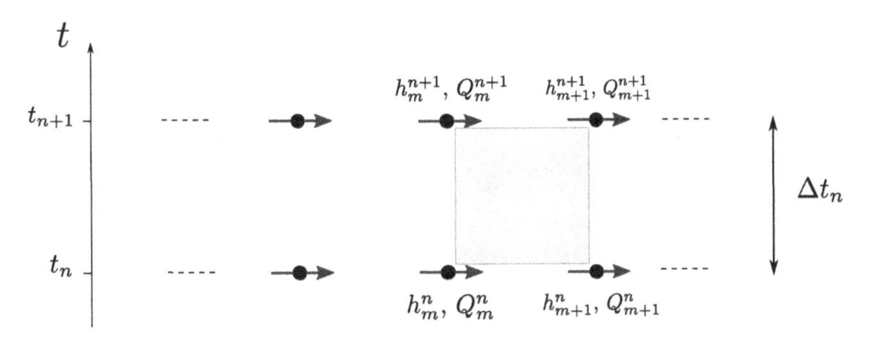

Fig. 11.8 Preissmann method

[1] In both physics and numerics skew-symmetry always hints at energy conservation. It is a desirable property of any numerical method used to simulate nondissipative systems.

Since the discrete equations are formulated sectionwise, the method generalizes easily to network systems, which is one of the reasons for its widespread use in commercial software packages.

Finite Element Method

We finally mention the *finite element method* where the problem variables Q and h are represented as linear combinations of predefined *trial functions*. The water level for instance is discretized as

$$h(s, t_n) = \sum_m h_m^n \varphi_m(s) \tag{11.29}$$

where the φ_m are the trial functions and h_m^n discrete coefficients. Generally, the trial functions are chosen to have compact support – i.e. they are zero in all nodes except the associated node m– and are interpolated between nodes. An example of such a set of trial functions and the corresponding interpolation of the water level is shown in Figure 11.9.

The coefficients h_m^n in Eq. (11.29), and similar coefficients for the discharge, are the unknowns of the resulting discrete problem. Before integrating them in time, the problem equations are multiplied by so-called *test functions* and integrated over the spatial domain. This is called a *variational formulation* of the problem. Helpful in this respect is that the representation of a function in terms of (continuous) trial functions also defines its gradient. If the number of test functions equals the number of trial functions, the procedure will give the same number of equations as there are unknowns. A choice commonly made is to use the trial functions as test functions as well, giving the *Galerkin method*. Since a wide range of function spaces can be used to define trial/test functions, an incredibly rich class of methods is obtained; see for instance Pironneau (1989) or Zienkiewicz et al. (2014).

11.3.5 Properties of the Semi-Implicit Method

This section discusses some basic properties of the previous semi-implicit method. Being aware of the numerical limitations and artefacts warrants a proper use of such modelling tools. Occasionally we will also touch on the properties of some of the alternative methods mentioned in the previous section.

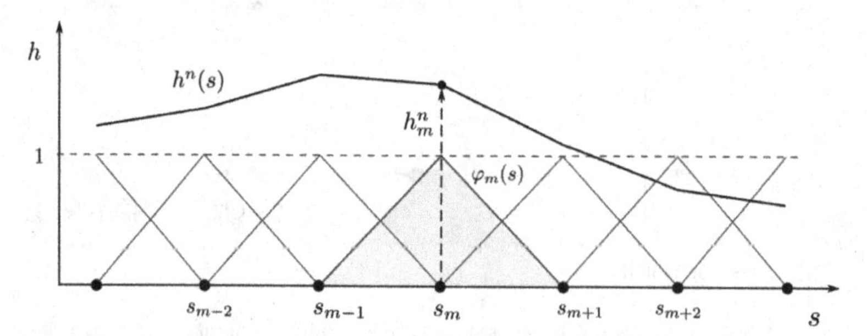

Fig. 11.9 Finite element method: interpolated function $h^n(s)$ using trial functions

Accuracy and Consistency

The forward and backward Euler methods, used for time integration of the continuity and momentum equations, respectively, involve a *truncation error* in the time derivatives that is proportional to the time step size. The local error in the approximation of the solution is then proportional to the time step size squared. Already demonstrated for the semi-implicit tidal basin model (see Eq. (11.9)), this holds invariably for the semi-implicit shallow water model.

The error due to the approximation of the various spatial derivatives by central differences can be quantified in a similar way, using Taylor series expansions. The discrete continuity equation, for instance, uses the discharges at the midpoints of sections m and $m+1$ to estimate the discharge gradient at node m; see Eq. (11.17). Developing these discharges in terms of function values and derivatives at point s_m and assuming a uniform section length Δs gives, respectively,

$$Q_m = Q(s_m) - \tfrac{1}{2}\Delta s \left.\frac{\mathrm{d}Q}{\mathrm{d}s}\right|_{s_m} + \tfrac{1}{8}\Delta s^2 \left.\frac{\mathrm{d}^2Q}{\mathrm{d}s^2}\right|_{s_m} - \tfrac{1}{48}\Delta s^3 \left.\frac{\mathrm{d}^3Q}{\mathrm{d}s^3}\right|_{s_m} + \cdots \tag{11.30}$$

$$Q_{m+1} = Q(s_m) + \tfrac{1}{2}\Delta s \left.\frac{\mathrm{d}Q}{\mathrm{d}s}\right|_{s_m} + \tfrac{1}{8}\Delta s^2 \left.\frac{\mathrm{d}^2Q}{\mathrm{d}s^2}\right|_{s_m} + \tfrac{1}{48}\Delta s^3 \left.\frac{\mathrm{d}^3Q}{\mathrm{d}s^3}\right|_{s_m} + \cdots \tag{11.31}$$

Subtraction of these equations and division by Δs leads to

$$\frac{Q_{m+1} - Q_m}{\Delta s} = \left.\frac{\mathrm{d}Q}{\mathrm{d}s}\right|_{s_m} + \tfrac{1}{24}\Delta s^2 \left.\frac{\mathrm{d}^3Q}{\mathrm{d}s^3}\right|_{s_m} + \cdots \tag{11.32}$$

Comparing this result with Eq. (11.17) reveals that, for a uniform partitioning with Δs sufficiently small, the discrete discharge gradient involves a truncation error that is proportional to Δs^2. (For a non-uniform partitioning the truncation error would be proportional to Δs.) This error can be made arbitrarily small by letting Δs go to zero. The approximation given in Eq. (11.17) is thus consistent with the continuous discharge gradient.

A similar result can be derived for the discrete water level gradient in Eq. (11.22), affirming its consistency with the corresponding continuous form. The one-sided differencing of the momentum flux gradient, due to the upwinding introduced in Eq. (11.25), is also consistent, but comes with a first-order truncation error. (This can be made plausible by noting its similarity to the forward and backward Euler methods for time differencing.) Overall, we can conclude that our semi-implicit method is consistent with the shallow water equations given in Eqs. (11.14) and (11.15).

In fact, all methods discussed in the previous section share this favourable property. Since they all employ some form of central differencing to approximate the spatial derivatives they are also equally accurate in space. Concerning the approximation of the time derivatives, the semi- and fully implicit methods involve an error proportional to the time step size. The Crank–Nicolson method, the Preissmann method of Eqs. (11.27) and (11.28) and the leapfrog method are second-order accurate in time.

Stability

Considering stability, we have to study the behaviour of small perturbations of the discrete solution. Such deviations will always occur during a computation; if not initiated by local errors, just discussed, they will be induced by the finite precision of computer arithmetic. These deviations should not grow in magnitude after repetitive application of the recurrence relations. Otherwise, the computed results will ultimately become useless.

Consider the recurrence relation in Eq. (11.18), which is used to compute h_m^{n+1} using the discrete discharge at time level n. We now express the required discharge in terms of the discrete water level at time level n by applying the momentum update, Eq. (11.23), from time level $n-1$ to time level n. For linear waves in a uniform domain, and ignoring resistance, the result can be written as follows:

$$h_m^{n+1}: \; = \sigma^2 \, h_{m-1}^n + \left(1 - 2\sigma^2\right) h_m^n + \sigma^2 \, h_{m+1}^n - \Delta t \frac{Q_{m+1}^{n-1} - Q_m^{n-1}}{B \, \Delta s} \tag{11.33}$$

where $\sigma = c \, \Delta t / \Delta s$ is the Courant number. This equation shows that an initial perturbation of the water level δh_m^n at time level n will cause deviations from the regular solution $\delta h_m^{n+1} = \left(1 - 2\sigma^2\right) \delta h_m^n$ and $\delta h_{m\pm1}^{n+1} = \sigma^2 \delta h_m^{n+1}$, respectively, at time level $n+1$. These new disturbances will then cause other ones after the next time step, and so on. The resulting sequence of deviations will decay if $|1 - 2\sigma^2| \leq 1$ and $\sigma^2 \leq 1$. This leads to the following sufficient stability criterion:

$$c \frac{\Delta t}{\Delta s} \leq 1 \tag{11.34}$$

In words, the distance travelled by a disturbance during one time step should not exceed the section length. The semi-implicit method is therefore conditionally stable. This restriction, which holds for the leapfrog method, too, is known as the *CFL condition* after Courant, Friedrichs and Lewy, who pioneered numerical analyses (Courant et al., 1928). This important result relies on a staggered spatial arrangement of the flow variables. Combining semi-implicit time stepping with a collocated grid, on the other hand, will lead to persisting disturbances, so-called *spurious modes*; see Box 11.3.

Box 11.3 — Spurious modes

An important feature of our semi-implicit method is the spatial staggering of the discrete variables. Without it unphysical fluctuations may persist in the solution. These so-called *spurious modes* are remnants of local errors that neither grow nor decay after repeated time stepping. To understand this, the collocated grid may be regarded as a superposition of two staggered grids, where one is shifted over a distance of $\frac{1}{2}\Delta s$ relative to the other. Using central differencing to approximate the spatial derivatives, two discrete realizations of the solution will emerge that are essentially decoupled from each other. These two sets of variables will behave independently. Aliasing of the differences between them is manifested as spurious modes. By formulating the discrete equations at locations between the nodes, the coupling is restored, which also eliminates the spurious part of the solution. This is the conceptual background of the Preissmann method.

It can be shown that the implicit θ-method is stable for $\theta \geq \frac{1}{2}$, which includes the fully implicit method and the Crank–Nicolson method. From the perspective of stability these methods do not require a restriction on the time step size; they are unconditionally stable. However, a large time step size will increase the truncation error in the time derivative. This may compromise the quality of the solution if the chosen time partitioning is too coarse with respect to the time scale of the wave phenomenon to be simulated. Using the fully implicit method ($\theta = 1$), increasing the time step size will enhance the damping of the simulated waves, which can be an advantage if time stepping is merely used to obtain a steady-state solution.

Convergence

Involving time differencing and multiplication with Δt, the updates in Eqs. (11.18) and (11.23) finally lead to a local error in the discrete solution of $\mathcal{O}\left(\Delta s^2, \Delta t^2\right)$. Having confirmed consistency and stability, it is guaranteed that after performing a number of time steps the sum of these local errors remains bounded. The associated global error, which is the accumulated error after a number of time steps, will not exceed the sum of the individual local errors. It can be made arbitrarily small by decreasing the section lengths and time step size. The important property of convergence is thus established; we may get as close to the exact solution as desired by increasing the resolution of our simulation.

Since the number of sections and time steps is inversely proportional to the section length and time step size, respectively, the convergence order of the global error is equal to the approximation order of the underlying partial differential equation. For the (staggered) semi- and fully implicit method this implies an $\mathcal{O}\left(\Delta s^2, \Delta t\right)$ behaviour of the global error. The Crank–Nicolson method, leapfrog method and the central Preissmann method involve an $\mathcal{O}\left(\Delta s^2, \Delta t^2\right)$ global error and are thus more accurate in the time domain.

All of this does not yet provide us with an estimate of the actual error; it merely states that the global error is bounded and shows how its convergence depends on the section length and time step size. In practice, to guarantee a sufficiently accurate solution, the spatial resolution and the number of time steps will be increased until the solution does not change significantly anymore.

11.3.6 Python Implementation

We will next code the semi-implicit method for shallow water waves into a simple Python program. Obviously, the core of this program will be the repeated execution of the recurrence relations, Eqs. (11.18) and (11.23). Besides, a number of administrative tasks are to be performed to provide the model with the required input (domain, initial and boundary conditions) and to transfer the computed results as data files or plots.

The resulting script is given in Box 11.4. The first lines loading the `pylab` module and the final lines where the solution is plotted to screen are omitted for brevity; these lines are essentially similar to those of the script in Box 11.1.

Box 11.4 **Python script Canal.py**

```
# physical constant(s)
g = 9.81                                   # gravitation                    [m/s2]

# domain partitioning
L  = 5000                                  # length                         [m]
M  = 200                                   # number of sections             [-]
s  = linspace(0,L,M+1)                     # nodal coordinates              [m]

# section lengths
ds = s[1:M+1] - s[0:M]                     # section length                 [m]
dS = zeros(M+1);
dS[0:M] += ds/2; dS[1:M+1] += ds/2         # section length at nodes        [m]

# function defining cross-section: s,h -> A,B,R
def geometry(s,h):
    d = 3.5 + (h[0:M]+h[1:M+1])/2          # midpoint depth in section      [m]
    A = 10*d                               # conveyance area section        [m2]
    B = 10 + 20*(s<L/2)                    # nodal storage width            [m]
    R = A/(10+2*d)                         # hydraulic radius section       [m]
    return A,B,R

# resistance coefficient
cf=.003                                    #                                [-]

# time stepping
t0,tN = 0,1500                             # start and end time simulation  [s]
N = 500                                    # number of time steps           [-]
t = linspace(t0,tN,N+1)                    # time instances                 [s]

# initial conditions
h = zeros(M+1)                             # initial water level            [m]
Q = zeros(M)                               # initial discharge              [m3/s]
A,B,R = geometry(s,h)                      # initial geometry

# time loop
for n in range(N):
    dt = t[n+1] - t[n]                     # time step size                 [s]
    # boundary conditions
    QL = 3*(1+tanh((t[n]-150.)/50.))       # discharge left boundary        [m3/s]
    QR = 0                                 # discharge right boundary       [m3/s]
    # continuity equation
    dh = append(Q,QR) - append(QL,Q)       # discharge difference
    h -= dt*dh/(B*dS)                      # water level update
    # momentum flux (explicit)
    F = zeros(M+1)                         # initialise momentum flux
    F[0:M]   += .5*(Q-abs(Q))*Q/A          # left contribution
    F[1:M+1] += .5*(Q+abs(Q))*Q/A          # right contribution
    F[0]     += max(QL,0)*QL/A[0]          # inward flux left boundary
    F[M]     += min(QR,0)*QR/A[M-1]        # inward flux right boundary
    # new geometry & resistance
    A,B,R = geometry(s,h)                  # geometry at time level n+1
    chi = cf*ds/R                          # resistance coefficient section [-]
    kappa = chi*abs(Q)/A                   # linear resistance factor       [m/s]
    # momentum equation
    dQ = g*A*(h[1:M+1] - h[0:M]) + kappa*Q      # surface slope & resistance
    dQ += F[1:M+1] - F[0:M]                # momentum flux difference
    Q  -= dQ/(ds/dt + 2*kappa)             # discharge update

# write final results to file
outfile = open('output.dat','w')
for m in range(M): print >> outfile, 2*'%10.5f' % ((h[m]+h[m+1])/2,Q[m])
```

Input and Initialization

Gravity is specified in line 2, the length `L` of the canal in line 5 and the number of sections `M` in line 6. Using a uniform partitioning of the domain, the sequence holding the spatial coordinates `s` is defined in line 7. The `linspace` function, used to define `s`, is available from `numpy`. It creates an array with a specified number of elements ($M + 1$ in this case), whose values are evenly distributed between a specified lower bound and an upper bound (being the first and second argument, respectively). The nodal coordinates also define the section lengths `ds`, computed in line 10, and the average lengths `dS` of sections adjacent to each node, which is computed in lines 11 and 12.

The function `geometry`, defined in lines 15–20, computes the required cross-sectional parameters as a function of the spatial coordinate s and the water level h, where the latter may vary in time t as well. Whenever this function is called it will return several arrays containing the geometrical parameters, using the current value of the nodal water levels. In respective order it computes the flow depth `d` and conveying area `A` for each section, the storage width `B` for each node, and the hydraulic radius `R`, which is also defined sectionwise. The given example features a rectangular cross section where the storage width changes abruptly halfway down the canal. The function can be easily adapted to define other geometries. Inclusion of the dimensionless resistance coefficient `cf` in this function is an option, but instead a constant value is assumed here, which is specified in line 23.

Lines 26–28 define the time interval of interest and its partitioning. Initial conditions for the water level and discharge are specified in lines 31 and 32, respectively. The `numpy` function `zero` is used here to create an array filled with zeros, of specified length. In line 33 the `geometry` function is called for the first time, to compute the canal geometry at the initial time `t0`. This concludes the input section.

Time Stepping

An indexed loop using the `for` statement starts the time-stepping sequence in line 36. In line 37 the new time step size `dt` is determined from the array `t` of time instances.

Each time step starts with the continuity equation. The discharge boundary conditions at the left and right end of the canal are defined in lines 39 and 40, respectively. In this example a time-varying discharge is imposed at the left boundary, while the discharge at the right boundary is set to zero. The discharge difference over each node is computed in line 41, using the temporary variable `dh`. For convenience the `numpy` function `append` is used to construct discharge arrays including the boundary values. The continuity equation is completed in line 42 by computing the new water level `h`. If water level boundary conditions are to be used, the corresponding array element of `h` is simply replaced with the imposed value of the water level after executing the general update.

For an elegant treatment of the momentum equation, we rewrite it in a form similar to Eq. (11.13) as used in the tidal basin script,

$$\left(\Delta s_m/\Delta t_n + 2\kappa_m^{n+1}\right)\Delta Q_m = -gA_m^{n+1}\left(h_m^{n+1} - h_{m-1}^{n+1}\right) - \left(\tilde{F}_m^n - \tilde{F}_{m-1}^n\right) - \kappa_m^{n+1}Q_m^n \tag{11.35}$$

where ΔQ_m is the discharge increment in section m, and $\kappa_m^{n+1} = \chi_m^{n+1}|Q_m^n|/A_m^{n+1}$ is the corresponding linear resistance factor; see also Eq. (11.23). Note that one Newton–Raphson step has been used to linearize the resistance term.

Momentum advection is computed explicitly, at time level n, by determining the momentum flux towards each node. After specifying an array of zeros having the required size (line 45), the flux from the left section (line 46) and the right section (line 47) are added subsequently. In a similar way the fluxes corresponding to the (imposed) discharge at the boundaries are added in lines 48 and 49.

Once the (explicit) momentum fluxes have been computed, the geometry can be updated by calling the `geometry` function in line 51. After calculating the new resistance parameters `chi` and `kappa`, in lines 52 and 53, respectively, the water level difference and the explicit part of the resistance term are computed in line 55 and stored as a temporary variable `dQ`. Line 56 adds the momentum flux difference, which completes the right-hand side of Eq. (11.35). Division by the factor $(\Delta s/\Delta t + 2\kappa)$ and adding the result to the discharge `Q` finalizes the solution update in line 57.

Output of Results

The above sequence is repeated `N` times. At each time step, the old values for water level and discharge are overwritten by their most recent values. This warrants an efficient use of internal memory, although in the presented set-up, with its relatively small number of sections and time steps, this is not a strict necessity. During the recursion, intermediate results may be plotted to screen or written to an output file. For some basic plotting commands, see Box 11.1.

As an example lines 61 and 62 contain some basic printing commands. First, an output file `output.dat` is opened in write mode (`'w'`), after which the midpoint water level and discharge per section are written to this file using a `for` loop. The `print` statement contains the format specifier `2*'%10.5f'`; two floating point numbers are written having ten digits and five decimal places each. The variables to be printed are enclosed in parentheses, which in Python means a tuple, an immutable sequence.

The script in Box 11.4 contains just the bare minimum of output commands. Programming examples containing sophisticated output formatting and interactive manipulation of the model (using widgets) can be found in Langtangen (2009).

11.3.7 Verification

For preliminary testing of the model, the propagation of low translatory waves in a canal of varying cross section will be simulated. The computed results will be compared with the corresponding linearized solutions derived in Chapter 4, giving an impression of the model's accuracy. The script in Box 11.4 already contains the most important parts to perform the tests that will be presented next.

Example 11.3 The computational domain is a 5 km long, horizontal canal having a rectangular cross section with a conveyance width B_c of 10 m and an initial depth of 3.5 m. Halfway down the canal the storage width B changes abruptly from $B_1 = 30$ m to $B_2 = 10$ m, in the positive s-direction. The corresponding wave speeds in the respective canal reaches are $c_1 = 3.4$ m/s and $c_2 = 5.9$ m/s. The bed resistance coefficient c_f has been set to 0.003, uniformly.

At the left boundary a discharge is imposed that gradually increases from 0 to 6 m^3/s after which it remains constant. The time variation of the discharge has a time scale of 50 s, i.e. the denominator in the hyperbolic tangent function in line 39 of the script Canal.py in Box 11.4. The right boundary is closed.

The canal is partitioned into 200 sections with a length of 25 m each. The chosen time step equals 3 s, leading to a maximum Courant number (the narrow part of the canal is normative in this respect): $\sigma_2 = c_2\Delta t/\Delta s = (5.9\,\text{m/s}) \times (3\,\text{s}) / (25\,\text{m}) \approx 0.7$, which satisfies the CFL condition.

Result

Figure 11.10 shows the computed water level in the canal at various time instances. The plots for $t = 300$ s and $t = 700$ s show the incident wave approaching the transition from the left (the wider canal reach). A slight surface slope is noticeable at the trailing edge of the wave, which is caused by the bed resistance, as mentioned in Section 4.4. At time

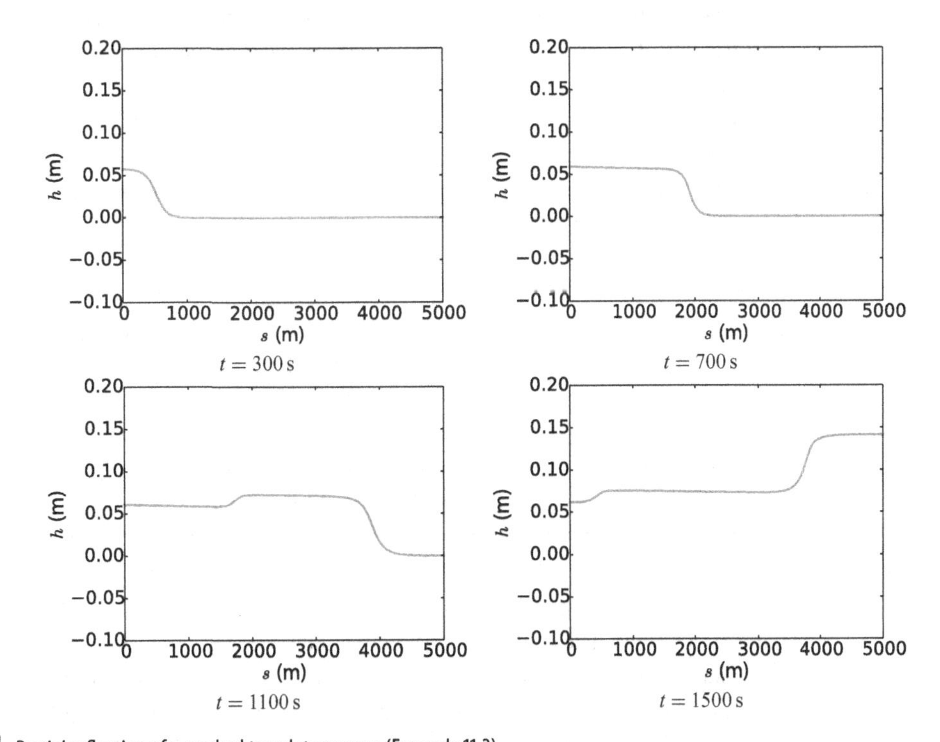

Fig. 11.10 Partial reflection of a gradual translatory wave (Example 11.3)

$t = 1100$ s, following partial reflection at the transition, a reflected wave is travelling back towards the left boundary and a transmitted wave continues travelling in the positive direction. At time $t = 1500$ s the transmitted wave has been reflected fully from the right (closed) boundary. All is in accordance with the theory of Chapter 4.

Let us next examine the results more closely. Between $t = 300$ s and $t = 700$ s (see Figure 11.10) the incident wave travels over a distance of about 1400 m. This gives an estimated wave speed of about 3.5 m/s, which agrees with the theoretical wave speed c_1 of 3.4 m/s. The computed height of the incident wave (0.06 m) agrees with the theoretical height $\delta h = Q/(B_1 c_1) = (6\,\mathrm{m}^3/\mathrm{s}) / ((30\,\mathrm{m}) \times (3.4\,\mathrm{m/s})) = 0.059$ m. This wave height is sufficiently small, relative to the water depth, to justify the use of linear wave theory in the above.

For the incident wave, the parameter γ characterizing the transition equals $\sqrt{1/3}$, giving a reflection coefficient $r = (1-\gamma)/(1+\gamma) \approx 0.27$. The theoretical height of the reflected wave is therefore $0.27 \times 0.059\,\mathrm{m} = 0.018$ m, which matches the computed result exactly. Since the computed water level is continuous at the transition it follows that the height of the transmitted wave is also reproduced correctly. Furthermore, the increase of the surface elevation at the right boundary after reflection of the transmitted wave equals twice the transmitted wave height, in accordance with theory.

Example 11.4 The previous results show that the semi-implicit method can provide accurate solutions. However, the assignments in Eqs. (11.18) and (11.23) involve truncation errors that will become more pronounced as the time and length scales of the wave decrease with respect to the chosen time step and section length, respectively. In the example, the partitioning of the space–time domain and the scaling parameters of the wave problem were chosen such that the resulting numerical error was hardly noticeable.

As a counterexample, a simulation is performed for the same channel where now the discharge at the left boundary increases more rapidly. To this end the denominator in the hyperbolic tangent function prescribing the inflow (line 39 of canal.py) has been adjusted from 50 s to 10 s. All other parameters and computational settings remain the same.

Result

Figure 11.11 shows the results for the steeper wave, resulting from the more rapid increase of the discharge at the left boundary. While the overall wave properties (speed and net height) are reproduced correctly, the result is plagued by oscillations that appear on the trailing edges of the waves. This is a signature of high wave number components travelling at a speed that is too low, an artefact referred to as *numerical dispersion*; the computed propagation speed depends on the wavelength relative to the section length.

The numerical results are in agreement with linear wave theory, for suitably chosen parameter values. The spatial discretization and time step size should be selected carefully, considering the time and length scales of the particular wave problem at hand. More elaborate analyses of the semi-implicit method will reveal the dispersion error as a function of the wavelength relative to the section length; see for instance Vreugdenhil (1994).

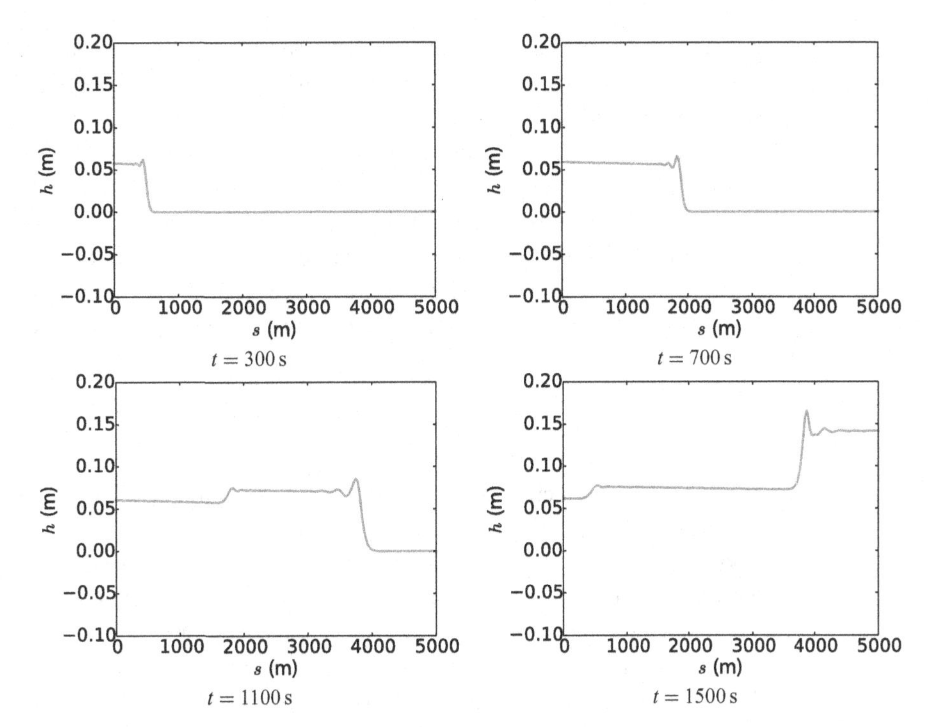

Fig. 11.11 Partial reflection of a steep translatory wave (Example 11.4)

11.4 Characteristics-Based Methods

The computation of compression waves and shocks requires numerical techniques dedicated specifically to these nonlinear problems. The methods from Section 11.3 cannot be used for these types of wave since they tend to generate artificial oscillations once the computed wave steepens; see also Example 11.4. Exact solutions of the homogeneous shallow water equations will never develop such maxima or minima, which is a consequence of the characteristic structure of the problem. Along each characteristic the associated Riemann invariant cannot exceed the value that is assumed initially or imposed at the boundary, which is known as the *maximum principle*. Numerical methods preserving this property are called *monotone*. They can be constructed by using the characteristic structure of the problem in a discrete manner.

11.4.1 Characteristic Equations

For convenience, the characteristic equations are repeated here. Assuming a rectangular canal of uniform width, the shallow water equations can be transformed into (see Section 5.2),

$$\frac{\partial R^{\pm}}{\partial t} + V^{\pm}\frac{\partial R^{\pm}}{\partial s} = g\left(i_b - i_f\right) \tag{11.36}$$

where $R^{\pm} = U \pm 2c$ are the *characteristic variables* (Riemannn invariants), and $V^{\pm} = U \pm c$ are the characteristic velocities, in which $c = \sqrt{gd}$ denotes the wave speed. The right-hand side of Eq. (11.36) contains the external forcing terms, which involve the bed slope i_b and the resistance; the latter is represented here using the friction slope $i_f \equiv (c_f/gR)\,|U|U$. Equation (11.36) is valid for canals of uniform width only. It can, however, be extended to the general case of non-uniform width; see Problem 11.8.

Equation (11.36) constitutes a pair of partial differential equations that can be solved independently of each other. Moreover, they are of a special mathematical form in that they can be cast into a set of coupled ordinary differential equations as follows:

$$\frac{\mathrm{d}R^{\pm}}{\mathrm{d}t} = g\left(i_b - i_f\right) \qquad \text{provided} \qquad \frac{\mathrm{d}s}{\mathrm{d}t} = V^{\pm} \tag{11.37}$$

The variation along a characteristic of the associated variable $R^{\pm}$ is determined by a first-order ordinary differential equation. Once $R^{\pm}$ is known for one point of the characteristic, it is known for all points of the characteristic. This initial value is naturally specified where a characteristic crosses a lateral boundary (boundary condition) or where it passes a time level where the solution is already known (initial condition).

This principle can be exploited numerically. From computational nodes at which the solution has to be computed, a pair of characteristics is traced back in time until they enter a part of the discrete (s, t)-plane where the solution is already known. This can be one of the lateral boundaries or it can be the previous discrete time level. This fixes the solution along the entire characteristic, which determines the solution in the target node. Repeating this process advances the solution in time.

11.4.2 Space–Time Discretization

To set out we construct a discrete (s, t)-plane first; see Figure 11.12. The spatial domain of interest S is partitioned into M sections by defining a sequence of $M + 1$ nodes; $S = [s_0, s_1, \ldots, s_M]$. The section lengths may vary where Δs_m denotes the length of section m that is situated between nodes s_m and s_{m-1}. The time interval of interest I is partitioned into N time steps, leading to a sequence of $N + 1$ discrete time levels: $I = [t_0, t_1, \ldots, t_N]$. The intervals between individual time levels may vary and will be denoted by $\Delta t_n = t_{n+1} - t_n$, as before.

We next prescribe the bed level and resistance of the canal. The bed level z must be prescribed at the nodes. This leads to a bed slope i_b, which is sectionwise constant. Also the friction slope i_f is defined to be sectionwise constant. It may depend on time through its dependence on the local water depth and flow velocity.

We use the water depth d and flow velocity U as flow variables. The discrete Riemann invariants are defined uniquely if d and U reside in the same computational nodes. For this reason characteristics-based methods commonly employ a *collocated* arrangement of the flow variables. We will allocate them in the nodes of the discrete (s, t)-plane, shown in Figure 11.12.

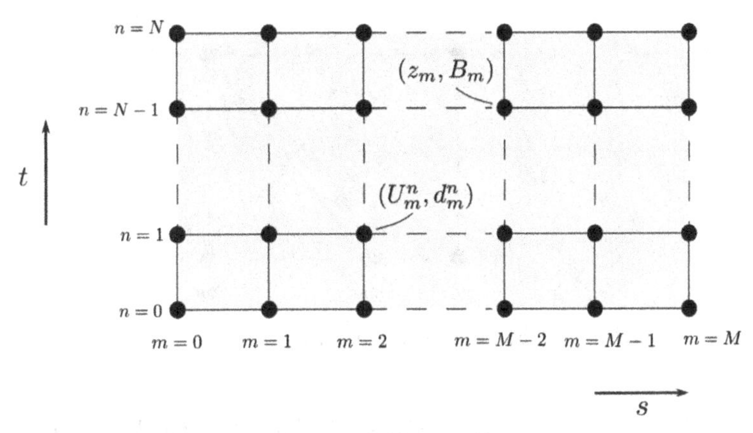

Fig. 11.12 Characteristic method: discrete (s, t)-plane and allocation of flow variables

The collocated positioning of variables leads to sequences of water depths $\left[d_m^n\right]$ and flow velocities $\left[U_m^n\right]$ that are coded as arrays. The array elements at time level $n = 0$ can be specified from the initial conditions.

11.4.3 Forward Time Backward Space Method

The recurrence relation we aim for must advance the numerical solution from one time level t_n, say, to the next (t_{n+1}) along characteristics. The ensuing procedure will be quite similar to the graphical solution method outlined in Section 5.4. We will first consider the homogeneous problem, omitting the right-hand side of Eq. (11.36). It will be reintroduced later on.

Interpolation

Consider a region in the discrete (s, t)-plane comprising the sections between the points s_{m-1} and s_{m+1} and the time interval from t_n to t_{n+1}; see Figure 11.13. Let K_P^+ denote the positive characteristic through node $T = (s_m, t_{n+1})$ issued from point $P = (s_P, t_n)$. Supposing that the flow is subcritical, s_P will be situated to the left of s_m.

Furthermore, with P being situated between s_m and s_{m-1}, an interpolation parameter $\alpha_P \in [0, 1]$ can be defined expressing s_P in terms of s_{m-1} and s_m:

$$s_P = \alpha_P s_{m-1} + (1 - \alpha_P)\, s_m \tag{11.38}$$

By approximation, the positive characteristic velocity V_P^+ in P will be expressed as the average of the characteristic velocities in nodes (s_{m-1}, t_n) and (s_m, t_n), as follows:

$$V_P^+ = \frac{\left(V^+\right)_{m-1}^n + \left(V^+\right)_m^n}{2} \tag{11.39}$$

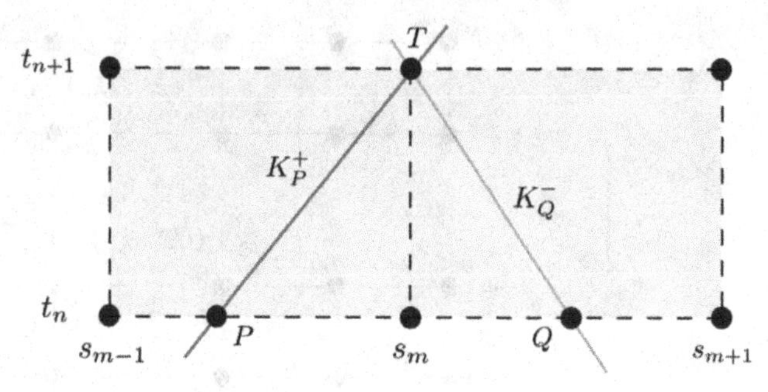

Fig. 11.13 Interpolation along characteristics: subcritical flow

If it is assumed that the characteristic K_P^+ is a straight line (which is exact for simple waves and holds approximately otherwise), the following relation holds:

$$V_P^+ = \frac{s_m - s_P}{\Delta t_n} = \frac{\alpha_P \Delta s_m}{\Delta t_n} \tag{11.40}$$

giving

$$\alpha_P = V_P^+ \frac{\Delta t_n}{\Delta s_m} \tag{11.41}$$

For a homogeneous problem the Riemann invariant R^+ is constant along K_P^+. Its value in target node T is therefore equal to the known value at point P, leading to

$$\left(R^+\right)_m^{n+1} := R_T^+ = R_P^+ = \alpha_P \left(R^+\right)_{m-1}^n + (1 - \alpha_P) \left(R^+\right)_m^n \tag{11.42}$$

A similar procedure applied to the negative characteristic K_Q^- issued from a point $Q = \left(s_Q, t_n\right)$ and going through target node T (see Figure 11.13) yields the Riemann invariant R^- at point s_m and time t_{n+1}:

$$\left(R^-\right)_m^{n+1} := R_T^- = R_Q^- = \alpha_Q \left(R^-\right)_{m+1}^n + \left(1 - \alpha_Q\right) \left(R^-\right)_m^n \tag{11.43}$$

in which the parameter α_Q is given by

$$\alpha_Q = -V_Q^- \frac{\Delta t_n}{\Delta s_{m+1}} \tag{11.44}$$

assuming once again that the flow is subcritical. Importantly, the procedure outlined above is valid only if the spatial interval $[s_{m-1}, s_{m+1}]$ comprises the *domain of dependence* of the target node $T = (s_m, t_{n+1})$ at time level t_n. After the assignments in Eqs. (11.42) and (11.43) both Riemann invariants are known at interior nodes of the domain, which also determines the corresponding flow states. Note that the advective acceleration does not receive a special treatment; it is naturally incorporated in the interpolation procedure.

Boundary Conditions

At the boundary nodes, only the Riemann invariants corresponding to the respective outgoing characteristic have been updated. It remains to complete the solution in these nodes by prescribing the boundary conditions, which effectively determines the Riemann invariant along the ingoing characteristic.

For subcritical flow, the ingoing characteristic at the left boundary ($m = 0$) specifies R^+, the value of which must be prescribed such that the resulting solution for U and c satisfies the local boundary condition at time $t = t_{n+1}$. For an imposed velocity U_0^{n+1} or depth d_0^{n+1} this gives, respectively,

$$\left(R^+\right)_0^{n+1} = 2U_0^{n+1} - \left(R^-\right)_0^{n+1} \qquad \text{or} \qquad \left(R^+\right)_0^{n+1} = 4\sqrt{gd_0^{n+1}} + \left(R^-\right)_0^{n+1} \qquad (11.45)$$

Likewise, at the right boundary node ($m = M$) the negative characteristic is ingoing (provided the flow is subcritical), implying that the local value of R^- must be prescribed, leading to

$$\left(R^-\right)_M^{n+1} = 2U_M^{n+1} - \left(R^+\right)_M^{n+1} \quad \text{or} \quad \left(R^-\right)_M^{n+1} = -4\sqrt{gd_M^{n+1}} + \left(R^+\right)_M^{n+1} \qquad (11.46)$$

imposing the flow velocity (U_M^{n+1}) or water depth (d_M^{n+1}), respectively.

It is also possible to compute the value of the incoming Riemann invariant at the boundary from an equation relating the local water depth to the flow velocity (e.g. a discharge relation of a control structure) or to impose its value directly, the latter giving a nonreflective boundary.

Forcing Terms

To account for nonhorizontal bed level and bed resistance, forcing terms must be added to the right-hand sides of the assignments in Eqs. (11.42) and (11.43). The Riemann invariants are no longer constant along characteristics.[2] The characteristics themselves and the corresponding interpolation procedure are not influenced by the forcing terms.

Having prescribed the bed level z at the nodes, the bed slope term $g\,i_b$ is sectionwise constant. It is independent of the solution and can be computed in advance. We now pick the value belonging to the section in which point P is situated and add this term, after multiplication with Δt_n, to the interpolated value of $\left(R^+\right)_m^{n+1}$ obtained from Eq. (11.42). In a similar way the value of $\left(R^-\right)_m^{n+1}$ is adjusted using the bed slope term in the section holding point Q. See also the remarks in Box 11.5.

Likewise, the resistance term can be computed sectionwise, using the average flow velocity and hydraulic radius in a section, after which the corresponding Riemann variable is augmented with this term. Since the resistance term is nonlinear, it must be recomputed at every time step, either explicitly or by using an implicit linearization step.

[2] In the presence of forcing terms Riemann invariants cease to be 'invariant', for which reason they are also named *characteristic variables*.

Box 11.5 Consistent implementation of forcing terms

The presented implementation of the bed slope term might seem pretty straightforward; however, it has been designed carefully to match with the chosen interpolation procedure. Otherwise, simple balance principles may be violated at the discrete level, resulting in numerical artefacts. To illustrate this, consider a state of rest in section $m-1$. Averaging the characteristic velocities sectionwise and using Eqs. (11.42) and (11.43) then leads to $\left(c_m^n + c_{m-1}^n\right)\left(c_m^n - c_{m-1}^n\right)/\Delta s_m = -g\left(z_m - z_{m-1}\right)/\Delta s_m$. It is easily verified that this implies $z_m + d_m^n = z_{m-1} + d_{m-1}^n$, or $h_m^n = h_{m-1}^n$, which is consistent with the rest state. Balancing the forcing terms consistently is a nontrivial issue in characteristics-based methods; see for instance Toro (2001).

Supercritical Flow

The assignments in Eqs. (11.42) and (11.43) rely on the flow being subcritical. For supercritical flow one of the characteristic velocities will change sign and the interpolation procedure to determine either R_P^+ or R_Q^- will fail. Assuming that the domain of dependence of the target node T at time t_n remains enclosed by the spatial interval $[s_{m-1}, s_{m+1}]$, the solution procedure can be adapted easily to account for supercritical conditions.

Consider supercritical flow with $U > c$, for which the characteristics[3] going through the target node T are shown in Figure 11.14. The intersection point P is situated in the same section as for subcritical flow, and the interpolation procedure formulated in Eqs. (11.38)–(11.42) remains valid. Point Q, however, has moved in upstream direction ending up in the same section as point P. Therefore, a similar procedure as for point P must be followed to determine R^- by interpolating between nodes $m-1$ and m, using an interpolation parameter

$$\alpha_Q = V_Q^- \frac{\Delta t_n}{\Delta s_m} \tag{11.47}$$

The velocity V_Q^- is now based on the state in section $m-1$, using the average of the characteristic velocities in nodes m and $m-1$.

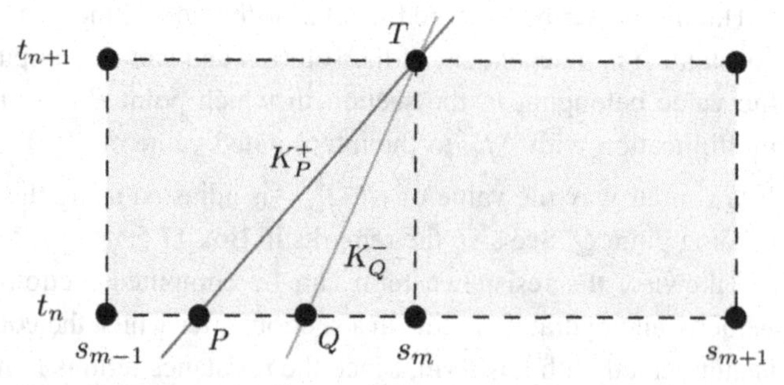

Fig. 11.14 Interpolation along characteristics: supercritical flow with $U > c$

[3] The notation using a '−' superscript (as in K^-, R^- etc.) suggests a negative direction of propagation, which is not true if $U > c$. For the same reason the '+' superscript is perhaps confusing if $U < -c$.

For supercritical flow with $U < -c$, the situation reverses. Point P will now move to another section, while Q stays in the same section as for subcritical flow. The procedure to obtain R_Q^- is therefore the same as for subcritical conditions, using Eqs. (11.43) and (11.44), while R_P^+ must now be interpolated using the same sequence of steps as for R_Q^-.

A practical way to implement this in a model is to compute the increments of the Riemann invariants (i.e. the terms $-\Delta t_n\, V^{\pm}\partial R^{\pm}/\partial s$) sectionwise, after which the result is assigned to either the right node, if $V^{\pm} \geq 0$, or to the left node, if $V^{\pm} < 0$. Summation of the contributions from all sections yields the nodal increments of $R^{\pm}$, which are then simply added to the respective values at the previous time level to obtain an updated solution.

11.4.4 Some Other Characteristics-Based Methods

Characteristics-based methods can be used in many alternative ways, of which we present a brief summary. A more complete overview of possible solution strategies exploiting characteristics is provided in LeVeque (1990).

Higher-Order Interpolation

The solution procedure explained so far is based on linear interpolation of the characteristic variables using the end nodes of the section in which the intersection points P and Q are located. Without altering this solution principle the accuracy can be improved by using *higher-order interpolation*. This requires the addition of supporting nodes to the left and/or right of the respective intersection point. After fitting a polynomial of the desired order through the corresponding nodal values of $R^{\pm}$, the values of R_P^+ and R_Q^- can be computed. The properties of the discrete solution will depend on the polynomial order and the location of supporting nodes. As a rule of thumb, they should be centered around the intersection points P and Q.

The interpolation domain must enclose the domain of dependence of the target node T to obtain stable solutions; see also Section 11.4.5. Increasing the number of supporting nodes also allows a larger domain of dependence of T, which, in principle, increases the maximum allowable time step size. Polynomial orders that are too high should be avoided, however, since the interpolation polynomial will be susceptible to oscillations. This may generate new maxima and minima of the discrete solution and may even compromise the stability of the algorithm. For this reason moderate polynomial orders of one to three, say, are commonly used. Higher-order interpolation methods can be made viable by suppressing the extrema. An example of the latter is given in Priestly (1993).

Implicit Interpolation

If the distance travelled along a characteristic during one time step exceeds the section length, the intersection point with the previous time level t_n will end up in a section that is not adjacent to the target point T. This is the case for the characteristic K_P^+ in Figure 11.15, having an intersection point P' at time level n. The associated variable $R_{P'}^+$ may be found

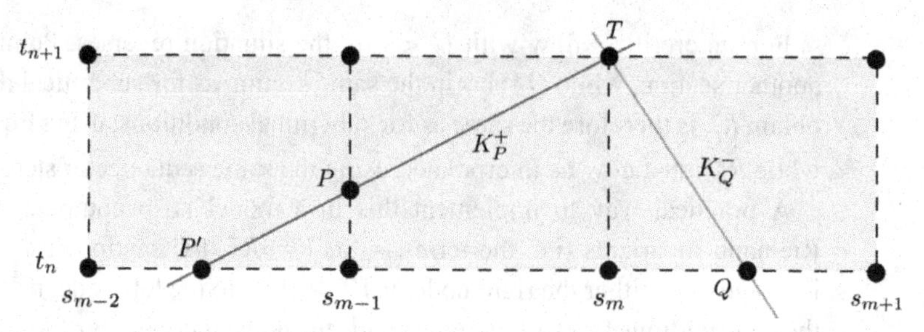

Fig. 11.15 Implicit interpolation (point P) and explicit interpolation (point Q)

by fitting a linear or higher-order interpolation polynomial using at least nodes $m-2$ and $m-1$. The solution procedure remains explicit but is complicated by the fact that the type of interpolation needed depends on the section (possibly varying) in which P' is situated.

As an alternative, we may consider intersection point P having a spatial coordinate s_{m-1} and situated between time levels t_n and t_{n+1}. The time level t_P of this point may be expressed in terms of an interpolation parameter β_P as $t_P = \beta_P t_n + (1-\beta_P)\, t_{n+1}$. To determine β_P, considering the slope of the characteristic K_P^+ leads to

$$V_P^+ = \frac{s_T - s_{m-1}}{t_{n+1} - t_P} = \frac{\Delta s_m}{\beta_P \Delta t_n} \tag{11.48}$$

from which it follows that

$$\beta_P = \frac{\Delta s_m}{V_P^+ \Delta t_n} \tag{11.49}$$

In absence of forcing terms, the Riemann variable R_T^+ at the target point T equals R_P^+ at intersection point P, leading to

$$\left(R^+\right)_m^{n+1} := R_T^+ = R_P^+ = \beta_P \left(R^+\right)_{m-1}^{n} + (1-\beta_P)\left(R^+\right)_{m-1}^{n+1} \tag{11.50}$$

after which the forcing terms can be included in the same way as for the explicit method. Since the variable $\left(R^+\right)_{m-1}^{n+1}$ appearing in Eq. (11.50) is unknown at time level t_n, Eq. (11.50) leads to an implicit solution procedure. That is, a system of equations has to be solved. The intersection point Q in Figure 11.15 is still situated in the section adjacent to the target node T and there is no need to abandon the explicit interpolation procedure for the characteristic K_Q^-. These types of method are coined IMEX (IMplicit-EXplicit), switching between implicit formulations where needed and explicit ones where possible.

Flux Difference Splitting

Instead of tracing the solution back along characteristics, the method of characteristics can also be used to compute fluxes, which are then used to update the solution. To this end the flow variables and channel geometry are defined section wise thus representing piecewise constant functions. They remain collocated. The evolution of these variables in time is governed by the associated fluxes at the interfaces between sections. Since the flow state is discontinuous at interfaces, the interface fluxes are not defined uniquely but must be

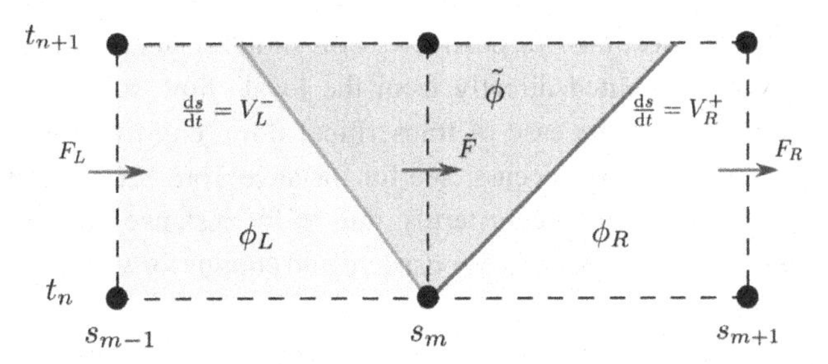

Fig. 11.16 Structure of the local characteristics problem

blended from the two states adjacent to each interface. Simple averaging will not do here. All information will then propagate in two directions, whether or not this is supported by a characteristic, which will lead to instabilities.

Each discontinuity in fact constitutes the initial condition of a local characteristics problem, the solution of which is given by an intermediate state $\tilde{\phi}$, say, separated from the initial states by a pair of characteristics; see Figure 11.16. For subcritical flow, and ignoring forcing terms, the flow state at an interface is the intermediate state, which can be computed from the initial states to the left and right of an interface, respectively, using the solution method from Chapter 5. Once the intermediate state is known, the associated fluxes can be computed, which are then used to update the solution in the sections adjacent to the considered interface, in a way similar to that of Eq. (11.24).

Alternatively, the flux at the interface may be computed directly, by applying conservation laws over portions of the right and left sections containing a characteristic. This approach was initially conceived by Godunov (1959) and developed further in Harten et al. (1983). Equating the increment of a conserved flow variable (d or q) in a section to the associated flux difference over the same section (see Figure 11.16) gives for the left and right sections, respectively,

$$V_R^+ \left(\phi_R - \tilde{\phi}\right) = F_R - \tilde{F} \tag{11.51}$$

$$V_L^- \left(\tilde{\phi} - \phi_L\right) = \tilde{F} - F_L \tag{11.52}$$

where ϕ denotes either the water depth or the specific discharge, and F denotes the associated flux, that is, the specific discharge or specific momentum flux, respectively. Subscripts L and R refer to the sections to the left and right of an interface, and a tilde denotes the interface itself. Eliminating $\tilde{\phi}$ from Eqs. (11.51) and (11.52) gives for the flux at the interface

$$\tilde{F} = \frac{V_R^+ F_L - V_L^- F_R + V_R^+ V_L^- \left(\phi_R - \phi_L\right)}{V_R^+ - V_L^-} \tag{11.53}$$

While derived for subcritical flow, Eq. (11.53) also holds in the case of supercritical flow. This unified treatment of both flow regimes is a clear advantage over interpolation-based methods.

Once the characteristic speeds V_L^- and V_R^+ are known, the interface flux can be computed using Eq. (11.53), and the resulting update of the flow state can be assigned to the left and right sections, respectively. For simple, subcritical flows the characteristic speeds may be computed directly from the local flow state. More sophisticated formulations are needed in the case of transcritical flow (e.g. hydraulic jump) or when flooding and drying of sections occurs; see for instance Toro (2001). An advantage of the flux-based approach is that forcing terms, due to for instance non-uniform width or nonhorizontal bottom, can be added in a consistent and straightforward manner, as for instance in Glaister (1993).

The above procedure is explicit; implicit versions can be developed from similar principles. Accuracy may be increased by determining the initial flow states at an interface using higher-order interpolation. The state within a section is reconstructed to higher order using the information from neighbouring sections. To remain monotone, oscillations that may arise must be suppressed, usually by *limiting* the contribution from higher-order terms where these would otherwise lead to new extrema; see for instance LeVeque (1990).

11.4.5 Properties

In the following some basic properties of the explicit characteristics-based interpolation method will be discussed. We will focus on the homogenous form of the equations.

Accuracy

Substituting the expression for α_P given in Eq. (11.41) into Eq. (11.42), the following discrete equation is satisfied by R^+:

$$\frac{\left(R^+\right)_m^{n+1} - \left(R^+\right)_m^n}{\Delta t_n} + V_P^+ \frac{\left(R^+\right)_m^n - \left(R^+\right)_{m-1}^n}{\Delta s_m} = 0 \tag{11.54}$$

A numerical method of the form of Eq. (11.54), which can be used to discretize an advection equation, is commonly referred to as the *forward time backward space* (FTBS) method, originally conceived by Courant et al. (1952). This method is also used in numerical transport models to discretize the advection term, in which case it is often combined with a term representing diffusion; see Box 11.6.

Since one-sided differencing is used, both in time and in space, the truncation error of Eq. (11.54) behaves as $\mathcal{O}(\Delta s, \Delta t)$ for Δs and Δt sufficiently small. This can be proven by developing $\left(R^+\right)_m^{n+1}$ and $\left(R^+\right)_{m-1}^n$, respectively, as Taylor series from the function properties at (s_m, t_n), following the same procedure as used in Sections 11.2.5 and 11.3.5. See also Problem 11.7.

The same error estimate holds for the Riemann variable R^-, since it satisfies the same type of equation. It follows that the characteristics-based interpolation method is consistent with the characteristic equations given in Eq. (11.36); the numerical error vanishes in the limit of Δt and Δs going to zero.

Box 11.6 Inclusion of a diffusion term

Addition of a diffusion term to Eq. (11.36) results in an advection-diffusion equation, such as Eq. (8.17) in the context of river flood waves, and Eq. (10.70) in the context of transport problems. The advective part of these equations has a characteristic structure and can be integrated numerically using the FTBS-method. The diffusion term has the form $-K\partial^2\phi/\partial s^2$, where K is a diffusion constant and ϕ is the modelled variable (e.g. the water depth in a flood wave model or the species concentration in a transport model). Using central differences, the diffusion term can be discretized as $-K\left(\phi_{m+1} - 2\phi_m + \phi_{m-1}\right)/\Delta s^2$. It can be treated explicitly, using the discrete variable ϕ at the known time level, or implicitly, leading to a system of equations in the unknown ϕ at the new time level. The importance of the discrete diffusion term relative to the advection term is measured by the *mesh Péclet number*, defined as $\text{Pe} = |V|\Delta s/2K$. This determines to a large extent the behaviour of the numerical solution. In particular, when $\text{Pe} > 1$ the solution may become oscillatory if the advection term is discretized by central differences as well, e.g. $V\left(\phi_{m+1} - \phi_{m-1}\right)/2\Delta s$. Backward differencing, as used in the characteristics-based interpolation method, will then cure this problem, at the price of introducing a larger error, acting as a diffusion term.

Monotonicity and Stability

Inspection of Eq. (11.42) reveals that the solution for $\left(R^+\right)_m^{n+1}$ will be bounded by the values of $\left(R^+\right)_{m-1}^n$ and $\left(R^+\right)_m^n$ if $\alpha_P \in [0, 1]$. If this condition is satisfied, the repeated use of Eq. (11.42) will never cause any of the R_T^+ to fall outside the range of previous values of R^+ anywhere in the domain. The discrete solution will be *monotone*.

This desirable property does not depend on the direction of the characteristic. It holds invariably for R^- if $\alpha_Q \in [0, 1]$ and also in the case of supercritical flow. It can therefore be concluded that our characteristics-based interpolation method is monotonicity preserving provided that

$$\frac{|U \pm c|\,\Delta t}{\Delta s} \leq 1 \tag{11.55}$$

which is reminiscent of the CFL condition for stability, Eq. (11.34).

A consequence of monotonicity is that disturbances of the discrete solution will not grow as the recursion proceeds. *A monotone method is therefore also a stable method.* The characteristics-based interpolation method is therefore stable, provided the condition in Eq. (11.55) is respected.

Forcing terms were omitted in the preceding. If they are included, monotonicity of the discrete solution is no longer guaranteed, which is in accordance with physics though. For stability, the bed slope term does not pose any problem since this (constant) term, cancels from the equations describing the behaviour of disturbances. The bed resistance term, on the other hand, is (quadratically) dependent on the flow velocity, which may compromise stability if it is treated in an explicit manner. Implicit treatment of this term must be considered if the resulting maximum time step size is much smaller than the value imposed by Eq. (11.34).

Convergence

Being both consistent and stable, the interpolation method is also convergent. The exact solution of the characteristic equations will be approached when Δt and Δs go to zero, provided they are chosen to meet Eq. (11.34). The global error of the obtained solution, that is, the accumulated error after performing a number of recursions, will be proportional to $\mathcal{O}(\Delta s, \Delta t)$.

It can be shown using Taylor series that this error has the appearance of a diffusion term. This effectively smoothens sharp wave fronts. The lower order accuracy of the FTBS method, compared with for instance the semi-implicit method, is thus the price to be paid for keeping the solution monotone. It was Godunov (1959) who proved that a linear monotone method is at most first-order accurate, known as the *order barrier theorem*.

11.4.6 Python Implementation

A Python implementation of the characteristics-based interpolation method is given in Box 11.7. The first line importing the `pylab` module has been omitted, as are the final lines containing plot functions and write statements.

Initialization

The computation contains a number of steps in which the differences between the discrete variables in neighbouring nodes or midpoint values in sections have to be computed; an example of this is the computation of the characteristic velocity in Eq. (11.39). As a general programming rule, repeated use of elementary program structures should be coded by means of predefined functions or subroutines. For this purpose the functions `dif` and `avg` are defined in lines 5 and 8, respectively, for differencing and averaging arrays. They take an arbitrary array and return a new array containing, respectively, the increments between and averages of successive elements of the input array.

The spatial partitioning is specified in lines 11–14, followed by assigning the variable bed level `z` (line 17) and the resistance coefficient `cf` (line 20), which is assumed constant. The time interval of interest and its partitioning are defined in lines 23–25.

The initial fields for the flow velocity (`U0`) and water level (`h0`) are prescribed in lines 28 and 29, which together with the bed level define the initial wave celerity `c0` (line 30). From these variables the initial solution arrays for U and c are constructed in lines 33 and 34. Next, the corresponding arrays for the initial Riemann invariants R^{+} and R^{-} (denoted by, respectively, `R1` and `R2`) are constructed in lines 37 and 38. The bed slope term gi_b, which is constant, is computed in line 41, which completes the input section.

Box 11.7 Python script Characteristics.py

```
# physical parameters
g = 9.81                                       # gravitation                 [m/s2]

# differencing function
def dif(f): return f[1:M+1] - f[0:M]

# averaging function
def avg(f): return (f[1:M+1] + f[0:M])/2

# spatial domain
L = 3000                                       # length                      [m]
M = 100                                        # number of sections
s = linspace(0,L,M+1)                          # spatial coordinates         [m]
ds = dif(s)                                    # section lengths             [m]

# bed level
z = -5 + (s>L/2)*4                             # bed level                   [m]

# bed resistance
cf = .004                                      # resistance coefficient      [-]

# time stepping
t0,tN = 0,450                                  # start and end time          [s]
N = 200                                        # number of time steps        [-]
t = linspace(t0,tN,N+1)                        # time instances              [t]

# initial conditions
U0 = 0                                         # initial flow velocity       [m/s]
h0 = 0                                         # initial water level         [m]
c0 = sqrt(g*(h0-z))                            # initial wave celerity       [m/s]

# create initial arrays
U = U0*ones(M+1)                               # flow velocity               [m/s]
c = c0                                         # wave speed                  [m/s]

# characteristic variables
R1 = U + 2*c                                   # Riemann variable R+         [m/s]
R2 = U - 2*c                                   # Riemann variable R-         [m/s]

# bed slope term
g_ib = -g*dif(z)/ds                            # bed slope term              [m/s2]

# time loop
for n in range(N):
    dt = t[n+1] - t[n]                         # time step size              [s]
    # forcing term
    R  = avg(c*c/g)                            # hydraulic radius            [m]
    Um = avg(U)                                # flow velocity mid section   [m/s]
    F  = g_ib - (cf/R)*abs(Um)*Um              # forcing term                [m/s2]
    # positive characteristic
    V1 = avg(U + c)                            # characteristic speed V+     [m/s]
    alpha_P = dt*V1/ds                         # interpolation parameter     [-]
    R1[1:M+1] -= alpha_P*dif(R1) - dt*F        # new Riemann variable R+     [m/s]
    # negative characteristic
    V2 = avg(U - c)                            # characteristic speed V-     [m/s]
    alpha_Q = dt*V2/ds                         # interpolation parameter     [-]
    R2[0:M] -= alpha_Q*dif(R2) - dt*F          # new Riemann variable R-     [m/s]
    # left (velocity) boundary condition
    UL = .25*(1+tanh((t[n]-20)/2.))            # imposed velocity            [m/s]
    R1[0] = 2*UL - R2[0]                       # ingoing Riemann variable    [m/s]
    # right (absorbing) boundary condition
    R2[M] = -2*c0[M]                           # ingoing Riemann variable    [m/s]
    # new flow variables
    U = (R1 + R2)/2                            # new flow velocity           [m/s]
    c = (R1 - R2)/4                            # new wave speed              [m/s]
```

Time Stepping

The recurrent loop is defined in line 44 after which the current time step size is calculated first in line 45. The recursion proceeds in three subsequent steps.

First, the forcing term, which is the right-hand side of Eq. (11.36), is computed using the midpoint value of the flow velocity and the hydraulic radius in the respective sections (lines 47–49) and the previously computed bed slope term.

Next, the interpolation procedure is performed by computing for the positive and negative characteristics, respectively, the midpoint values of the characteristic velocity (lines 51 and 55), and the interpolation parameters α_P (line 52) and α_Q (line 56). The increments for the Riemann invariants are computed in lines 53 and 57 by adding the contributions from the interpolation procedure, using the `dif` function, and the forcing terms.

Finally, the boundary conditions are implemented assuming subcritical conditions. In the example script the (left) boundary condition at node $m = 0$ has a prescribed velocity `UL`, which is specified as a function of time in line 59. Together with the already computed value for R_0^-, this defines the value of R_0^+, which is assigned in line 60. At node M, the ingoing characteristic determines R_M^-, the value of which must be computed from the current value of R_M^+, and the prescribed boundary condition. The example has a nonreflective boundary condition here, specifying the value of R_M^- directly.

The recursion ends by computing from the Riemann variables the updated values for the flow velocity `U` (line 64) and the wave celerity `c` (line 65). This is essentially similar to the construction of the intersection point in the state plane, used in the graphical procedure of Chapter 5.

11.4.7 Verification

It is easily verified, by computing for instance the propagation of a positive translatory wave over an abrupt depth discontinuity (see Example 11.4), that computed wave fronts remain monotone, even if they are high and steep. Here, we will demonstrate the capability of the method to reproduce nonlinear effects by studying the deformation of simple waves, for which analytical solutions were derived in Section 5.5.

Example 11.5 The domain is a horizontal canal of uniform width, having an initial depth d_0 of 3.65 m and a length of 8 km. It is partitioned into 80 uniform sections with a length Δs of 100 m each. At the left boundary of the canal a gradually increasing outflow is imposed, with the corresponding velocity U_L prescribed as

$$U_L(t) = U_{\max} \frac{1 + \tanh\left(\left(t - t'\right)/\tau\right)}{2} \tag{11.56}$$

in which the maximum velocity $U_{\max} = -2$ m/s, the time scale $\tau = 25$ s, and the time shift $t' = 250$ s. At the right boundary a nonreflective boundary condition is applied by prescribing the ingoing Riemannn invariant as $R_M^- = -2\sqrt{gd_0}$. A time step size Δt of 10 s is used, giving a Courant number $\sigma \approx 0.6$ based on the initial state, which satisfies the CFL condition.

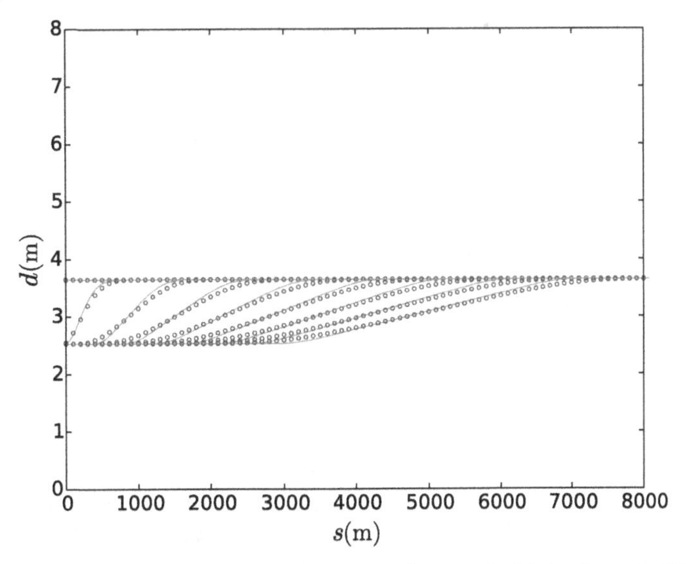

Fig. 11.17 Expansion wave (Example 11.5): progression of wave front starting from $t = 0$ with time intervals of 150 s; analytical solution (solid lines) and computed solution (circles)

Result

The solution is a right-travelling expansion wave that is initially steep, flattening out as it propagates into the canal. Figure 11.17 shows the sequence of computed wave fronts together with the corresponding analytical solution, obtained from Eqs. (5.23) and (5.24). The computed and analytical wave profiles match closely, indicating that the nonlinear effects causing the deformation of the wave are reproduced adequately by the model. The numerical wave deforms slightly faster than predicted by the analytical solution, which is due to the first-order accuracy of the FTBS method.

Example 11.6 The computational set-up of Example 11.6 is used now to simulate a compression wave. To this end, the maximum inflow velocity at the left boundary is set to $U_{\max} =$ 2 m/s and the time scale τ is increased to 125 s. This generates a compression wave that is initially flat and will steepen as it moves into the channel. In reality such a wave will finally break, after which it continues as a bore.

In the initial stages, the deformation of the wave can be computed analytically from Eqs. (5.23) and (5.24). After some time, however, a bore is formed and the characteristic solution ceases to be valid. Reverting to Eqs. (4.19) and (4.21), it can be shown that the bore has a height Δd of 1.3 m and a speed c of 7.6 m/s.

Result

The numerical solution is plotted in Figure 11.18 together with the analytical solution, which is valid during the early stages of the wave deformation. The computed and analytical wave profiles are in good agreement.

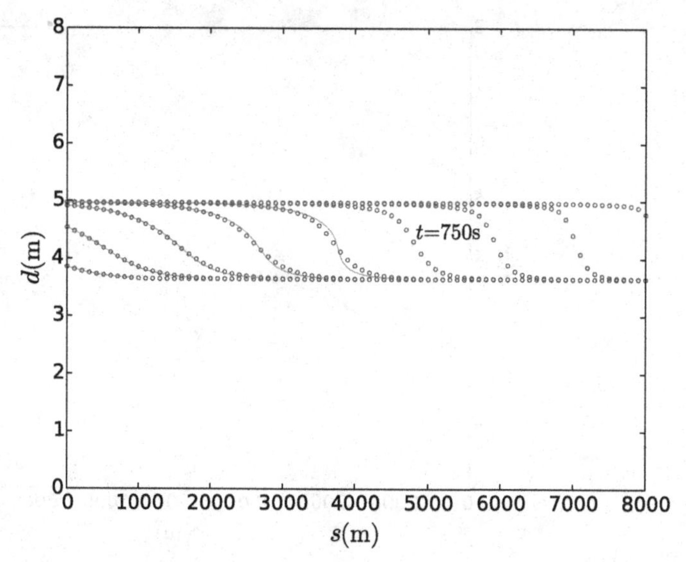

Fig. 11.18 Compression wave (Example 11.6): progression of wave fronts starting from $t = 0$ with time intervals of 150 s; analytical solution (solid lines) and computed solution (circles)

At $t \approx 750\,\mathrm{s}$ the analytical compression wave transforms into a bore. In the model, though, further steepening of the wave front ceases before the actual bore is formed. This is a signature of the first-order interpolation of the FTBS method, which manifests itself as a diffusion term in the discrete equations. This tends to flatten the computed wave and ultimately balances its further steepening, capturing the bore over typically a few nodes. However, the overall properties of the bore, its height and speed of propagation, are reproduced correctly.

The example illustrates that the propagation and deformation of nonlinear waves is modelled adequately by our characteristics-based method. Numerical dissipation influences the rate at which the wave deforms, but also warrants a correct overall behaviour of the computed bore. Should wave fronts have to be resolved more accurately, then (limited) higher-order characteristics-based methods must be used.

Problems

11.1 Prove that the fully implicit method for the discrete basin–canal system is unconditionally stable.

11.2 Using Taylor series expansion, prove that the Crank–Nicolson method for time stepping has a second-order truncation error in the time derivative.

11.3 Use the fully implicit tidal basin model (available online as supplementary material) to compute the natural oscillation of a canal–basin system. Compare the results with those obtained from the semi-implicit method.

11.4 Perform some tests with the script Canal.py in order to verify the CFL condition for the semi-implicit model for long waves.

11.5 Use the script Canal.py to simulate a small amplitude tidal wave in an estuary for different values of the relative resistance σ. Verify the results using analytical solutions from Chapter 7.

11.6 Use the script Canal.py to simulate a river flood wave. Verify the result using the analytical solutions in Chapter 8.

11.7 Prove that the FTBS method has a local truncation error that is first order in time and first order in space.

11.8 Show that a non-uniform canal width can be accommodated in the characteristic equations by adding terms $\mp\left(U\sqrt{gd}/B\right)\mathrm{d}B/\mathrm{d}s$ to the right-hand side of Eq. (11.37). Modify the script Characteristics.py to include this extension and perform some tests to demonstrate its validity.

11.9 Use the script Characteristics.py to simulate a low standing wave in a semi-closed basin without resistance. Perform the same simulation using the script Canal.py. Compare both computed results with the theoretical solution (Chapter 7) and explain the differences.

11.10 Modify the script Characteristics.py to handle supercritical conditions as well. Verify the algorithm by computing a backwater curve for a mildly sloping canal with supercritical inflow and subcritical outflow, as in the river reach downstream of the underflow gate in Figure 9.10.

A

Appendix A Pressurized Flow in Closed Conduits

In the preceding chapters, we have seen that the one-dimensional shallow-water equations (named after De Saint-Venant) form a set of partial differential equations of hyperbolic type. Their mathematical structure is defined by the underlying characteristics, which property is used in the technique of integration. Pressurized flow of a liquid (water, oil, etc.) in closed conduits is mathematically described by a similar set of equations. Therefore, with relatively little extra effort, a solution method for pressurized flow can be obtained. This fact in itself would not be a sufficient justification for the inclusion of this subject in the present book, but since pressurized flow frequently occurs in conjunction with free-surface flow in the context of hydro-engineering projects, it was decided to include the subject, albeit in abbreviated form and in an appendix in order to not interrupt the line of development of the principal subject of this book. The presentation below rests heavily on Chapter 5, to which the reader is referred for more extensive background information. A more detailed account of the subject than is appropriate here can be found in Streeter and Wylie (1967), Jaeger (1977), Fox (1989) and Thorley (1991).

A.1 Introduction

In free-surface flows, storage takes place through variations of the free-surface elevation. This is accompanied by pressure variations of a few metres of water column at most, too small to cause appreciable changes in density. The water can therefore be treated as incompressible. Pressurized flows do not have a free surface, so that a corresponding storage cannot occur. In these cases, storage can take place only through *elasticity of the pipe wall*, allowing profile variations, and *compression of the liquid*, allowing variations in mass in a given volume.

The abrupt closure or opening of a valve or the abrupt switching on or off of a pump in a pipeline for irrigation, hydropower, drinking water supply, etc., either purposefully or as the result of a failure or an accident, results in rapid variations in flow velocity, accompanied by large pressure variations. This phenomenon is called *water hammer* because it can sound as if the pipe wall is struck by a hammer. Too large pressures should be avoided, or at least reduced in view of the limited strength of the materials. Modelling of these effects requires the use of the *constitutive equations* for the elasticity of the pipe wall and the compression of the liquid in addition to the equations of conservation of mass and momentum. This is elaborated in the following.

Water hammer induces negative pressure variations as well. When the pressure reduces to the vapour pressure of the water, vapour bubbles are formed, the so-called process of *cavitation*, resulting in a two-phase system of water and bubbles. This mixture is far more compressible than pure water, so that the speed of propagation of the pressure waves through the pipe/liquid/bubbles system is drastically reduced. Locally, a zone with a free surface of the liquid can develop, even to the point of column separation. These processes are not considered in this chapter.

The actual magnitudes of the variations in cross-sectional area and density are quite small, resulting in an almost rigid response. In fact, if the flow varies gradually, the pressure variations are mild, and these storage effects can be neglected, leading to the so-called *rigid-column approximation*, in which the liquid moves axially as a rigid body. In this case the conservation of mass is fulfilled *a priori* so that we have to deal with the conservation of momentum only.

A.2 Governing Equations

We restrict ourselves to liquids (water, oil, etc.) for which the density varies exclusively as a result of compression, ignoring possible variations of the density due to differences in salinity or temperature.

A.2.1 Constitutive Equations

We need to establish so-called constitutive equations for the liquid and for the pipe wall, providing the connection between the pressure p and

- the liquid density (ρ) and
- the cross-sectional area (A).

We will use linear, elastic models for this purpose.

Liquid Compressibility

The modulus of compression (K) of a liquid is defined through the relation

$$\frac{d\rho}{dp} = \frac{\rho}{K} \tag{A.1}$$

Under normal operating conditions, the modulus of compressibility of water is $K = 2.2$ GPa approximately, virtually independent of pressure or temperature. (In the case where the liquid contains gas or vapour bubbles, even in minute amounts, the bulk value of K is reduced drastically because of the high compressibility of the gas or vapour in the bubbles.)

We will need the partial derivatives of ρ with respect to t and s in the conservation equations. Using Eq. (A.1), these can be expressed as follows in terms of the derivatives with respect to the pressure p:

$$\frac{\partial \rho}{\partial t} = \frac{\mathrm{d}\rho}{\mathrm{d}p}\frac{\partial p}{\partial t} = \frac{\rho}{K}\frac{\partial p}{\partial t} \tag{A.2}$$

$$\frac{\partial \rho}{\partial s} = \frac{\mathrm{d}\rho}{\mathrm{d}p}\frac{\partial p}{\partial s} = \frac{\rho}{K}\frac{\partial p}{\partial s} \tag{A.3}$$

Pipe Elasticity

Consider a pipeline with a circular cross section with inner diameter D and a uniform, relatively thin wall thickness δ (so $\delta \ll D$; see Figure A.1, in which the relative wall thickness has been exaggerated). Suppose now that a small increase in pressure ($\mathrm{d}p$) causes an increase in hoop stress in the pipe wall equal to $\mathrm{d}\sigma$. Neglecting the inertia of the fluid (radially) and of the wall, equilibrium relations can be used, from which it follows that

$$2\delta \times \mathrm{d}\sigma = D \times \mathrm{d}p \tag{A.4}$$

Because of the elasticity of the pipe wall, with modulus E, an increase in hoop stress $\mathrm{d}\sigma$ causes an increase in the circumference ($P = \pi D$) and therefore also of the pipe diameter, which according to Hooke's law can be expressed by

$$\frac{\mathrm{d}D}{D} = \frac{\mathrm{d}P}{P} = \frac{\mathrm{d}\sigma}{E} \tag{A.5}$$

Since the cross-sectional area A is proportional to D^2, and using Eq. (A.4), it follows that

$$\frac{\mathrm{d}A}{A} = 2\frac{\mathrm{d}D}{D} = \frac{D}{\delta}\frac{\mathrm{d}p}{E} \tag{A.6}$$

so that

$$\frac{\mathrm{d}A}{\mathrm{d}p} = \frac{D}{\delta E}A \tag{A.7}$$

Using this, the partial derivatives of A with respect to t and s can be expressed as

$$\frac{\partial A}{\partial t} = \frac{\mathrm{d}A}{\mathrm{d}p}\frac{\partial p}{\partial t} = \frac{D}{\delta E}A\frac{\partial p}{\partial t} \tag{A.8}$$

$$\frac{\partial A}{\partial s} = \frac{\mathrm{d}A}{\mathrm{d}p}\frac{\partial p}{\partial s} = \frac{D}{\delta E}A\frac{\partial p}{\partial s} \tag{A.9}$$

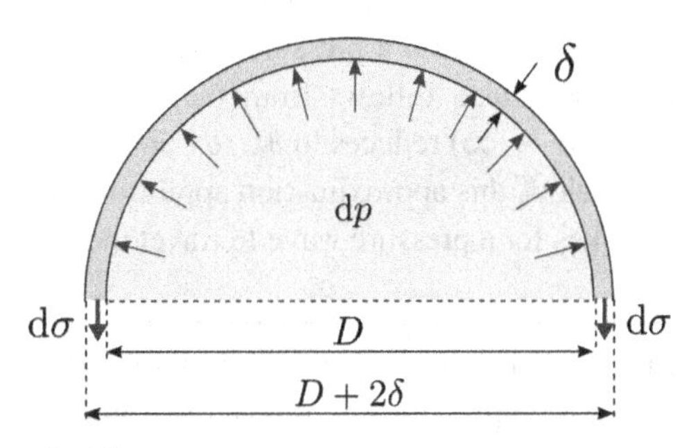

Fig. A.1 Cross section of pipe for pressurized flow

A.2.2 Conservation of Mass

The mass balance for the liquid under pressure in a pipeline reads

$$\frac{\partial}{\partial t}(\rho A) + \frac{\partial}{\partial s}(\rho A U) = 0 \qquad \text{(A.10)}$$

This can be expanded into

$$A\frac{\partial \rho}{\partial t} + \rho\frac{\partial A}{\partial t} + \rho U\frac{\partial A}{\partial s} + \rho A\frac{\partial U}{\partial s} + UA\frac{\partial \rho}{\partial s} = 0 \qquad \text{(A.11)}$$

We substitute Eqs. (A.2), (A.3), (A.8) and (A.9), and divide by A, with the result

$$\left(\frac{\rho}{K} + \frac{\rho D}{E\delta}\right)\frac{\partial p}{\partial t} + \left(\frac{\rho}{K} + \frac{\rho D}{E\delta}\right)U\frac{\partial p}{\partial s} + \rho\frac{\partial U}{\partial s} = 0 \qquad \text{(A.12)}$$

Defining a quantity c through

$$\frac{1}{c^2} = \frac{\rho}{K} + \frac{\rho D}{E\delta} \qquad \text{(A.13)}$$

and substituting this into Eq. (A.12) brings the latter in the following compact form that will be used in water-hammer computations:

$$\frac{\partial p}{\partial t} + U\frac{\partial p}{\partial s} + \rho c^2\frac{\partial U}{\partial s} = 0 \qquad \text{(A.14)}$$

We will see below that c represents the propagation speed of axial pressure waves through the pipeline with the pressurized liquid. In an infinitely rigid pipe ($E \to \infty$), we have $c = \sqrt{K/\rho}$, which is the classical expression for the propagation speed of compression waves (the speed of sound) in a liquid, which for water (without bubbles!) is about 1500 m/s. The elasticity of the pipe wall causes the actual speed in the coupled system to be less than this, often on the order of 1000 m/s in the case of steel pipes; see also Example A.1.

In the approximation of an incompressible liquid and a rigid pipe, $c \to \infty$. This implies that in this approximation a pressure perturbation would be felt instantly over the entire pipe length. This also follows from the mass balance Eq. (A.12), which in this case ($K \to \infty$ and $E \to \infty$) reduces to $\partial U/\partial s = 0$; i.e. the fluid behaves as a rigid column. As we will see below, this approximation applies when the flow varies slowly compared with the time it takes for a pressure wave to travel the length of the pipe.

Example A.1 Consider the pressurized flow of water ($\rho = 1000\,\text{kg/m}^3$) in a pipeline. The bulk modulus (compressibility) of water (K) amounts to 2.2×10^9 Pa. Compute the speed of pressure waves in the pipeline in the case of:

1. a pipeline with an infinitely rigid wall
2. a steel pipeline (tensile modulus $E = 220 \times 10^9$ Pa) with a pipe diameter of 50 times the wall thickness
3. a glass-reinforced plastic pipeline ($E = 17 \times 10^9$ Pa), also with a pipe diameter of 50 times the wall thickness.

Solution

To calculate the speed of pressure waves in a pipeline use Eq. (A.13).

1. For an infinitely rigid pipe wall $E \to \infty$ reducing Eq. A.13 to $1/c^2 = \rho/K$ (second right-hand side term is zero), from which it follows that $c = \sqrt{K/\rho} = 1483$ m/s.
2. Using the full expression for c and setting $D/\delta = 50$ gives $c = 1211$ m/s.
3. Carrying out the same steps as in question 2 leads to $c = 543$ m/s.

Provided the wall thickness does not change relative to the pipe diameter, increasing elasticity of the pipe wall slows down the speed of pressure waves. For synthetic materials frequently used in civil engineering, such as high-density polyethylene, this decrease can be considerable.

A.2.3 Conservation of Momentum

The mass balance must be supplemented with an expression for conservation of momentum. Since we are primarily interested in relatively rapid variations, we will initially neglect flow resistance and use Euler's equation for the streamwise motion:

$$\frac{\partial U}{\partial t} + U\frac{\partial U}{\partial s} + \frac{1}{\rho}\frac{\partial(p + \rho g z)}{\partial s} = 0 \tag{A.15}$$

In a state of equilibrium, the sum $(p + \rho g z)$ is constant (hydrostatic pressure distribution), whose value we set at zero for convenience. (This is like setting the pressure gauges at zero prior to the onset of flow.) With this convention, the pressure p is actually the deviation from this initial hydrostatic value, and we can omit the term $\rho g z$ in Eq. (A.15).

A.3 Pressure Waves in Pipelines

This section deals with applications of the method of characteristics to pressure waves in closed conduits, in particular, pipelines, for which it is highly suitable, as we will see.

A.3.1 Characteristic Equations

The characteristic relations for U and p can be derived from Eqs. (A.14) and (A.15) through the same kind of procedure as was used in Chapter 5 for open water, with the result

$$\frac{dR_p^{\pm}}{dt} = 0 \quad \text{provided} \quad \frac{ds}{dt} = U \pm c \tag{A.16}$$

in which

$$R_p{}^{\pm} = p \pm \rho c U \tag{A.17}$$

Using the so-called piezometric level $h \equiv p/\rho g$ and U as the two state variables, we define

$$R_h{}^{\pm} = h \pm \frac{c}{g} U \tag{A.18}$$

and obtain

$$\frac{dR_h^{\pm}}{dt} = 0 \quad \text{provided} \quad \frac{ds}{dt} = U \pm c \tag{A.19}$$

It follows from these characteristic relations that c is the speed of longitudinal propagation of pressure waves in the coupled fluid–pipe system, relative to the fluid.

Although the pressure-induced variations of the mass density ρ have been accounted for in the mass balance, and so in the expression for the wave speed, the relative magnitude of these is always quite small and can be neglected where ρ appears in the equations as a multiplying factor. The same applies to the diameter D and the wall thickness δ. In this approximation, the wave speed c is independent of the actual state of motion, and lines of constant values of $R_p{}^{\pm}$ in the (U, p) state diagram are straight; so are lines of constant values of $R_h{}^{\pm}$ in the (U, h)-plane. This means that variations in U and p or h are proportional, not only for infinitesimal variations, but also for finite values. Written in finite difference form, the characteristic relations can be expressed as

$$\Delta p = \mp \rho c \Delta U \quad \text{and} \quad \Delta h = \mp \frac{c}{g} \Delta U \quad \text{provided} \quad \frac{ds}{dt} = U \pm c \tag{A.20}$$

These relations are valid for disturbances of arbitrary magnitude, not just infinitesimal ones.

For low disturbances in free-surface flows, the relation $\delta d = \mp (c_0/g) \delta U$ was derived (Eq. (5.22)), valid along the $\pm$ characteristics. At the free surface, $\delta p = 0$, so that $\delta d = \delta h$ and $\delta h = \mp (c_0/g) \delta U$ along the $\pm$ characteristics, the same as for pressurized flow in a pipe. The difference is that for free-surface flows these proportionalities are valid only for weak disturbances, whereas such restriction does not apply in pressurized flows.

A.3.2 Physical Behaviour

As a result of the very limited storage available in pressurized flow (in contrast to free-surface flow where mass can be stored by a rise in the free surface), the speed of pressure waves in pipe flow is quite high, of the order of 1000 m/s; see Example A.1. Relative to this, the flow velocity U can be neglected in the characteristic velocities $U \pm c$, which therefore can be approximated as $ds/dt = \pm c$, in which moreover c can be considered to be a constant (for pipes of constant cross section and elastic properties), independent of the actual pressure. This means that the characteristic velocities to a very good approximation can be considered as constant; i.e. the characteristics are straight lines, independent of the state of motion.

It follows from Eq. (A.20) that large variations in pressure can result from even moderate changes in flow velocity, because of the high values of c. For example, if $c = 1000$ m/s, a pressure variation of 100 m water column results from a change in flow velocity of 1 m/s.

The large pressure variations associated with water hammer can be positive as well as negative. When and where the fluid pressure tends to become lower than the vapour pressure, *cavitation* occurs; i.e. vapour bubbles or even complete cavities are formed. When these collapse, an intense sound is generated (which in some houses can be heard when closing the kitchen tap too rapidly), not unlike the sound caused by hammering on steel.

Water hammer and the associated cavitation can cause serious damage, even fracture, to the pipeline system including its appurtenances such as pumps and valves. Therefore, water hammer requires careful consideration in the design, and it imposes restrictions on the allowable operation of pipeline systems.

A.4 Closure Procedures

Next, some examples are given of application of the method of characteristics to pipe flow involving the operation of valves in pipeline systems. In each case, the solution is determined graphically. It is important to study the example problems and their solutions closely and to rework them. Reference is made to Chapter 5 for more information about the use of the state diagram (the (U, h)-plane) and the characteristics diagram (the (s, t)-plane).

A.4.1 Abrupt Closure

Consider a pipe with length ℓ between two reservoirs. Initially, the flow in the pipe is uniform with velocity U_0. Wall friction and velocity head effects at the upstream end are ignored, so the piezometric level in the pipe at that end is set equal to the free-surface level in the adjacent reservoir, which we take as our reference level $h = 0$ (boundary condition in $s = 0$), which is also the initial level in the whole pipe. At $t = 0$, a valve at the downstream end is suddenly closed completely (boundary condition $U = 0$ in $s = \ell$ for $t > 0$). The solution is given in Figure A.2, showing the (s, t)-diagram and the (U, h) state diagram.

At the location of the closed valve, the initial flow (state I) is suddenly brought to a halt and the pressure rises steeply (state II). The front of this transition travels upstream and reaches the open end at time $t = \ell/c$, where it is 100% negatively reflected (state III) because $h =$ constant as a result of the presence of the reservoir. The reflected negative wave arrives at time $t = 2\ell/c$ at the end of the closed valve and is reflected there by 100% (state IV).

At time $t = 4\ell/c$, the front has traveled up and down the pipe once more, after which the original state is restored and the process repeats itself with a period $T = 4\ell/c$. This goes on

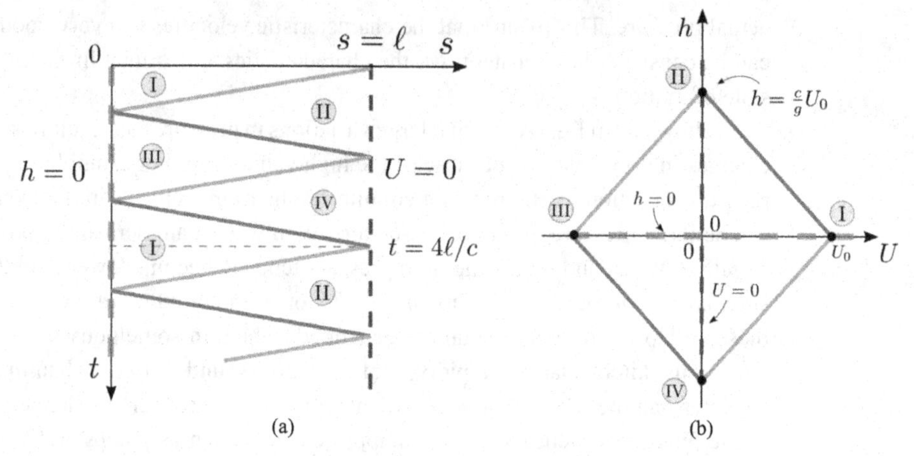

Fig. A.2 Abrupt closure at downstream end of a pipe; (s, t)-diagram (a) and (U, h)-state diagram (b)

'forever' because energy losses (due to wall friction and expansion at the exit) have been neglected.

Ignoring cavitation, the maximum and minimum piezometric level are $\pm c\,U_0/g$ higher/lower than the undisturbed value of zero, corresponding to pressure variations of $\pm\rho c U_0$. The states I through IV (Figure A.2b) are indicated in the (s, t)-diagram (Figure A.2a) as well as in the graphs in Figure A.3 where they are visible as a sequence of longitudinal profiles.

At a fixed point, the maximum and minimum pressures alternate as time goes on. Near either end, the durations of maximum and of minimum pressure are unequal (consider some sections s = constant in the (s, t)-plane), but halfway down the length of the pipe they last equally long.

Figure A.4 shows a time record of absolute pressure head measured at a fixed point in the middle of the pipe for the situation considered presently. The initial pressure was sufficiently high to prevent the occurrence of cavitation (see the minimum pressure head of a little more than -10 m water column). The pattern of the pressure variations and the values agree with the theory. The most notable deviation is the gradual decay of the measured oscillations, which is not predicted by this theory in which all losses were neglected. (For a quantitative check, the experimental data listed in Problems A.9 and A.10 at the end of this chapter can be used.)

A.4.2 Gradual Closure

The large pressure variations associated with abrupt closure of a valve (or the sudden start or shut-off of a pump) are undesirable. They can be avoided to a controllable extent by a more gradual closure. This can be seen as follows.

The preceding example shows that the high-pressure wave, originated at the location of the sudden closure, is negatively reflected at the other, open end of the pipe. When this reflected, negative wave arrives at the closed end, it can compensate for the pressure build-

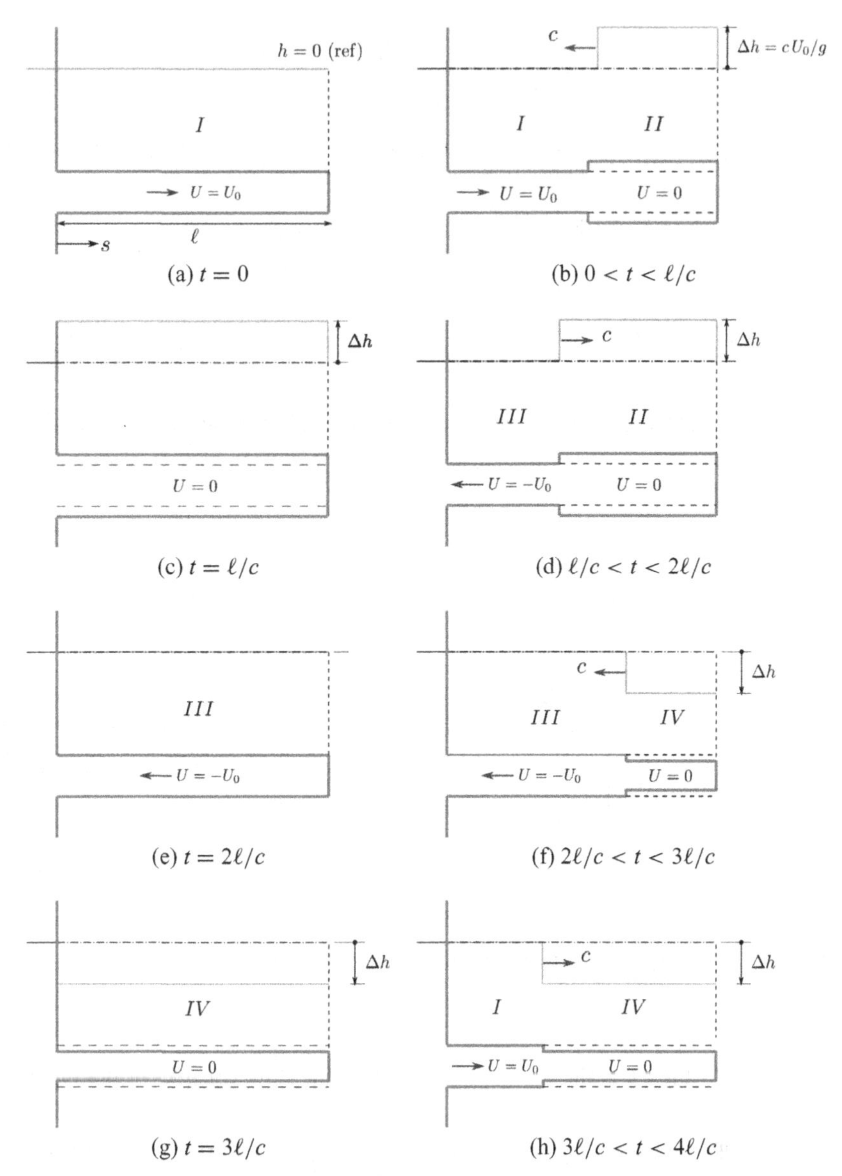

Fig. A.3 Abrupt closure: snapshots of pressure variation and flow velocity in a pipeline; state at $t = 4\ell/c$ equals that at $t = 0$ (a)

up there, provided the closure was not yet complete by the time of arrival of the reflected, negative wave. This means that the closure should take longer than $2\ell/c$.

In the following elaboration of this idea we assume the same situation and approximations as in the preceding example, except for the presence of a valve at the downstream end. Initially, the valve is fully open ($h = 0$). We assume the following relationship to describe the effect of partial closure of the valve:

$$h = \left(\frac{1}{\mu^2} - 1\right)\frac{|U|U}{2g} \tag{A.21}$$

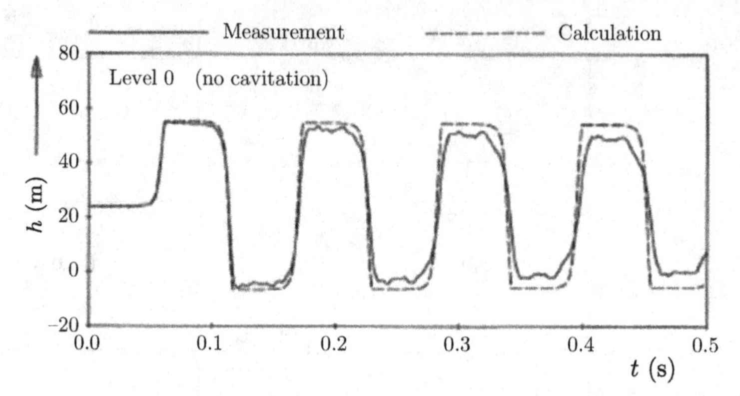

Fig. A.4 Abrupt closure: measured (solid line) and computed (dashed line) pressures; from Tijsseling (1993)

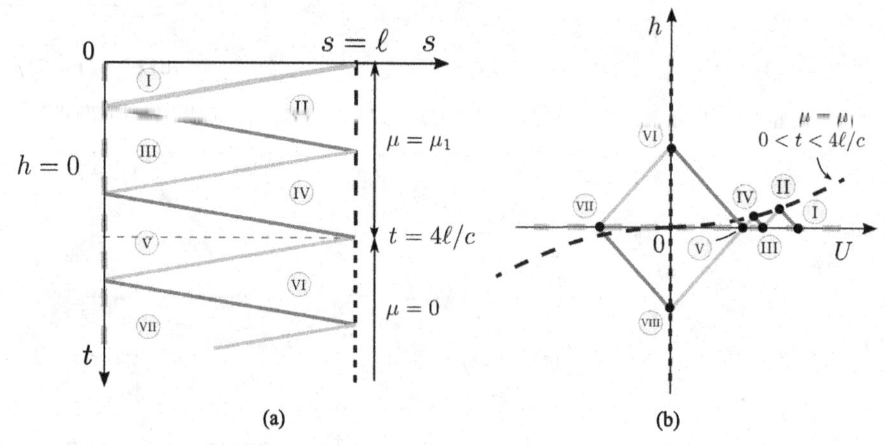

Fig. A.5 Gradual closure at downstream end of a pipe; (s, t)-diagram (a) and (U, h)-state diagram (b)

in which h is the piezometric level; U the flow velocity in the pipe near the valve, taken positive for outflow; and μ is the ratio of the effective cross-sectional area of the valve opening to that of the pipe. This relation is shown graphically in the state diagram of Figure A.5 as two mirrored half-parabolas. For a fully open valve, $\mu = 1$ ($h = 0$ for finite U), and for a fully closed valve, $\mu = 0$ ($U = 0$ for finite h).

Instead of using a truly gradual closure we approximate it as a two-step process, assuming that at $t = 0$ the valve is abruptly but partially closed, such that $\mu = \text{constant} = \mu_1$ for $0 < t < 4\ell/c$, with $0 < \mu_1 < 1$; it is abruptly fully closed ($\mu = 0$) at time $t = 4\ell/c$.

The solution is shown in Figure A.5. It can be seen that the partial closure causes a moderate pressure rise (state II). In the end, after the valve has been closed completely, a periodic state is established (states V-VI-VII-VIII-V), as in the preceding example, but the maximum and minimum pressures are smaller in absolute magnitude due to the effect of the reflected negative wave arriving at the valve at $t = 2\ell/c$, when the valve was not yet fully closed.

It is obvious from the above that closing the valve in a sequence of small steps, or gradually, can reduce the maximum and minimum pressures (in absolute magnitude) at will, provided the duration is sufficiently long compared with the basic travel time of $2\ell/c$.

A.4.3 Influence of Exit Losses and/or Wall Friction

We return to the situation first considered, of initially uniform flow in a pipe that at the downstream end is abruptly closed. The only difference is in the boundary condition at $s = 0$, where we now take velocity head effects into account, leading to $h = 0$ during outflow and $h = -U^2/2g$ during (streamlined) inflow. This results in the following relation to be imposed in $s = 0$:

$$h = \begin{cases} 0 & \text{if } U \leq 0 \\ -U^2/2g & \text{if } U > 0 \end{cases} \tag{A.22}$$

The solution is presented in Figure A.6. Again, an oscillation develops with a period equal to $4\ell/c$, but this time the extreme values decrease in time as a result of the assumed exit losses.

So far, we neglected wall resistance. It can in principle be taken into account with the method described in Chapter 5, using numerous points distributed along the length of the pipe. The overall effect, i.e. the gradual decay of the oscillations, can be obtained more simply by lumping the overall resistance in one or two end points, accounting for it through a modification of the boundary conditions.

A.4.4 Influence of Time Scales

The examples presented above clearly show that two time scales are important: the time scale τ_e of external influences, e.g. the duration of closure of a valve on the one hand, and the internal system time scale of the travel time of pressure waves over a pipe length (ℓ/c) on the other. Their ratio determines the dynamics of the system:

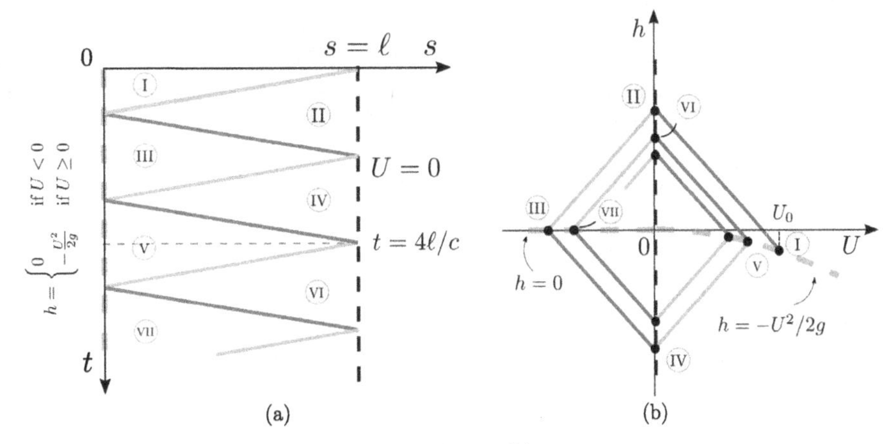

Fig. A.6 Influence of exit losses; (s, t)-diagram (a) and (U, h)-state diagram (b)

- For $\tau_e \ll \ell/c$ (relatively fast excitation), dynamic effects are important and compression waves must be taken into account.
- For $\tau_e \gg \ell/c$, the influence of (slowly varying) boundary conditions is quasi-instantaneously present throughout the pipe length via frequent pressure waves travelling back and forth.

In the latter case, by approximation, the fluid reacts as a rigid column; compression and expansion waves need not be considered.

Problems

A.1 Where is storage taking place in pressurized flow?

A.2 Describe an important consequence of the difference in character and magnitude of the storage as compared with free-surface flows.

A.3 Derive the mass balance equation for pressurized flow in a pipe.

A.4 Considering the characteristic velocities of pressure waves in a closed pipe, what can be said about the relative importance of the flow velocity compared with the wave speed?

A.5 Verify step by step the graphical solutions for the examples of Section A.4 by critical analyses of the corresponding (s, t)-diagram and the state diagram.

A.6 Elaborate the examples of Section A.4 by making sketches of the corresponding (s, t)-diagram and the state diagram.

A.7 Choose some instants in the solutions of the examples of Section A.4 and sketch the corresponding longitudinal profiles of the state variables.

A.8 Choose some fixed locations in the solutions of the examples of Section A.4 and sketch the corresponding time variations of the state variables at those points.

A.9 The following data apply to the experiments of Figure A.4: $E = 120$ GPa, $D = 19.05$ mm, $\delta = 1.588$ mm, $\ell = 36$ m, $K = 1.95$ GPa, $\rho = 1000$ kg/m^3. Calculate c and ℓ/c and compare the latter value with the experimental data in the figure.

A.10 The pressure shown in Figure A.4 was measured after the sudden closure of a valve in a flow with initial velocity $U_0 = 0.239$ m/s. Calculate the difference between the maximum and the minimum piezometric levels and compare it with the experimental result in the figure.

Appendix B Summary of Formulas

One-Dimensional Shallow Water Equations

Continuity equation

$$B\frac{\partial h}{\partial t}+\frac{\partial Q}{\partial s}=0$$

Momentum equation

$$\frac{\partial Q}{\partial t}+\frac{\partial}{\partial s}\left(\frac{Q^2}{A_c}\right)+gA_c\frac{\partial h}{\partial s}+c_f\frac{|Q|Q}{A_cR}=0$$

Elementary Wave Equation

Partial differential equation

$$\frac{\partial^2 h}{\partial t^2}-c^2\frac{\partial^2 h}{\partial s^2}=0$$

where $c=\sqrt{gA_c/B}$

General solution (d'Alembert)

$$h\,(s,t)=h^+\,(s-ct)+h^-\,(s+ct)$$

Corresponding discharge, flow velocity

$$\delta Q=Bc\left(\delta h^+-\delta h^-\right),\quad \delta U=\frac{g}{c}\left(\delta h^+-\delta h^-\right)$$

Translatory Waves

Reflection and transmission (low wave)

Abrupt change of cross section ($1 \rightarrow 2$):

$$r_r \equiv \frac{\delta h_r}{\delta h_i} = \frac{1-\gamma}{1+\gamma}, \quad r_t \equiv \frac{\delta h_t}{\delta h_i} = \frac{2}{1+\gamma}$$

where $\gamma = \dfrac{B_2 c_2}{B_1 c_1}$

Gradual change of cross section (Green's law)

$$Bc\,\zeta^2 = \text{constant}$$

Wave damping (of pulse with height $\hat{\zeta}$)

$$\hat{\zeta}(s) = \frac{\hat{\zeta}_0}{1 + s/S}$$

where $S = \dfrac{2\alpha\, d^2}{c_f\, \hat{\zeta}_0}$, $\alpha = 1$ (block pulse), $\alpha = \frac{3}{8}\pi$ (half-sinusoidal pulse)

Bore entering still water

$$c = \sqrt{g \frac{d_0 + d_1}{2} \frac{d_1}{d_0}}$$

Characteristic Method

Characteristic relations for prismatic conduits (horizontal bottom, without friction)

$$U \pm 2c = \text{constant} \quad \text{along} \quad \frac{\mathrm{d}s}{\mathrm{d}t} = U \pm c$$

Solution for small disturbances, superimposed on rest state (d_0, c_0)

$$\delta d = \mp \frac{c_0}{g} \delta U = \mp \sqrt{\frac{d_0}{g}}\, \delta U \quad \text{along} \quad \frac{\mathrm{d}s}{\mathrm{d}t} = \pm c_0$$

Linearized Channel-Basin System

Differential equation

$$\omega_0^{-2}\frac{\mathrm{d}^2\zeta_b}{\mathrm{d}t^2}+\tau\frac{\mathrm{d}\zeta_b}{\mathrm{d}t}+\zeta_b=\zeta_s$$

where $\omega_0=\sqrt{\dfrac{gA_c}{\ell A_b}}$, $\tau=\dfrac{8}{3\pi}\chi\dfrac{\hat{Q}A_b}{gA_c^2}$ and $\chi=\frac{1}{2}+c_f\dfrac{\ell}{R}$

Complex response factor

$$\tilde{r}\equiv\frac{\tilde{\zeta}_b}{\tilde{\zeta}_s}=\frac{1}{1-\omega^2/\omega_0^2+i\omega\tau}$$

Amplitude ratio

$$\hat{r}\equiv\frac{\hat{\zeta}_b}{\hat{\zeta}_s}=\frac{1}{\sqrt{2}\,\Gamma}\sqrt{-\left(1-\omega^2/\omega_0^2\right)^2+\sqrt{\left(1-\omega^2/\omega_0^2\right)^4+4\Gamma^2}}$$

where $\Gamma\equiv\dfrac{\omega\tau}{r}=\dfrac{8}{3\pi}\chi\left(\dfrac{A_b}{A_c}\right)^2\dfrac{\omega^2\hat{\zeta}_s}{g}$

Phase lag

$$\tan\theta=\frac{\omega\tau}{1-\omega^2/\omega_0^2}=\frac{\Gamma r}{1-\omega^2/\omega_0^2}$$

Harmonic Method

Differential equation

$$\frac{\mathrm{d}^2\tilde{\zeta}}{\mathrm{d}s^2}+k_0^2\left(1-i\sigma\right)\tilde{\zeta}=0$$

where $k_0=\dfrac{c_0}{\omega}$, $c_0=\sqrt{gA_c/B}$, $\sigma=\kappa/\omega$, and $\kappa=\dfrac{8}{3\pi}c_f\dfrac{\hat{Q}}{A_cR}$

General solution

$$\tilde{\zeta}(s)=C^+\exp\left(-ps\right)+C^-\exp\left(ps\right)=\tilde{\zeta}^++\tilde{\zeta}^-$$

where $p\equiv\mu+ik$, $k=\dfrac{k_0}{\sqrt{1-\tan^2\delta}}$, $\mu=k\tan\delta$, $\delta=\frac{1}{2}\arctan\sigma$

Corresponding discharge

$$\tilde{Q}(s) = \frac{i\omega B}{p}\left(C^{+}\exp(-ps) - C^{-}\exp(ps)\right) = Bc\cos\delta\exp(i\delta)\left(\tilde{\zeta}^{+} - \tilde{\zeta}^{-}\right)$$

Basin closed at one side ($s = \ell$)

$$\tilde{\zeta}_{\ell} = \frac{\tilde{\zeta}_0}{\cosh p\ell}$$

$$\tilde{Q}_0 = \frac{i\omega B}{p}(\tanh p\ell)\,\tilde{\zeta}_0$$

where $\sinh p\ell = \cos k\ell \sinh \mu\ell + i \sin k\ell \cosh \mu\ell$
$\cosh p\ell = \cos k\ell \cosh \mu\ell + i \sin k\ell \sinh \mu\ell$

Response factor

$$r \equiv \frac{\hat{\zeta}_{\ell}}{\hat{\zeta}_0} = \frac{1}{|\cosh p\ell|} = \frac{1}{\sqrt{\sinh^2 \mu\ell + \cos^2 k\ell}}$$

Discharge amplitude

$$\hat{Q}_0 = Bc\cos\delta\,|\tanh p\ell|\,\hat{\zeta}_0 = Bc\cos\delta\sqrt{\frac{\sinh^2\mu\ell + \sin^2 k\ell}{\sinh^2\mu\ell + \cos^2 k\ell}}\,\hat{\zeta}_0$$

Flood Waves

Quasi-steady approximation

$$B\frac{\partial d}{\partial t} + \frac{\partial Q}{\partial s} = 0$$

where $Q = Q_u\sqrt{1 - i_b^{-1}\partial d/\partial s}, \quad Q_u = A_c\sqrt{gR\,i_b/c_f}$

Kinematic wave model, prismatic conduit

$$\frac{\mathrm{d}d}{\mathrm{d}t} = 0 \quad \text{along} \quad \frac{\mathrm{d}s}{\mathrm{d}t} = c_{\mathrm{HW}}$$

where $c_{\mathrm{HW}} = \dfrac{1}{B}\dfrac{\mathrm{d}Q_u}{\mathrm{d}d} = \dfrac{3}{2}\dfrac{1}{B}\dfrac{Q_u}{d} = \dfrac{3}{2}\dfrac{B_c}{B}U_u$

Diffusion wave model, prismatic conduit (small deviations from uniform flow state)

$$\frac{\partial d}{\partial t} + c_{\mathrm{HW}} \frac{\partial d}{\partial s} - K \frac{\partial^2 d}{\partial s^2} = 0$$

where $K = \dfrac{Q_u}{2 i_b B}$

Response following abrupt release of volume V in $s = 0$ at time $t = 0$

$$d(s,t) = d_u + \frac{V/B}{\sqrt{2\pi}\,\sigma_s} \exp\left(-\frac{(s - c_{\mathrm{HW}} t)^2}{2\sigma_s^2}\right)$$

where $\sigma_s = \sqrt{2Kt}$

Steady Flow

Discharge relations

Underflow gate (free outflow)

$$q = \sqrt{2g \frac{d_1^2 d_2^2}{d_1 + d_2}}$$

Weir (free outflow)

$$q = m\tfrac{2}{3} h \sqrt{g \tfrac{2}{3} h}$$

where h is the upstream water level with respect to the crest of the weir

Hydraulic jump

$$q = \sqrt{g d_1 d_2 \frac{d_1 + d_2}{2}}$$

Backwater curves

Backwater equation

$$\frac{\mathrm{d}d}{\mathrm{d}s} = \frac{i_b - i_f}{1 - Fr^2} \cong i_b \frac{d^3 - d_u^3}{d^3 - d_{cr}^3} \quad (\text{for} \quad R \cong d)$$

where $d_{cr} = \left(\dfrac{q^2}{g}\right)^{1/3}, \quad d_u = \left(\dfrac{c_f}{i_b} \dfrac{q^2}{g}\right)^{1/3} = \left(\dfrac{c_f}{i_b}\right)^{1/3} d_{cr}$

Solution for small deviations from uniform flow state, boundary condition $d = d_0$ at $s = s_0$

$$d(s) = d_u + (d_0 - d_u) \exp\left(\frac{s - s_0}{L}\right)$$

where $L = \dfrac{1 - i_b/c_f}{3 i_b} d_u$ (adaptation length)

Solution for horizontal bed

$$\tfrac{1}{4} d^4 - d_{cr}^3 d + c_f d_{cr}^3 s = \text{const}$$

Uniform flow

Equilibrium relation

$$\tau_b = \rho g R i_b$$

Shear velocity

$$u_* \equiv \sqrt{\tau_b/\rho} = \sqrt{c_f}\, U$$

Uniform flow velocity, discharge

$$U = \sqrt{g R i_b / c_f}, \quad Q = A_c \sqrt{g R i_b / c_f}$$

where $\dfrac{1}{\sqrt{c_f}} = 5.75 \log\left(\dfrac{12R}{k}\right)$

Transport

Vertical diffusion

Turbulence diffusivity

$$\epsilon_t = \kappa u_* z \left(1 - \frac{z}{d}\right)$$

Vertical distribution of horizontal velocity

$$u(z) = \frac{u_*}{\kappa} \ln \frac{z}{z_0}, \quad z_0 = k/30 \quad \text{(Nikuradse)}$$

Advection and two-dimensional dispersion (zone II)

Depth-averaged advection–diffusion equation

$$\frac{\partial \bar{c}}{\partial t} + U \frac{\partial \bar{c}}{\partial x} - \frac{1}{d} \frac{\partial}{\partial x}\left(K_x d \frac{\partial \bar{c}}{\partial x}\right) - \frac{1}{d} \frac{\partial}{\partial y}\left(K_y d \frac{\partial \bar{c}}{\partial y}\right) = 0$$

$$K_x \cong 6\, u_* d, \quad K_y \cong 0.6\, u_* d \pm 50\%$$

Solution for instantaneous release of mass M in $(x, y) = (0, 0)$ at time $t = 0$, uniform depth

$$\bar{c}(x,y,t) = \frac{M/d}{2\pi\,\sigma_x\sigma_y}\exp\left(-\frac{(x-Ut)^2}{2\sigma_x^2} - \frac{y^2}{2\sigma_y^2}\right)$$

where $\sigma_x = \sqrt{2K_x t}$ and $\sigma_y = \sqrt{2K_y t}$

Solution for continuous release $\dot{M}$ in $(x, y) = (0, 0)$, uniform depth

$$\bar{c}(x,y) = \frac{\dot{M}/q}{\sqrt{2\pi}\,\sigma_y}\exp\left(-\frac{y^2}{2\sigma_y^2}\right) \quad (x > 0)$$

where $\sigma_y = \sqrt{2K_y x/U}$

Advection and one-dimensional dispersion (zone III)

Cross-sectionally averaged advection–diffusion equation

$$\frac{\partial\bar{\bar{c}}}{\partial t} + U\frac{\partial\bar{\bar{c}}}{\partial s} - K\frac{\partial^2\bar{\bar{c}}}{\partial s^2} = 0$$

$$K = \frac{\gamma}{\sqrt{c_f}}\frac{QB_c}{d^2}$$ in which $\gamma \cong 10^{-2}$, with an uncertainty of a factor of about 4

Solution for instantaneous release M in $s = 0$ at time $t = 0$, uniform cross section

$$\bar{\bar{c}}(s,t) = \frac{M/A}{\sqrt{2\pi}\,\sigma_s}\exp\left(-\frac{(s-Ut)^2}{2\sigma_s^2}\right)$$

where $\sigma_s = \sqrt{2Kt}$

Pressurized Pipe Flow

Continuity equation

$$\frac{\partial p}{\partial t} + U\frac{\partial p}{\partial s} + \rho c^2\frac{\partial U}{\partial s} = 0$$

where $c = 1/\sqrt{\rho/K + \rho\delta/(ED)}$ (wave speed)

Momentum equation

$$\frac{\partial U}{\partial t} + U\frac{\partial U}{\partial s} + \frac{1}{\rho}\frac{\partial\,(p+\rho g z)}{\partial s} + c_f\frac{|U|U}{R} = 0$$

Characteristic relations, absence of forcing terms (resistance, slope)

$$\Delta p = \mp\rho c\,\Delta U \quad \text{and} \quad \Delta h = \mp\frac{c}{g}\,\Delta U \quad \text{along} \quad \frac{ds}{dt} = U \pm c$$

References

Abbott, M. B., and Basco, D. R. 1997. *Computational fluid mechanics: an introduction for engineers*. Water Science and Technology Library. Longman Scientific & Technical.

Bertrand, G., and Hiver, J. M. 1998. *Quatrième Écluse de Lanaye Études des ondes de sassement*. Tech. rept. Rapport Intermédiaire 2: Annexes. Laboratoires de Recherches Hydrauliques. Chatelet, Belgium. In French.

Bos, M. G. (ed). 1989. *Discharge measurement structures*. 3rd edn. International Institute for Land Reclamation and Improvement, Wageningen, The Netherlands.

Brickhill, P. 1951. *The dam busters*. Evans Bros., London.

Chanson, H. 2012. *Tidal bores*. World Scientific, Singapore.

Chézy, A. 1768. Cited in Rouse and Ince, 1957. Pages 118–119.

Chaudry, M. H. 1993. Open-Channel Flow. Prentice Hall, Englewood Cliffs, N.J.

Courant, R., Friedrichs, K. O., and Lewy, H. 1928. Über die partiellen Differenzengleichungen der mathematischen Physik. *Mathematische Annalen*, **100**(1–2). In German.

Courant, R., Isaacson, E., and Rees, M. 1952. On the solution of nonlinear hyperbolic differential equations by finite differences. *Communications on Pure and Applied Mathematics*, **5**(3), 243–255.

De Jong, M. P. C., and Battjes, J. A. 2004. Low-frequency sea waves generated by atmospheric convection cells. *Journal of Geophysical Research*, **109**, C01011. doi:10.1029/2003JC001931.

De Saint-Venant, M. 1871. Théorie du mouvement non permanent des eaux, avec application aux crues des rivières et à l'introduction des marées dans leur lit. *Comptes Rendus Hebdomadaires des Séances de l'Académie des Sciences*, **73**, 147–154. In French.

Dronkers, J. J. 1964. *Tidal computations in rivers and coastal waters*. North-Holland, Amsterdam.

Dubs, R. 1909. Stollen und Wasserschloß. Pages 219–221 of: Dubs, R., Bataillard, V., and Alliévi, L. (eds), *Allgemeine Theorie über die veränderliche Bewegung des Wassers in Leitungen*. Verlag von Julius Springer, Berlin. In German.

Elder, J. W. 1959. The dispersion of marked fluid particles in turbulent shear flow. *Journal of Fluid Mechanics*, **5**, 544–560.

Fischer, H. B., List, C., Koh, C., Imberger, J., and Brooks, N. 1979. *Mixing in inland and coastal waters*. Academic Press. San Diego, CA.

Forel, F. A. 1875. Les Seiches – Vagues d'oscillation fixe des lacs. *Verhandlungen der Naturforschenden Gesellschaft*, **58**, 157–168. In French.

Fox, J. A. 1989. *Transient flow in pipes, open channels and sewers*. Ellis Horwood, Chichester.

French, R. H. 1985. Open-Channel Hydraulics. McGraw-Hill, New York.

Glaister, P. 1993. Flux difference splitting for open-channel flows. *International Journal for Numerical Methods in Fluids*, **16**(7), 629–654.

Godunov, S. K. 1959. Difference scheme for numerical solution of discontinuous solution of hydrodynamic equations. *Math Sbornik*, 271–306. Translated US Joint Publ. Res. Service, JPRS 7226, 1969.

Green, G. 1837. On the motion of waves in a variable canal of small depth and width. In: Ferrers, N. M. (ed), *Mathematical papers of the late George Green*. Cambridge University Press, Cambridge, 2014.

Harten, A., Lax, P. D., and Van Leer, B. 1983. On upstream differencing and Godunov-type schemes for hyperbolic conservation laws. *SIAM Review*, **25**(1), 35–61.

Henderson, F. M. 1966. *Open channel flow*. Macmillan, New York.

Hetland, M. L. 2005. *Beginning Python*. Apress, Berkeley, CA.

Ippen, A. T., and Harleman, D. R. F. 1966. Tidal dynamics in estuaries. Pages 493–545 of: Ippen, A. T. (ed), *Estuary and coastline hydrodynamics*. McGraw-Hill, New York.

Jaeger, C. 1977. *Fluid transients (in hydro-electric practice)*. Blackie & Son, Glasgow.

Keulegan, G. H. 1951. *Water level fluctuations of basins in communication with seas*. Tech rept. 1146. National Bureau of Standards, Washington, DC.

Lamb, H. 1932. *Hydrodynamics*. Dover, New York.

Langtangen, H. P. 2009. *Python scripting for computational science*. Springer Verlag, Berlin.

LeVeque, R. J. 1990. *Numerical methods for conservation laws*. Birkhauser Verlag, Basel.

Lorentz, H. A. 1926. *Verslag Staatscommissie Zuiderzee*. Tech. rept. In Dutch.

Manning, R. 1889. On the flow of water in open channels and pipes. *Transactions of the Institution of Civil Engineers of Ireland*, **20**, 166–195.

Massau, M. J. 1878. Mémoires sur l'intégration graphique et ses applications. *Extrait des Annales de l'Association des ingénieurs sortis des écoles spéciales de Gand*. Imprimerie Félix Callewaert Pére, Bruxelles. In French.

Mehta, A. J., and Özsoy, E. 1978. Flow dynamics and nearshore transport. In: Bruun, P. (ed), *Stability of tidal inlets*. Developments in geotechnical engineering, vol. 23. Elsevier, Amsterdam.

Merrifield, M. A., Firing, Y. L., Aarup, T., Agricole, W., Brundrit, G., Chang-Seng, D., Farre, R., Kilonsky, B., Knight, W., Kong, L., Magori, C., Manurung, P., McCreery, C., Mitchell, W., Pillay, S., Schindele, F., Shillington, F., Testut, L., Wijeratne, E. M. S., Caldwell, P., Jardin, J., Nakahara, S., Porter, F.-Y., and Turetsky, N. 2005. Tide gauge observations of the Indian Ocean tsunami, December 26, 2004. *Geophysical Research Letters*, **32**, L09603, doi:10.1029/2005GL022610.

Miles, W., and Munk, J. 1961. Harbor paradox. *ASCE Journal of the Waterways and Harbor Division*, **87**(3), 111–132.

Nikuradse, J. 1933. Strömungsgezetze in rauhen Rohren. *VDI Forschungsheft*, **361**, 237. In German.

Parsons, W. B. 1918. The Cape Cod Canal. *Transactions of the American Society of Civil Engineers*, **82**, 1–143.

PIANC, Inland Navigation Commission. 2015. *Ship behaviour in lock approaches*. Tech. rept. 155.

Pironneau, O. 1989. *Finite element methods for fluids*. Wiley, Chichester.

Prandtl, L. 1925. Bericht über Untersuchungen zur ausgebildeten Turbulenz. *Zeitschrift für Angewandte Mathematik und Mechanik*, **5**(2), 136. In German.

Priestly, A. 1993. Quasi-Riemannian method for the solution of one-dimensional shallow water flow. *Journal of Computational Physics*, **106**(1), 139–146.

Rehbock, Th. 1929. Wassermessung mit scharfkantigen Überfallwehren. *Z. des Vereines Deutscher Ingenieure*, **73**(24), 817–823. In German.

Reynolds, O. 1895. On the dynamical theory of incompressible viscous fluids and the determination of the criterion. *Philosophical Transactions of the Royal Society of London A: Mathematical, Physical and Engineering Sciences*, **186**, 123–164.

Rouse, H., and Ince, S. 1957. *History of hydraulics*. Dover, New York.

Rowbotham, F. 1983. *The Severn bore*. David & Charles, Newton Abbot.

Rutherford, J. C. 1994. *River mixing*. Wiley & Sons, New York.

Sanz-Serna, J. M., and Calvo, M. P. 1994. *Numerical Hamiltonian problems*. Chapman & Hall, London.

Savenije, H. H. G. 2005. *Salinity and tides in alluvial estuaries*. Elsevier, Amsterdam.

Schönfeld, J. C. 1948. Voortplanting en verzwakking van hoogwatergolven op een rivier. *De Ingenieur*, **4**, B1–B17. In Dutch.

Seelig, W. N., and Sorensen, R. M. 1977. *Hydraulics of Great Lakes Inlets*. Tech. rept. 77-8. Coastal Engineering Research Center, U.S. Army Corps of Engineers.

Stoker, J. J. 1957. *Water waves*. Interscience Publishers, New York.

Strauss, W. A. 1992. *Partial differential equations*. John Wiley & Sons, New York.

Streeter, V. L., and Wylie, E. B. 1967. *Hydraulic transients*. McGraw-Hill, New York.

Strickler, A. 1923. Beiträge zur Frage der Geschwindigkeitsformel und der Rauhigkeitszahlen für Ströme, Kanäle und geschlossene Leitungen. *Mitteilungen des Eidg. Amtes für Wasserwirtschaft*, **16**. In German.

Sturm, T. W. 2001. *Open-channel hydraulics*. McGraw-Hill, New York.

Terra, G. M. 2005. *Nonlinear tidal resonance*. Ph.D. thesis, University of Amsterdam, The Netherlands.

Thijsse, J. Th. 1935. A study of the effect upon navigation and upon the upkeep of the banks and bed of canalized rivers of undulatory movements and of currents in the reaches adjacent to locks with high lift. In: *Proceedings XVI International Congress of Navigation*. PIANC, Brussels.

Thorley, A. R. D. 1991. *Fluid transients in pipeline systems*. D & L George, Barnet, England.

Tijsseling, A. S. 1993. *Fluid-structure interaction in case of waterhammer with cavitation*. Ph.D. thesis, Delft University of Technology.

Toffolon, M., and Savenije, H. H. G. 2011. Revisiting linearized one-dimensional tidal propagation. *J. Geophys. Res.*, **116**, C07007, doi:10.1029/2010JC006616.

Toro, E. F. 2001. *Shock-capturing methods for free-surface shallow flows*. Wiley, New York.

Van Mazijk, A. 1996. *One-dimensional approach of transport phenomena of dissolved matter in rivers*. Ph.D. thesis, Delft University of Technology.

Ven Te Chow. 1959. *Open-channel hydraulics*. McGraw-Hill, New York.

Vreugdenhil, C. B. 1994. *Numerical methods for shallow-water flow*. Water Science and Technology Library. Springer Netherlands.

Wemelsfelder, P. J. 1947. Hoogwatergolf doorbraak Möhnetalsperre. *De Ingenieur*, **42**, B103–B105. In Dutch.

Wesseling, P. 2001. *Principles of computational fluid mechanics*. Springer Verlag, Berlin.

Williamson, J. 1951. The laws of flow in rough pipes. *La Houille Blanche*, **6**(5), 738–748.

Zienkiewicz, O. C., Taylor, R. L., and Nithiarasu, P. 2014. *The finite element method for fluid dynamics*. 7th edn. Elsevier, Amsterdam.

Author Index

Subject Index